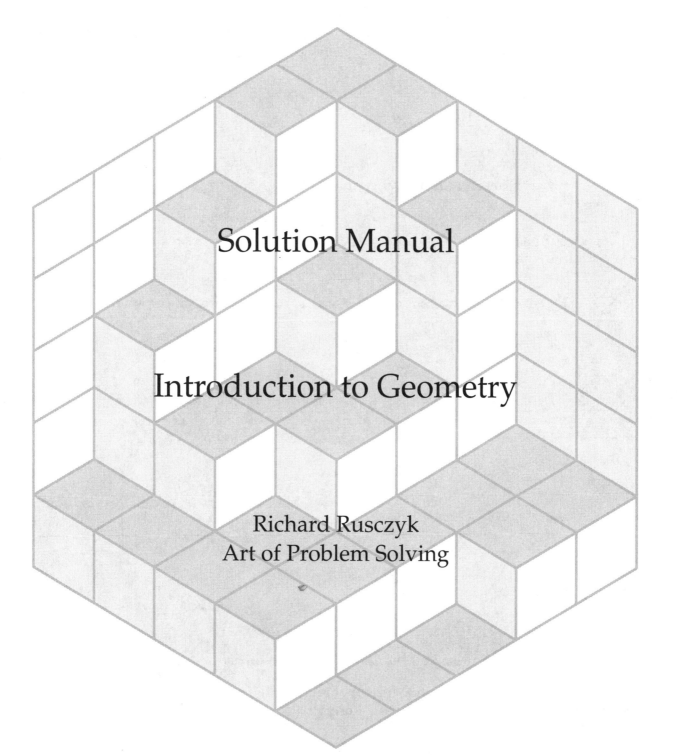

Solution Manual

Introduction to Geometry

Richard Rusczyk
Art of Problem Solving

Published by: AoPS Incorporated
                P.O. Box 2185
                Alpine, CA 91903-2185
                (619) 659-1612
                books@artofproblemsolving.com

ISBN: 978-1-934124-09-3

Visit the Art of Problem Solving website at http://www.artofproblemsolving.com

Cover image designed by Vanessa Rusczyk using KaleidoTile software.

Cover includes a satellite image of the Mississippi River Delta from NASA Earth Observatory and a photo of protractor and compasses by Vanessa Rusczyk.

Printed in the United States of America.

Second Edition Second Printing 2009.

Contents

19 Problem Solving Strategies in Geometry

CHAPTER 1

Exercises for Section 1.2

1.2.1 If Alice just gives Bob one point, he can't tell which line Alice is thinking about because there are infinitely many lines that could go through that point. Bob can't tell which one is Alice's line. However, once Alice gives Bob a second point on the line, he knows for sure which line is hers because there is only one line through any two given points – the continuation of the segment that connects the two points. Therefore, Bob needs $\boxed{\text{two}}$ points to determine Alice's line.

1.2.2 Since N is the midpoint of $\overline{BM}$ and $BN = 4$, we have $BM = 2BN = 8$. Similarly, since M is the midpoint of $\overline{AB}$ and $BM = 8$, we have $AB = 2BM = \boxed{16}$.

1.2.3 Our midpoints tell us that $ST = RT/2$, $RT = QT/2$, and $QT = PT/2$. Therefore, $ST = RT/2 = (QT/2)/2 = QT/4 = (PT/2)/4 = PT/8$. Since $ST + PS = PT$, we have $PS = PT - ST = PT - PT/8 = 7PT/8$. Therefore, $PT = 8PS/7 = \boxed{72/7}$.

1.2.4 Since $AB = AE/2$, B is the midpoint of $\overline{AE}$ and $AB = BE = AE/2 = 6$. Since $CD = AB/2$, $CD = 3$. Since $BC = CD/2$, $BC = 3/2$. We are given that C is between B and E, so $AC = AB + BC = 6 + 3/2 = 15/2$. Finally, we are told D is between C and E, so $AD = AC + CD = 15/2 + 3 = \boxed{21/2}$.

Exercises for Section 1.3

1.3.1

(a) $\overline{CO}$ is a $\boxed{\text{radius}}$.

(b) $\overleftrightarrow{EF}$ is a $\boxed{\text{tangent line}}$.

(c) $\overline{CD}$ is a $\boxed{\text{chord}}$.

(d) $\overline{AB}$ is a $\boxed{\text{diameter and a chord}}$.

(e) $\overleftrightarrow{CD}$ is a $\boxed{\text{secant line}}$.

1.3.2 $\boxed{\text{Yes}}$. Later in the text we learn exactly how to construct this line with basic geometric tools.

1.3.3 A line segment can intersect a circle in at most two points. Since a triangle consists of three line segments, the maximum number of intersections between a triangle and a circle is $3 \times 2 = 6$. The diagram at right shows such a configuration with $\boxed{6}$ intersections.

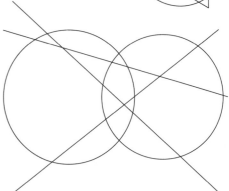

1.3.4 Two different circles can intersect in at most 2 points (see if you can figure out why). Each line can meet each circle in at most two points, for a total of at most $3 \times 2 \times 2 = 12$ possible intersections between a line and a circle. Each pair of lines can meet in at most 1 point of intersection. This gives us 3 more intersection points, one for each of the 3 ways we can choose two of the lines to intersect. This gives us a total of at most $2 + 12 + 3 = 17$ intersections. The diagram at right shows such a configuration with $\boxed{17}$ intersection points.

Exercises for Section 1.4

1.4.1

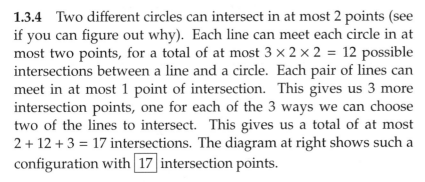

(a) We draw a line, pick a starting point X, then copy segment $\overline{AB}$ onto the line to find point Y such that $XY = AB$. Then we find point Z beyond Y on $\overrightarrow{XY}$ such that $YZ = CD$. Finally, we find point W such that $WZ = EF$, but notice that we locate W between X and Z, so that $XW = AB + CD - EF$.

(b) We draw a line, pick a starting point X, then copy segment $\overline{AB}$ onto the line to find point Y such that $XY = AB$. Then, we find point Z beyond Y on $\overrightarrow{XY}$ such that $YZ = AB$. Therefore, $XZ = 2AB$.

(c) Again, we start with a line and copy $\overline{AB}$ onto the line to make $\overline{XY}$ such that $XY = AB$. Then, we copy $\overline{CD}$ three times onto the end of $\overline{XY}$ (after Y on $\overrightarrow{XY}$) to make $\overline{XZ}$ such that $XZ = AB + 3CD$. Finally, we go back towards X, twice copying $\overline{EF}$ onto our line starting from Z. Thus, we find V such that $XV = AB - 2EF + 3CD$ as desired.

CHAPTER 2

Angles

Exercises for Section 2.2

2.2.1

We can easily make angles of 45°, 90°, and 135° from a given ray $\overrightarrow{OA}$.

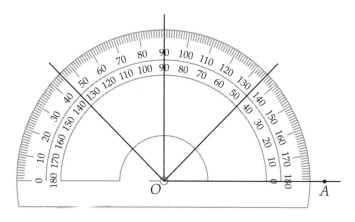

In order to create an angle of 220°, we first note that 220° = 360° − 140°. Our angle will be the reflex angle of a 140° angle. Therefore, we just make a 140° angle on the other side of the ray.

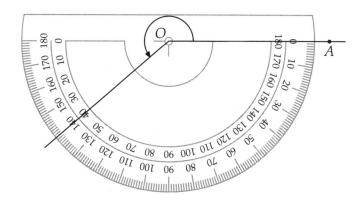

2.2.2

(a) $\angle X = \boxed{32°}$. $\angle X$ is $\boxed{\text{acute}}$.

(b) $\angle ABC = \boxed{120°}$ and $\angle DBC = \boxed{60°}$. $\angle ABC$ is $\boxed{\text{obtuse}}$ and $\angle DBC$ is $\boxed{\text{acute}}$.

(c) $\angle PQR = \boxed{70°}$, $\angle PRQ = \boxed{73°}$, and $\angle RPQ = \boxed{37°}$. All three angles are $\boxed{\text{acute}}$.

Exercises for Section 2.3

2.3.1 The given angle, together with the $x°$ angle, forms a straight angle, so $115° + x = 180°$. Therefore, $x = \boxed{65°}$. Since the angle with measure y and the given angle are vertical angles, $y = \boxed{115°}$. Finally, since the angles with measures x and z are vertical, $z = x = \boxed{65°}$.

2.3.2 Two angles are supplementary if their measures add to $180°$. Therefore, to find the measures of the angles that are supplementary to the given angles, we subtract each of our original angles from $180°$:

(a) $180° - 120° = \boxed{60°}$

(b) $180° - 45° = \boxed{135°}$

(c) $180° - 90° = \boxed{90°}$

2.3.3 Two angles are complementary if their measures add to $90°$. Therefore, to find the measures of the angles that are complementary to the given angles, we subtract each of our original angles from $90°$:

(a) $90° - 30° = \boxed{60°}$

(b) $90° - 45° = \boxed{45°}$

(c) $90° - 75° = \boxed{15°}$

Exercises for Section 2.4

2.4.1 We notice that the $118°$ angle and the angle with measure y are alternate exterior angles, so they are equal and $y = \boxed{118°}$. Then, since the angle with measure y and the angle with measure x form a straight angle, we have $x + y = 180°$ and $x = \boxed{62°}$.

2.4.2 Since $\overline{AB}$ is a transversal cutting the parallel segments $\overline{DA}$ and $\overline{CB}$, we know that $\angle A + \angle B = 180°$. Since $\overline{DC}$ also cuts $\overline{DA}$ and $\overline{CB}$, we know that $\angle D + \angle C = 180°$. Because $\overline{CB}$ cuts the parallel segments $\overline{AB}$ and $\overline{DC}$, we know that $\angle B + \angle C = 180°$. Finally, we combine these equations: $\angle A + \angle B = 180° = \angle B + \angle C$, so $\angle A = \angle C$, and $\angle B + \angle C = 180° = \angle D + \angle C$, so $\angle B = \angle D$.

2.4.3 We fill in corresponding angles as shown in the diagram at the top left of the next page. Then, since the marked angles at point K add up to a straight angle, we deduce that $(3x + y) + (2x - y) = 180°$. This reduces to $5x = 180°$, which gives $x = \boxed{36°}$.

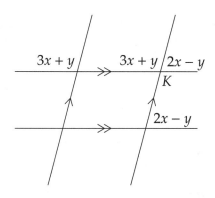

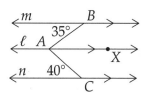

Figure 2.1: Diagram for Problem 2.4.3   Figure 2.2: Diagram for Problem 2.4.4

2.4.4 We draw line ℓ parallel to m and n through the point A. This divides $\angle BAC$ into the two angles $\angle BAX$ and $\angle XAC$. Now $\angle BAX = 35°$ since $\overline{BA}$ cuts the parallel lines m and ℓ, and $\angle XAC = 40°$ since $\overline{AC}$ cuts the parallel lines ℓ and n. Finally, we know that $x = \angle BAC = \angle BAX + \angle XAC = 35° + 40° = \boxed{75°}$.

2.4.5 We use our knowledge of parallel lines and equal angles to find the angle measures as shown to the right. We can then see from line n that the angles with measures $x - y$ and $x + y$ are supplementary, so $(x+y)+(x-y) = 180°$. Thus $x = 90°$. Then from line m we can conclude that $x - y$ and $2x - 2y$ are supplementary, so $(x - y) + (2x - 2y) = 180°$, or $3x - 3y = 180°$. Substituting our value for x, we obtain $y = 30°$. Therefore, our answer is $x = \boxed{90°}$ and $y = \boxed{30°}$.

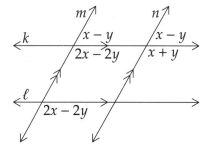

Exercises for Section 2.5

2.5.1 Call the measure of the unknown angle x. We know that the sum of the angles in a triangle is equal to 180°, so $30° + 57° + x = 180°$. Solving for x, we find that $x = \boxed{93°}$.

2.5.2 We let the smallest angle in the triangle have measure x. Then the problem says that the three angles have measures x, $2x$, and $3x$. Since the three angles in a triangle add up to 180°, we have $x + 2x + 3x = 180°$, so $x = 30°$. Therefore, the three angles have measures $\boxed{30°}$, $\boxed{60°}$, and $\boxed{90°}$.

2.5.3 *Solution 1:* Since the angles of $\triangle XYZ$ add up to 180°, we have $180° = \angle ZXY + \angle XYZ + \angle YZX = 58° + 52° + \angle YZX$. Therefore, $\angle YZX = 70°$. Since $\angle PZX$ is a straight angle, $\angle YZX$ and $\angle PZY$ are supplementary, so $\angle PZY = 180° - \angle YZX = 110°$. Finally, since the angles of $\triangle YZP$ add up to 180°, we have $180° = \angle ZYP + \angle YPZ + \angle PZY = \angle ZYP + 29° + 110°$. Solving, we get $\angle ZYP = \boxed{41°}$.

Solution 2: The angles of $\triangle XYP$ add up to 180°, so we have $180° = \angle XYP + \angle YPX + \angle PXY = \angle XYP + 29° + 58°$. Thus, we have $\angle XYP = 93°$. We can then split $\angle XYP$ into $\angle XYZ + \angle ZYP$: $93° = \angle XYP = \angle XYZ + \angle ZYP = 52° + \angle ZYP$. Solving, we find $\angle ZYP = \boxed{41°}$.

2.5.4 Call the vertex of the right angle A, and let the other two vertices be B and C. Since the angles of a triangle add up to $180°$, we know that $\angle A + \angle B + \angle C = 180°$. We also know that $\angle A$ is a right angle, so $\angle A = 90°$. Therefore, $90° + \angle B + \angle C = 180°$, or $\angle B + \angle C = 90°$. Therefore, angles $\angle B$ and $\angle C$ are complementary.

2.5.5 Extend $\overline{AB}$ to meet line m at H, as shown in the diagram to the right. Since $\angle ABH$ is a straight angle, $\angle HBC = 180° - \angle CBA = 90°$. Also, since $\overline{AH}$ cuts the parallel lines ℓ and m, $\angle BHC = 180° - 130° = 50°$ (same-side interior angles). Finally, since the angles of $\triangle HBC$ add up to $180°$, we have $180° = \angle CHB + \angle HBC + \angle BCH = 50° + 90° + \angle BCH$. Hence $\angle BCH = 40°$, so $x = \boxed{40°}$.

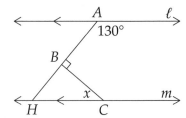

Exercises for Section 2.6

2.6.1 Since an exterior angle is equal to the sum of its remote interior angles, we know that $y = 62° + 38° = \boxed{100°}$.

2.6.2 Using the fact that an exterior angle is equal to the sum of its remote interior angles, we see that $110° = \angle B + \angle C = 38° + \angle C$. Solving, we find $\angle C = \boxed{72°}$.

2.6.3 *Solution 1:* Using the fact that an exterior angle is equal to the sum of its remote interior angles, we can write $123° = \angle XZY + \angle Y$ and $97° = \angle YXZ + \angle Y$. Since the angles of a triangle sum to $180°$, we can write $\angle YXZ + \angle XZY + \angle Y = 180°$. We can add our first two equations to find that $220° = \angle XZY + \angle Y + \angle YXZ + \angle Y = (\angle XZY + \angle Y + \angle YXZ) + \angle Y = 180° + \angle Y$. Therefore, $\angle Y = \boxed{40°}$.

Solution 2: We use the straight angles at X and Z to see that $123° + \angle YXZ = 180°$ and $97° + \angle XZY = 180°$. Therefore, $\angle YXZ = 57°$ and $\angle XZY = 83°$. Since the angles of a triangle sum to $180°$, we have $\angle Y = 180° - \angle YXZ - \angle XZY = 180° - 57° - 83° = \boxed{40°}$.

2.6.4 $\boxed{\text{No!}}$ Let $\triangle ABC$ have $\angle B = \angle C = 30°$. Then $\angle A = 120°$, and an exterior angle at A is equal to $60°$, which is clearly less than $90°$.

Exercises for Section 2.7

2.7.1 Suppose that the lines met to the left of $\overleftrightarrow{PQ}$, as shown in the diagram to the right. Then, in $\triangle PQR$, we have $\angle QPR + \angle PQR + \angle PRQ = 180°$. We have $\angle P = 40°$ and $\angle PQR = 180° - 40° = 140°$, so this equation is $40° + 140° + \angle PRQ = 180°$. Therefore, we have $\angle PRQ = 0°$, which is impossible. Therefore, we conclude that it is impossible for the two lines to meet to the left of $\overleftrightarrow{PQ}$.

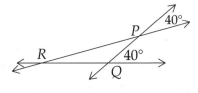

2.7.2 Suppose that $y = z$. The angles with measures z and x are vertical angles, so they are equal. Thus, $z = x$. Hence, $y = z = x$, and so by the previous result mentioned in the problem statement, lines k and m are parallel.

2.7.3 We notice that $\angle ABD = \angle BDE$. Since these are alternate interior angles, we conclude by the result of the previous problem that $\overleftrightarrow{AB} \parallel \overleftrightarrow{DE}$. Then, since the given 122° angle and our desired angle are same-side exterior angles, we have $x + 122° = 180°$, so $x = \boxed{58°}$.

2.7.4

(a) "If two teams are playing soccer, then they must be playing in the World Cup Finals." This is false.

(b) "If one angle of a triangle is 100°, then two angles of the triangle must add to 80°." This is true.

(c) "If a river is the Nile, then it must be the longest river in the world." This is true.

(d) "If an animal is a bird, then it must be a duck." This is false.

Review Problems

2.26

We first place the protractor as shown in the diagram below.

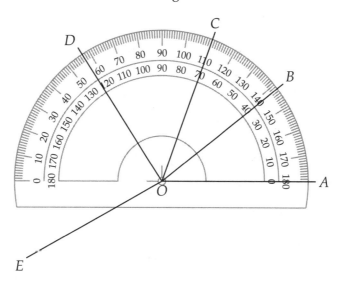

Since $\angle AOB$ and $\angle AOC$ are both less than 90°, we read the smaller number and find:

(a) $\angle AOB = \boxed{38°}$

(b) $\angle AOC = \boxed{70°}$

To find $\angle BOC$, we could move our protractor, or we can note that $\angle AOC = \angle AOB + \angle BOC$, so that

(c) $\angle BOC = \boxed{32°}$

Finally, to find $\angle DOE$, we rotate the protractor as shown on the next page.

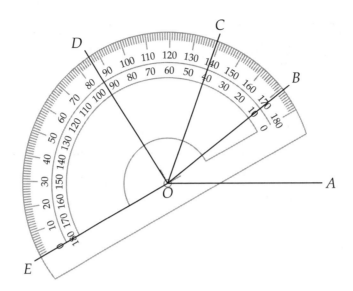

Since ∠*DOE* is less than 90°, we read the smaller number and find

(d) ∠*DOE* = $\boxed{86°}$

2.27   The second hand rotates through the entire circle, or 360°, in 60 seconds. Since 72° is $\frac{72}{360} = \frac{1}{5}$ of a circle, it will have rotated through an angle of 72° in $\frac{1}{5}$ the time, or $\boxed{12}$ seconds.

2.28

(a) ∠*CGE* is vertical to ∠*HGD*, so ∠*CGE* = $\boxed{58°}$.

(b) ∠*HGC* and ∠*HGD* form a straight angle, so ∠*HGC* = 180° − 58° = $\boxed{122°}$.

(c) ∠*FHB* and ∠*HGD* are corresponding angles, so ∠*FHB* = $\boxed{58°}$.

(d) ∠*BHG* and ∠*HGD* are same-side interior angles, so ∠*BHG* + ∠*HGD* = 180° and ∠*BHG* = 180° − 58° = $\boxed{122°}$.

2.29   Since the three angles of a triangle must add up to 180°, the third angle is 180° − 30° − 70° = $\boxed{80°}$.

2.30   We know that an exterior angle of a triangle is equal to the sum of its remote interior angles. Since the angles remote to the exterior angle at *A* are ∠*B* and ∠*C*, we have 170° = ∠*B* + ∠*C* = 60° + ∠*C*. Therefore, ∠*C* = $\boxed{110°}$.

2.31   We observe that $\overline{CB}$ cuts the parallel segments $\overline{AB}$ and $\overline{CD}$, so ∠*BCD* = ∠*ABC* = 3*x*. Furthermore, $\overline{CD}$ cuts the parallel segments $\overline{CB}$ and $\overline{DE}$, so the angles ∠*BCD* and ∠*CDE* are supplementary. Therefore, 8*x* = ∠*BCD* + ∠*CDE* = 180°, so *x* = $\boxed{22.5°}$.

2.32   We begin by drawing the diagram described in the problem at left atop the next page. From this diagram we can see that ∠*RYX* and ∠*PXY* are same-side interior angles. Since ∠*PXY* = 90°, ∠*RYX* = 180 − ∠*PXY* = $\boxed{90°}$.

Figure 2.3: Diagram for Problem 2.32   Figure 2.4: Diagram for Problem 2.33

2.33 We draw segment $\overline{HG}$ parallel to $\overline{AB}$ and $\overline{DE}$ as shown at right above. Then, since $\overline{BG}$ cuts parallel segments $\overline{AB}$ and $\overline{HG}$, we have $\angle ABG + \angle BGH = 180°$. Since $\overline{GE}$ cuts parallel segments $\overline{HG}$ and $\overline{DE}$, we also have $\angle HGE + \angle GED = 180°$. Now $\angle BGH + \angle HGE = \angle BGE = \angle FGC = 140°$, and combining this with the previous equations gives $(180° - \angle ABG) + (180° - \angle GED) = 140°$. So, $\angle ABG = x = 220° - \angle GED = \boxed{110°}$.

2.34 Since $\overline{AB} \parallel \overline{EF}$, we have $\angle DAB + \angle AEF = 180°$. Since $\angle DAB = \angle DAC + \angle CAB = 2x + 10°$, we have $2x + 10° + 4x - 10° = 180°$, so $x = \boxed{30°}$.

2.35 The fact that $\angle A$ and $\angle B$ are complementary can be written as $\angle A + \angle B = 90°$. Similarly, we find that $\angle B + \angle C = 90°$ and $\angle C + \angle A = 90°$. Adding these equations together and dividing by 2, we obtain $\angle A + \angle B + \angle C = 135°$. Subtracting each of the original equations from this one, we obtain $\angle A = \angle B = \angle C = 45°$, as required.

Alternatively, we can write $\angle A + \angle B = 90° = \angle B + \angle C$, so $\angle A = \angle C$. Similarly, $\angle B + \angle C = 90° = \angle A + \angle C$ gives us $\angle A = \angle B$, so $\angle A = \angle B = \angle C$.

2.36 Suppose that $\triangle ABC$ is labeled such that $\angle A \leq \angle B \leq \angle C$. Then, since the angles are in the ratio $3 : 4 : 5$, there exists some number x such that $\angle A = 3x$, $\angle B = 4x$, and $\angle C = 5x$. Using the fact that the three angles of a triangle add up to $180°$, we obtain $12x = \angle A + \angle B + \angle C = 180°$, so $x = 15°$. Since the exterior angles of the triangle are equal to the sums of their remote interior angles, the smallest exterior angle has measure equal to the least of $3x + 4x = 7x$, $3x + 5x = 8x$, and $4x + 5x = 9x$. Since the smallest of these is $7x$, the measure of the smallest exterior angle of the triangle is $7x = \boxed{105°}$.

Alternatively, we can use $x = 15°$ to note that the interior angles are $\angle A = 3x = 45°$, $\angle B = 4x = 60°$, and $\angle C = 5x = 75°$. Therefore, the exterior angles are $135°$, $120°$, and $105°$.

2.37 Let the exterior angles have measures $x \leq y \leq z$, and let the corresponding interior angles have measures a, b, and c. Since the exterior angles are in the ratio $2 : 3 : 4$, there is some number n such that $x = 2n$, $y = 3n$, $z = 4n$. We show two solutions from here.

Solution 1: Since the sum of the exterior angles of a triangle is $360°$ (as proved in the text), we have $2n + 3n + 4n = 360°$, so $n = 40°$. Therefore, the exterior angles are $80°$, $120°$, and $160°$. Since each interior angle is supplementary to an exterior angle, the interior angles are $\boxed{100°, 60°, \text{ and } 20°}$.

Solution 2: Since an exterior angle is equal to the sum of its remote interior angles, $b + c = x = 2n$, $a + c = y = 3n$, and $a + b = z = 4n$. Adding these together and using the fact that the sum of the angles of a triangle is $180°$, we find $360° = 2(a + b + c) = 9n$, so $n = 40°$. Thus, we have the equations $b + c = 80°$, $a + c = 120°$, and $a + b = 160°$. Solving this system, we get $\boxed{a = 100°}$, $\boxed{b = 60°}$, and $\boxed{c = 20°}$.

2.38 It is not possible. To see this, we draw segment $\overline{EF}$ parallel to $\overline{AB}$ and $\overline{CD}$ 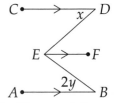 as shown. Then, since $\overline{DE}$ cuts parallel segments $\overline{CD}$ and $\overline{EF}$, and $\overline{EB}$ cuts parallel segments $\overline{EF}$ and $\overline{AB}$, we have $\angle DEB = \angle DEF + \angle FEB = \angle EDC + \angle EBA$. If we substitute the given measures into $\angle DEB = \angle EDC + \angle EBA$, we get $x + y = x + 2y$, or $y = 0$. But this is impossible, since $\angle EBA = 2y$ is not zero.

2.39 We proved in the text that the sum of the exterior angles of a triangle is $360°$. If two of these exterior angles are supplementary, then their sum is $180°$. Since all three exterior angles add to $360°$, this means the third exterior angle must be $180°$. The interior angle corresponding to this third exterior angle would then have measure $0°$. This is impossible, so it is impossible for two of the exterior angles to be supplementary.

2.40 Since the ratio of $\angle COB$ to $\angle BOF$ is $7:2$, there is some x such that $\angle COB = 7x$ and $\angle BOF = 2x$. Since $\angle COF$ is a straight angle, $9x = \angle COB + \angle BOF = \angle COF = 180°$, so $x = 20°$. Since $\angle BOF$ and $\angle COE$ are vertical angles, $\angle COE = \angle BOF = 2x = 40°$. Finally, since $\angle COD = \angle DOE$, we have $\angle COD = \frac{1}{2}\angle COE = \boxed{20°}$.

2.41 We draw lines m and n parallel to $\overleftrightarrow{BA}$ and $\overleftrightarrow{FE}$ through C and D, and label the angles as shown. Then, since $\overline{BC}$ cuts parallel lines $\overleftrightarrow{BA}$ and m, $r = \angle ABC = 45°$. Since $\overline{CD}$ cuts the parallel lines m and n, we have $s = t$. Since $\overline{DE}$ cuts the parallel lines n and $\overleftrightarrow{FE}$, $u = \angle DEF = 50°$. Also, we know from the original diagram that $r + s = 85°$, so $s = 85° - 45° = 40°$, and therefore $t = s = 40°$ as well from our parallel lines. Finally, since $w = t + u$, we have $w = 40° + 50° = \boxed{90°}$.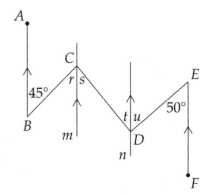

2.42 It seems at first that we do not have enough information: we don't know whether the $20°$ angle is the smallest angle, the largest angle, or in between. But rather than give up, we check each of these cases:

1. The $20°$ angle is the largest angle. Then the other two angles are both smaller than $20°$, so the sum of all of the angles is less than $3 \cdot 20° = 60°$. But this is impossible, since the sum of all of the angles of the triangle must be $180°$.

2. The $20°$ angle is neither the smallest nor the largest angle. Then the smallest angle is less than $20°$, so the largest angle is less than $6 \cdot 20° = 120°$. Then the sum of all of the angles is less than $20° + 20° + 120° = 160°$, again contradicting the fact that the sum of all of the angles of the triangle must be $180°$.

3. The $20°$ angle is the smallest angle of the triangle. Then the largest angle must be $6 \cdot 20° = 120°$, and the third angle must be $180° - 120° - 20° = 40°$, since the sum of all of the angles is $180°$. Since $20° \le 40° \le 120°$ and $6(20°) = 120°$, this triangle satisfies the conditions of the problem.

Since only the last case yields a triangle that satisfies the conditions of the problem, we conclude that this triangle is the only solution. The required angles are $\boxed{20°, 40°, \text{ and } 120°}$.

2.43 Since $\overline{AD}$ cuts the parallel segments $\overline{AC}$ and $\overline{EB}$, we have $3x = \angle CAB = \angle ABE$. Angle $\angle ABD$ is a straight line, so $5x = \angle ABE + \angle EBD = 180°$, giving $x = 36°$. Also, $\angle CBD$ is an exterior angle of $\triangle ABC$, so $4x = \angle CBD = \angle C + \angle CAB = \angle C + 3x$. Therefore, $\angle C = x = \boxed{36°}$.

2.44 Using the fact that the angles of a triangle sum to 180°, we determine that $180° = \angle A + \angle B + \angle C = (x - 2y) + (3x + 5y) + (5x - 3y) = 9x$. Solving for x, we find $x = \boxed{20°}$.

2.45 First, we note that $\angle A$ must be greater than 0°. Therefore, $x - 2y > 0°$, so $y < x/2$. In the previous problem, we found $x = 20°$, so we know $y < 10°$. Since $x = 20°$ and $y < 10°$, we know that $\angle C = 5x - 3y = 100° - 3y$ must be greater than 10°. Therefore, we must only consider the two cases $\angle A = 10°$ and $\angle B = 10°$. If $\angle A = 10°$, we have $x - 2y = 10°$ and $x = 20°$, so $y = 5°$. If $\angle B = 10°$, we have $3x + 5y = 10°$ and $x = 20°$, so $y = -10°$. We are told y is positive, so our only solution is $y = \boxed{5°}$.

2.46 We draw and label a diagram at right. The problem is then to show that if $\angle FBA + \angle BAD = 180°$, then $\overleftrightarrow{CD} \parallel \overleftrightarrow{EF}$. Since angles $\angle HBF$ and $\angle FBA$ together form a straight angle, we have $\angle HBF + \angle FBA = 180°$. Therefore, $\angle HBF = 180° - \angle FBA = \angle BAD$. But we have already shown in the text that $\angle HBF = \angle BAD$ implies that $\overleftrightarrow{CD} \parallel \overleftrightarrow{EF}$. Therefore, we conclude that if a transversal cuts two lines such that the same-side interior angles are supplementary, then the two lines are parallel.

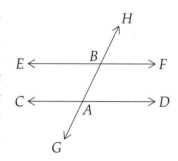

Challenge Problems

2.47 We label the points as in the diagram to the right. Using the fact that the sum of the angles of $\triangle BED$ is 180°, we find $\angle EDB + \angle DBE + \angle BED = 180°$. Since $\angle BED = 90°$ and $\angle DBE = \phi = 27°$, we have $\angle EDB = 180° - 90° - 27° = 63°$. Then, since $\angle ADE + \angle EDB = \angle ADB = 90°$, we have $\theta + 63° = 90°$, or $\theta = \boxed{27°}$.

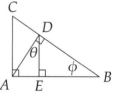

2.48 First, we find the position of the minute hand. The minute hand makes a 0° angle with 12 o'clock on the hour, and it travels a total of 360° in an hour, so in the 10 minutes after 11 o'clock, it moves $\frac{10}{60} \cdot 360° = 60°$ clockwise from the 12.

Next, we consider the hour hand. It travels all the way around the clock in 12 hours, so it travels at a rate of $\frac{1}{12} \cdot 360° = 30°$ per hour. Since it will be pointing at the 12 at 12 o'clock, and it will travel $(\frac{5}{6} \text{ hour})(30° \text{ per hour}) = 25°$ in the time between 11:10 PM and 12:00 midnight, it currently rests 25° counterclockwise from the 12. Therefore, since the angle made by the two hands is the sum of the two angles made between the hands and the ray pointing from the center of the clock towards the 12, the angle between the hands is $60° + 25° = \boxed{85°}$.

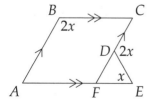

2.49 We extend $\overline{DC}$ past D to meet $\overline{AE}$ at F as shown. Then $\angle CDE$ is an exterior angle of $\triangle DEF$, so $\angle DEF + \angle EFD = \angle CDE$. Thus, $x + \angle EFD = 2x$, so $\angle EFD = x$. Since $\overline{FC}$ cuts the parallel segments $\overline{BC}$ and $\overline{AE}$, $\angle BCF = \angle DFE = x$. Finally, since $\overline{BC}$ cuts the parallel segments $\overline{AB}$ and $\overline{FC}$, angles $\angle ABC$ and $\angle BCF$ are supplementary, so $3x = \angle ABC + \angle BCF = 180°$, giving us $x = \boxed{60°}$.

2.50 From the fact that $\angle EAC = \angle DFC$, we can conclude that $\overline{AE} \parallel \overline{FD}$. Also, since the sum of the angles in $\triangle EGD$ is 180°, we have $\angle BDF = 180° - 90° - 60° = 30°$. Finally, since $\overline{BD}$ cuts the parallel segments $\overline{AE}$ and $\overline{FD}$, we deduce that $\angle BEA = \angle BDF = \boxed{30°}$.

2.51 Let the triangle be $\triangle ABC$, and suppose that $\angle A = \angle B + \angle C$. Then, since the exterior angles of a triangle are equal to the sum of their opposite interior angles, the three exterior angles have measures $\angle A + \angle B$, $\angle B + \angle C = \angle A$, and $\angle A + \angle C$. Adding the first and last of these gives a measure of $2\angle A + \angle B + \angle C = \angle A + (\angle A + \angle B + \angle C) = \angle A + 180°$ (since the angles of a triangle add to $180°$). Therefore, the sum of the exterior angles with measures $\angle A + \angle B$ and $\angle A + \angle C$ is $180°$ greater than the third (which we showed has measure $\angle A$).

2.52 Let the triangle be $\triangle ABC$, with $\angle A = \angle B + \angle C$. Then, using the fact that the angles of a triangle add up to $180°$, we have $180° = \angle A + \angle B + \angle C = 2\angle A$, so $\angle A = 90°$. Since one of the angles of the triangle is $40°$, the third angle must be $180° - 90° - 40° = 50°$. Thus, the two required angles have measures $\boxed{50°}$ and $\boxed{90°}$.

2.53 Let the middle angle have measure x. Since the difference between the largest and middle angle is the same as the difference between the middle and the smallest, we can let this common difference be y. Therefore, we can write the angle measures as $x - y$, x, and $x + y$ for some values of x and y. Then, using the fact that their sum is $180°$, we get $(x - y) + x + (x + y) = 3x = 180°$, or $x = 60°$. Therefore, one of the angles of the triangle must be $60°$. Since we are told that one angle is $100°$, it follows that the third angle is $180° - 100° - 60° = 20°$, and thus the two angle measures we are looking for are $\boxed{60°}$ and $\boxed{20°}$.

2.54 Suppose the smallest exterior angle has measure x. Then, the other two exterior angles have measures $2x$ and $6x$. We learned in the text that the sum of the exterior angles of a triangle is $360°$. Therefore, we have $x + 2x + 6x = 360°$, so $x = 40°$. But this means that one of the exterior angles of the triangle has measure $6x = 240°$, which is clearly impossible, since exterior angles must be less than $180°$. Therefore, the exterior angles cannot be in the ratio $1 : 2 : 6$.

2.55 Draw segment $\overline{AC}$. Then, we use the fact that the sum of the angles of a triangle is $180°$ to determine that $\angle CAB + \angle ABC + \angle BCA = 180°$ and $\angle ACD + \angle CDA + \angle DAC = 180°$. Adding these equations and rearranging, we find that $(\angle CAB + \angle DAC) + \angle ABC + (\angle BCA + \angle ACD) + \angle CDA = 360°$. But $\angle DAB = \angle CAB + \angle DAC$ and $\angle BCD = \angle BCA + \angle ACD$. Thus, we find that $\angle DAB + \angle ABC + \angle BCD + \angle CDA = \boxed{360°}$.

2.56 We use the result of the previous problem. Labeling the vertices A, B, C, and D clockwise around the figure as shown, we notice that $x = 180° - \angle DAB$, $w = 180° - \angle ABC$, $z = 180° - \angle BCD$, and $y = 180° - \angle CDA$. Adding these four equations, we find that $w + x + y + z = 720° - (\angle DAB + \angle ABC + \angle BCD + \angle CDA)$. Since the sum of the interior angles of the figure is equal to $360°$ by the previous problem, we find that $w + x + y + z = \boxed{360°}$.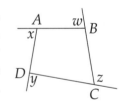

2.57 Since $\angle PZQ = \angle PQZ$, we have $\angle ZQR = \angle PQR - \angle PQZ = \angle PQR - \angle PZQ$. $\angle PZQ$ is an exterior angle of $\triangle ZQR$, so $\angle PZQ = \angle R + \angle ZQR$. Therefore, we have $\angle ZQR = \angle PQR - (\angle R + \angle ZQR)$, so $2\angle ZQR = \angle PQR - \angle R = 42°$. Thus, $\angle ZQR = \boxed{21°}$.

CHAPTER 3 Congruent Triangles

Exercises for Section 3.2

3.2.1

(a) We are given that $EF = GH$ and $FG = EH$. Also, triangles EFG and GHE share a common side $\overline{EG}$, so by SSS, they are congruent.

(b) Angles $\angle EGF$ and $\angle GEH$ are corresponding parts in triangles EFG and GHE, so they must be equal.

(c) Viewing $\overline{EG}$ as a transversal of lines $\overleftrightarrow{HE}$ and $\overleftrightarrow{FG}$, we realize $\overline{HE}$ is parallel to $\overline{FG}$ since $\angle HEG = \angle EGF$.

(d) Angles $\angle EGH$ and $\angle FEG$ are also corresponding parts in congruent triangles EFG and GHE, so $\angle EGH = \angle FEG$. Therefore, $\overline{HG} \parallel \overline{EF}$.

3.2.2

(a) We are given that $AB = AC$. Since M is the midpoint of $\overline{BC}$, we have $BM = CM$. Also, triangles ABM and ACM share a common side $\overline{AM}$, so by SSS, they are congruent. Then the corresponding parts of triangles ABM and ACM are congruent, so in particular, $\angle BAM = \angle CAM$. Hence, $\overline{AM}$ bisects $\angle BAC$.

(b) Angles $\angle AMB$ and $\angle AMC$ are also corresponding parts of congruent triangles ABM and ACM, so $\angle AMB = \angle AMC$. Angles $\angle AMB$ and $\angle AMC$ add up to $\angle BMC$, which is a straight angle, so $\angle AMB + \angle AMC = 180^\circ$. Therefore, $\angle AMB = \angle AMC = 90^\circ$. In other words, $\overline{AM} \perp \overline{BC}$.

3.2.3 Since $AD = CD$, $AB = BC$, and $BD = BD$, we have $\triangle ABD \cong \triangle CBD$ by SSS Congruence. Therefore, $\angle ABD = \angle CBD$.

Exercises for Section 3.3

3.3.1 In $\triangle DEF$, we have $\angle D = 180^\circ - \angle E - \angle F = 80^\circ$. Therefore, by SAS we have $\triangle DEF \cong \triangle GHI$. No other two triangles shown need be congruent.

3.3.2

(a) We are given that $AB = AC$. Since $\overline{AM}$ bisects $\angle BAC$, $\angle BAM = \angle CAM$. Also, triangles ABM and

ACM share a common side $\overline{AM}$, so by SAS, they are congruent. $\overline{BM}$ and $\overline{CM}$ are corresponding parts of these triangles, so $BM = CM$. In other words, M is the midpoint of $\overline{BC}$.

(b) Angles $\angle AMB$ and $\angle AMC$ are also corresponding parts of congruent triangles ABM and ACM, so $\angle AMB = \angle AMC$. But $\angle AMB$ and $\angle AMC$ add up to $\angle BMC$, which is a straight angle, so $\angle AMB + \angle AMC = 180°$. Therefore, $\angle AMB = \angle AMC = 90°$. In other words, $\overline{AM} \perp \overline{BC}$.

3.3.3 Since $XW = XV$, $\angle ZXV = \angle YWX$, and $WY = XZ$, we have $\triangle XWY \cong \triangle VXZ$. Therefore, $VZ = XY = 3$.

3.3.4 We are given that $AB = CD$ and $\angle BAC = \angle DCA$. Furthermore, triangles CAB and ACD share a common side $\overline{AC}$, so by SAS, they are congruent. Angles $\angle ACB$ and $\angle CAD$ are corresponding parts, so $\angle ACB = \angle CAD$. Therefore, $\overline{AD} \parallel \overline{BC}$.

Exercises for Section 3.4

3.4.1 In each triangle, the third angle is $180° - 55° - 50° = 75°$. To tell which triangles are congruent, we focus on where the $50°$ and $55°$ angles are relative to the sides of length 5 in each triangle. In $\triangle DEF$ and $\triangle GHI$, the side of length 5 is between angles with measures $50°$ and $55°$. Therefore, these triangles are congruent. We have to be careful in matching up the corresponding vertices: $\triangle DEF \cong \triangle HGI$ by ASA. In triangles $\triangle MNO$ and $\triangle ABC$, the side of length 5 is adjacent to a $55°$ angle and opposite a $50°$ angle. Therefore, $\triangle ABC \cong \triangle NMO$ by AAS Congruence.

3.4.2 If M is the midpoint of $\overline{AC}$, then $AM = MC$. Our parallel lines give $\angle C = \angle A$ and $\angle D = \angle B$. Therefore, we have $\triangle MAB \cong \triangle MCD$ by AAS Congruence. (We could also have used the vertical angles at M.) So, $BM = MD$ as corresponding parts of these triangles. Therefore, if M is the midpoint of $\overline{AC}$, then M must also be the midpoint of $\overline{BD}$.

3.4.3 Suppose that triangles $\triangle ABC$ and $\triangle DEF$ satisfy the conditions of AAS Congruence such that $\angle A = \angle D$, $\angle B = \angle E$, and $AC = DF$. (Make sure you see that this is covers all possible cases of two triangles satisfying the conditions of AAS Congruence!) We can use our two pairs of equal angles to show that the other pair of angles are equal:

$$\angle C = 180° - \angle A - \angle B = 180° - \angle D - \angle E = \angle F.$$

Since $\angle C = \angle F$, $\angle A = \angle D$, and $AC = DF$, we have $\triangle ABC \cong \triangle DEF$ by ASA Congruence. Therefore, we have used ASA Congruence to prove that AAS Congruence works.

3.4.4

(a) We are given that $PQ = PR$ and $\angle PQY = \angle PRX$. Triangles YQP and XRP also have angle QPR in common, so by ASA, they are congruent. Since $\overline{QY}$ are $\overline{RX}$ are corresponding parts of triangles YQP and XRP, we have $QY = RX$.

(b) Since $\triangle YQP \cong \triangle XRP$, we have $PX = PY$ because they are corresponding sides of the two triangles. We also have $PQ = PR$, and we can combine these two side length equalities to show $XQ = YR$:

$$XQ = PQ - PX = PR - PY = YR.$$

Our congruent triangles also give us $\angle Q = \angle R$. Also, $\angle XNQ = \angle YNR$ since they are vertical angles. $XQ = YR$, $\angle Q = \angle R$, and $\angle XNQ = \angle YNR$ give us $\triangle XNQ \cong \triangle YNR$ by AAS Congruence, so $XN = YN$.

Exercises for Section 3.6

3.6.1 We are given $\angle P = 43°$, and since $PQ = PR$, we know that $\angle Q = \angle R$. Therefore, $180° = \angle P + \angle Q + \angle R = 43° + 2\angle Q$, so $\angle Q = (180° - 43°)/2 = \boxed{68.5°}$.

3.6.2 We connect A to the midpoint, M, of $\overline{BC}$ as shown. Since $AB = AC$, $BM = CM$, and $AM = AM$, we have $\triangle ABM \cong \triangle ACM$ by SSS Congruence. Therefore, corresponding angles $\angle B$ and $\angle C$ of these triangles are equal.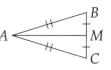

3.6.3

(a) Let $PQ = QR = PR$ in triangle $\triangle PQR$ as given. Since $PQ = QR$, we have $\angle P = \angle R$. Similarly, since $QR = RP$, we have $\angle P = \angle Q$. Therefore, all the angles are equal. Since they must sum to $180°$, each must have measure $180°/3 = 60°$.

(b) Let our triangle be $\triangle XYZ$, with $\angle X = \angle Y = \angle Z$. Since $\angle X = \angle Y$, we have $XZ = YZ$. Since $\angle Y = \angle Z$, we have $XZ = XY$. Therefore, $XY = XZ = YZ$, so all the sides of the triangle have the same length.

3.6.4 Since the triangle is equilateral, each angle is $60°$. Therefore, $3x + 27° = 60°$, so $x = 11°$. Also, $2y - 4° = 60°$, so $y = 32°$. So, $x + y = \boxed{43°}$.

3.6.5 Since O is the center of the circle, $\overline{OA}$ and $\overline{OB}$ are both radii of the circle. Therefore, $OB = OA$, so $\angle B = \angle A = 70°$. Hence $\angle O = 180° - \angle A - \angle B = \boxed{40°}$.

3.6.6 Let $\theta = \angle CAD = \angle ABE = \angle BCF$. Then

$$\angle BAE = \angle BAC - \angle CAD = 60° - \theta,$$
$$\angle CBF = \angle CBA - \angle ABE = 60° - \theta,$$
$$\angle ACD = \angle ACB - \angle BCF = 60° - \theta,$$

so $\angle BAE = \angle CBF = \angle ACD = 60° - \theta$. Since $\angle EDF$ is an exterior angle of triangle ACD, $\angle EDF = \angle DAC + \angle ACD = \theta + (60° - \theta) = 60°$. Similarly, $\angle FED = \angle DFE = 60°$. Therefore, triangle DEF is equilateral.

3.6.7

(a) Since $VW = VX$, we have $\angle VWX = \angle VXW$. Since $\overline{WX} \parallel \overline{YZ}$, we have $\angle VYZ = \angle VWX$ and $\angle VXW = \angle VZY$. Combining these with $\angle VWX = \angle VXW$ gives $\angle VYZ = \angle VZY$. Therefore, $VY = VZ$. Since $VW = VX$, we have

$$WY = VY - VW = VZ - VX = XZ,$$

as desired.

(b) Since $WY = XZ$, $YZ = YZ$, and $\angle WYZ = \angle XZY$, we have $\triangle WYZ \cong \triangle XZY$ by SAS Congruence. Corresponding sides $\overline{XY}$ and $\overline{WZ}$ therefore have the same length. (We could also have used SAS to show that $\triangle VXY \cong \triangle VWZ$.)

Exercises for Section 3.7

3.7.1 Construct the perpendicular bisector of $\overline{AB}$, and let it intersect $\overline{AB}$ at C. Then construct the perpendicular bisector of $\overline{AC}$ and let it intersect $\overline{AC}$ at D. Finally, construct the perpendicular bisector of $\overline{BC}$ and let it intersect $\overline{BC}$ at E. Then $AD = DC = CE = EB$, so points D, C, and E divide $\overline{AB}$ into four equal segments.

3.7.2 Suppose, as the problem states, we draw two circles centered at A and B, with radius r not necessarily equal to AB. Let the two circles intersect at C and D, so $AC = BC = AD = BD = r$. Since we also have $AB = AB$ and $CD = CD$, SSS Congruence gives us $\triangle ACB \cong \triangle ADB$ and $\triangle CAD \cong \triangle CBD$.

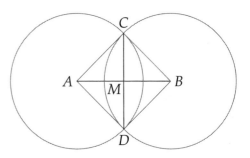

Let M be the point where $\overline{AB}$ and $\overline{CD}$ meet. $\angle ACM = \angle BCM$ (from $\triangle CAD \cong \triangle CBD$), so by SAS, triangles ACM and BCM are congruent. It follows that $AM = BM$, and $\angle AMC = \angle BMC$. But $\angle AMC + \angle BMC = 180°$, so $\angle AMC = \angle BMC = 90°$; in other words, $\overline{AB}$ is perpendicular to $\overline{CD}$. Hence, $\overline{CD}$ is the perpendicular bisector of $\overline{AB}$, and the construction still works.

3.7.3 Let $\overline{AB}$ be an arbitrary segment. Construct the perpendicular bisector of $\overline{AB}$. Call the perpendicular bisector $\overleftrightarrow{CD}$, and let it intersect $\overline{AB}$ at P. Then $\angle APC = 90°$.

3.7.4 In the text, we constructed equilateral triangles ABC and ABC', then drew $\overline{CC'}$ to create a segment perpendicular to $\overline{AB}$ as shown. We have $\angle CAC' = \angle CAB + \angle C'AB = 120°$ and $AC = AC'$, so $\angle ACC' = \angle AC'C = (180° - 120°)/2 = 30°$. Therefore, in constructing $\overline{CC'}$ we have constructed 30° angle $\angle ACC'$.

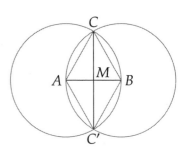

Review Problems

3.24

(a) $\triangle ABC \cong \triangle DCB$ by SSS Congruence.

(b) Since SSA is not a valid congruence theorem, we have no pair of triangles that must be congruent.

(c) Since $\overline{JK} \parallel \overline{MN}$, we have $\angle K = \angle M$ and $\angle J = \angle N$. Therefore, $\triangle JLK \cong \triangle NLM$ by either AAS or ASA Congruence.

(d) Since $EF = BC$, we have $FE + EB = EB + BC$, so $FB = EC$. Since we also have $\angle A = \angle D$ and $\angle ABF = \angle DEC$, we have $\triangle ABF \cong \triangle DEC$ by AAS Congruence.

(e) Our parallel lines give us $\angle HJI = \angle JHG$. However, SSA is not a valid congruence theorem, so we don't necessarily have a pair of congruent triangles.

(f) $\triangle KLM \cong \triangle NLM$ by SSS Congruence.

(g) Since $\angle BCA = \angle DCE$, $\angle ABC = \angle EDC$, and $AB = ED$, we have $\triangle BCA \cong \triangle DCE$ by AAS Congruence.

(h) Since $GH = GH$, we have $\triangle FGH \cong \triangle IGH$ by SAS Congruence.

(i) Since $\angle KLJ = \angle MLN$, $JL = LN$, and $\angle JKL = \angle NML$, we have $\triangle JKL \cong \triangle NML$ by AAS Congruence. Since $JL = NL$, we have $\angle NJL = \angle JNL$. Together with $JN = JN$ and $\angle K = \angle M$, this gives us $\triangle JKN \cong \triangle NMJ$ by AAS Congruence.

3.25 Let the measure of the vertex angle be x. Therefore, each base angle has measure $x/2 - 6°$ as given in the problem. Together, the three angles sum to $180°$, so we have $x + 2(x/2 - 6°) = 180°$. Therefore, $x = \boxed{96°}$.

3.26 Our parallel lines give us $\angle B = 2x$. Since $AB = AC$, we have $\angle C = \angle B$, so $\angle C = 2x$ as well. From $\triangle ABC$ we have $x + 2x + 2x = 180°$, so $x = \boxed{36°}$.

3.27 Since $\triangle BCD$ is equilateral, each of its angles is $60°$. Therefore, $\angle ABD = 180° - \angle CBD = 120°$. Since $AB = BD$, we have $\angle A = \angle ADB$. From triangle $\triangle ABD$, we have $180° = \angle A + \angle ABD + \angle ADB = 2\angle A + 120°$. Therefore, $\angle A = \boxed{30°}$.

3.28

(a) Since $AP = CP$, $\angle APD = \angle BPC$ (vertical angles), and $PD = BP$, we have $\triangle APD \cong \triangle CPB$ by SAS Congruence.

(b) Our diagram shows a case in which clearly $\overline{AD}$ and $\overline{BC}$ are not parallel. Our triangle congruence gives us $\angle A = \angle C$, not the $\angle A = \angle B$ needed to prove that $\overline{AD}$ and $\overline{BC}$ are parallel.

3.29

(a) Since $\triangle PTS$ and $\triangle RTS$ are congruent, we have $\angle PST = \angle RST$. These two angles together make a straight angle, so each angle must be $90°$.

(b) Since $\triangle PTQ \cong \triangle RTS$, we have $\angle PQT = \angle RST$, so $\angle PQT = 90°$, too. Angles $\angle QPT$, $\angle TPS$, and $\angle SRT$ are all equal due to the triangle congruences. Therefore, $\angle QPR = 2\angle PRQ$ and from triangle $\triangle PQR$, we have

$$
\begin{aligned}
180° &- \angle PQR + \angle QPR + \angle PQR \\
&= 90° + 3\angle PRQ.
\end{aligned}
$$

Therefore, $\angle PRQ = 30°$. Finally, we see that the angles of $\triangle PQR$ are $\boxed{\angle PQR = 90°}$, $\boxed{\angle PRQ = 30°}$, and $\boxed{\angle QPR = 60°}$.

3.30

(a) Since $KO = KN$, $\triangle KNO$ is isosceles with $\angle KON = \angle N$. Since $IA = IN$, we have $\angle IAN = \angle N$. Since $OI = ON$, we have $\angle N = \angle OIN$.

(b) Since $\angle KON = \angle N$ and $\angle N = \angle IAN$, we have $\angle KON = \angle IAN$. Therefore, $\overline{IA} \parallel \overline{KO}$.

3.31

(a) Since O is the center of the circle, we have $OA = OB = OC$. Therefore, we have $OC = OB$, $\angle BOA = \angle OCD$, and $CD = OA$, so $\triangle OCD \cong \triangle BOA$ by SAS Congruence.

(b) Our triangles from part (a) are congruent isosceles triangles, so $\angle ODC = \angle COD = \angle OAB = \angle OBA$. Since $\angle OAB = \angle COD$, we have $\overline{OC} \parallel \overline{AB}$.

(c) Since $\angle OCB$ is an exterior angle of $\triangle CDO$, we have $\angle OCB = \angle COD + \angle CDO$. We also have $\angle COD = \angle CDO = \angle OBA$, so $\angle OCB = \angle COD + \angle CDO = \angle OBA + \angle OBA = 2\angle OBA$.

(d) Let $x = \angle OBA$. Our previous part gives us $\angle OCB = 2x$. Since $OB = OC$, we have $\angle OBC = \angle OCB = 2x$ as well. Finally, our parallel lines give us $\angle OCB + \angle CBA = 180°$, so $2x + 3x = 180°$ and $x = \boxed{36°}$.

3.32 Let $x = \angle ADE$. Since $AD = AE$, $\angle AED = x$. Then $\angle BDE = 180° - \angle ADE = 180° - x$ and $\angle AEC = \angle 180° - \angle AED = 180° - x$, so by SAS, triangles BDE and AEC are congruent. Therefore, $AC = BE$.

3.33

(a) Since $OC = OB$, we have $\angle OCB = \angle OBC$. From $\triangle OBC$ we have $180° = 42° + \angle OCB + \angle OBC = 42° + 2\angle OCB$, so $\angle OCB = \angle CBO = (180° - 42°)/2 = \boxed{69°}$.

(b) Let $\angle BCD = x$, so $\angle OAB = 2x$. Since $OB = OA$, we have $\angle OBA = \angle OAB = 2x$. Therefore, $\triangle OAB$ gives us $\angle BOA = 180° - 4x = 180° - 4\angle BCD$.

(c) Again we let $\angle BCD = x$, so $\angle BOA = 180° - 4x$ from the previous part. Since $\overline{CD} \parallel \overline{OA}$, we have

$$\begin{aligned} 180° &= \angle COA + \angle OCD \\ &= \angle AOB + \angle BOC + \angle DCB + \angle BCO \\ &= 180° - 4x + 42° + x + 69°. \end{aligned}$$

Therefore, $x = 37°$, so $\angle BOA = 180° - 4x = \boxed{32°}$.

3.34 Since $\overline{AB}$ and $\overline{XY}$ are diameters, they intersect at O. Therefore, $AO = XO$ and $BO = YO$ since all four are radii of the circle. Finally, $\angle AOX = \angle BOY$ (vertical angles), so we have $\triangle AOX \cong \triangle BOY$ by SAS Congruence. As corresponding sides of these triangles, we have $AX = BY$.

3.35 Let $x = \angle ABC$ and $y = \angle ACB$. Since $MA = MB$, $\angle MAB = x$. Similarly, since $MA = MC$, $\angle MAC = y$.

The angles of a triangle sum to $180°$, so $\angle BAC + \angle ABC + \angle ACB = \angle BAM + \angle CAM + \angle B + \angle C = x + y + x + y = 2x + 2y = 180°$. Hence, $\angle BAC = x + y = 90°$.

3.36

(a) Since $\angle A = 90°$, $\angle AEH = 90° - \angle AHE$. Since $\angle AEH$, $\angle FEH$, and $\angle BEF$ together make a straight angle, we have

$$\angle AEH + \angle FEH + \angle BEF = 180°.$$

Since $EFGH$ is a square, we know $\angle FEH = 90°$. So, we have $\angle AEH = 90° - \angle BEF$. Equating this to our earlier expression for $\angle AEH$ gives $\angle BEF = \angle AHE$.

(b) From square $EFGH$, we have $EF = FG = GH = HE$. From square $ABCD$, we have $\angle A = \angle B = \angle C = \angle D$. From part (a) (and similar arguments), we have $\angle AHE = \angle BEF = \angle CFG = \angle DGH$. Therefore, $\triangle AHE \cong \triangle BEF \cong \triangle CFG \cong \triangle DGH$ by AAS Congruence.

3.37

(a) We follow the very similar problem in the text as a guide. Since $AE = DC = BF$, $EB = AD = CF$, and $AB = AC = CB$, we have $\triangle AEB \cong \triangle CDA \cong \triangle BFC$ by SSS Congruence. Therefore, $\angle EAB = \angle DCA = \angle FBC$ and $\angle EBA = \angle DAC = \angle FCB$. This, combined with $\angle BAC = \angle ABC = \angle ACB$, gives us

$$\angle EAD = 360° - \angle EAB - \angle BAC - \angle DAC = 360° - \angle ACD - \angle ACB - \angle BCF = \angle DCF.$$

Similarly, we can show that $\angle EAD = \angle EBF = \angle DCF$. This, combined with $AE = DC = BF$ and $EB = CF = AD$, gives us $\triangle EAD \cong \triangle DCF \cong \triangle FBE$ by SAS Congruence. Therefore, $ED = DF = FE$, so $\triangle DEF$ is equilateral.

(b) Even if ABC were not fully inside $\triangle DEF$, $\triangle DEF$ would still be equilateral. The situation is depicted at right. As we did earlier, we can still use our side equalities to show that $\triangle AEB \cong \triangle CDA \cong \triangle BFC$ by SSS Congruence, then use the resulting angle equalities to show that $\triangle EAD \cong \triangle DCF \cong \triangle FBE$ by SAS Congruence. Therefore, $DE = EF = FD$, so $\triangle DEF$ is equilateral.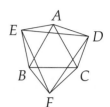

3.38 We are given that $DC = DE$ and $\angle CDA = \angle EDA$. We also have $AD = AD$, so $\triangle CDA \cong \triangle EDA$ by SAS. Therefore, $AC = AE$, and $\triangle ABC \cong \triangle AFE$ by SSS. Therefore, $\angle ABC = \angle EFA$.

3.39

(a) Since $\angle MNP = \angle ONP$, $NP = NP$, and $\angle MPN = \angle NPO$, we have $\triangle MNP \cong \triangle ONP$ by ASA Congruence. Therefore, $NO = MN$ since they are corresponding sides of these two triangles.

(b) Nothing in the problem implies $NO = OP$. Indeed, the given diagram shows a counterexample. We can easily make $\overline{NO}$ and $\overline{OP}$ have different lengths and still satisfy the problem.

(c) Since $\angle MPN$ and $\angle PNO$ do not need to have the same measure, we cannot conclude that $\overline{MP} \parallel \overline{NO}$.

(d) Let $\overline{MO}$ meet $\overline{NP}$ at X. Since $MP = PO$ (from our congruent triangles in the first part), $\angle MPX = \angle OPX$, and $PX = PX$, we have $\triangle MPX \cong \triangle OPX$. Therefore, $\angle MXP = \angle OXP$. Since these angles must also sum to $180°$, they must both be right angles. Therefore, we have $\overline{MO} \perp \overline{NP}$.

3.40

(a) Let $\angle DCB = x$. From our given information, we have $\angle ACD = 2x$. From right triangle $\triangle ADC$, we have $\angle DAC = 90° - 2x$. Since $\angle BAC = \angle DAC$, we have $\angle BAC = 90° - 2x$. Since $AB = AC$, we have $\angle ABC = \angle ACB = 2x + x = 3x$. We also could have noted that right triangle $\triangle DBC$ gives us $\angle ABC = \angle DBC = 180° - \angle BDC - \angle DCB = 90° - x$, and we could note that $\triangle ABC$ gives us $\angle BAC = 180° - \angle B - \angle ACB = 180° - 6x$.

(b) In the previous part, we found that $\angle ABC = 3x$ and $\angle ABC = 90° - x$. Equating these two expressions gives $4x = 90°$, so $x = 22.5°$. Therefore, $\angle ABC = 3x = \boxed{67.5°}$. (Alternatively, we could have used $\angle ABC = 3x$ and the angles of $\triangle ABC$ to write $3x + 3x + (90° - 2x) = 180°$, so $x = 22.5°$.)

Challenge Problems

3.41 There are 3 possible cases:

(a) *The two angles given are equal.* In this case, we have $3x + 4° = x + 17°$, so $x = 6\frac{1}{2}°$.

(b) *There are two angles with measure $3x + 4°$.* The sum of the three angles of the triangle must be $180°$, so we have $2(3x + 4°) + x + 17° = 180°$. Therefore, $x = 22\frac{1}{7}°$.

(c) *There are two angles with measure $x + 17°$.* The sum of the three angles of the triangle must be $180°$, so we have $3x + 4° + 2(x + 17°) = 180°$. Therefore, $x = 28\frac{2}{5}°$.

Therefore, the possible values for x are $\boxed{6\frac{1}{2}°, 22\frac{1}{7}°, 28\frac{2}{5}°}$.

3.42 We have $CA = AH = AB = BD = BC = CF$. Furthermore, $\angle CAH = \angle ABD = \angle BCF = 90°$, so by SAS, triangles CAH, ABD, and BCF are congruent. Therefore, $AD = BF = CH$.

Now, $\angle EAH = 360° - \angle EAB - \angle BAC - \angle CAH = 360° - 90° - 60° - 90° = 120°$, and $\angle DAE = 45°$ (from isosceles triangle $\triangle DAE$). Therefore, $\angle DAH = \angle DAE + \angle EAH = 45° + 120° = 165°$. In the same way, we can show that $\angle FBD = \angle HCF = 165°$.

Therefore, again by SAS, triangles DAH, FBD, and HCF are congruent. Then $DH = FD = HF$, so triangle DHF is equilateral.

3.43 Note that $\angle EAD = \angle EAB + \angle BAD = 60° + 90° = 150°$. We also have $\angle DCF = \angle DCB + \angle BCF = 60° + 90° = 150°$, and $\angle EBF = 360° - \angle ABE - \angle ABC - \angle CBF = 360° - 60° - 90° - 60° = 150°$.

Since $\angle DAB + \angle ABC = 180°$, $\overline{AD} \parallel \overline{BC}$. Therefore, when we draw $\overline{BD}$, we have $\angle ADB = \angle DBC$. Similarly, $\overline{AB} \parallel \overline{DC}$, so $\angle ABD = \angle CDB$. Therefore, $\triangle BAD \cong \triangle DCB$ by ASA Congruence, which gives us $AD = BC$ and $AB = CD$. (This property of rectangles will be explored in the chapter on quadrilaterals.) Now we have $EA = EB = AB = CD$, and $AD = BC = BF = CF$. Therefore, by SAS, triangles EAD, DCF, and EBF are congruent. So, $ED = DF = FE$, which implies that triangle DEF is equilateral.

3.44 Since $AD = AC$, $\angle ACD = \angle ADC$. Also, $\angle ACD + \angle ADC + \angle CAD = 180°$, so $2\angle ACD = 180° - \angle CAD = 180° - \angle CAB$. Therefore, $\angle ACD = 90° - (\angle CAB)/2$. We also have $\angle CAB = 180° - \angle ACB - \angle ABC = 90° - \angle B$, so $\angle ACD = \angle ADC = 45° + \frac{\angle B}{2}$. Then, $\angle BCD = \angle BCA - \angle ACD = 90° - \angle ACD = 45° - \frac{\angle B}{2}$. From isosceles triangle $\triangle BCE$, we have $\angle BCE = \angle BEC = (180° - \angle B)/2 = 90° - \frac{\angle B}{2}$, so $\angle ACE = 90° - \angle BCE = \frac{\angle B}{2}$. Therefore,

$$\angle DCE = 90° - \angle ACE - \angle BCD = 90° - \frac{\angle B}{2} - \left(45° - \frac{\angle B}{2}\right) = \boxed{45°}.$$

3.45 First, $\angle ADE = \angle EDF = \angle FDC$, and $\angle ADE + \angle EDF + \angle FDC = \angle ADC = 90°$. Therefore, $\angle ADE = \angle EDF = \angle FDC = 30°$. By SAS, $\triangle ADE \cong \triangle EDF \cong \triangle FDC$.

Since $DA = DE$, $\angle DAE = \angle DEA$. But $\angle DAE + \angle DEA = 180° - \angle ADE = 150°$, so $\angle DAE = \angle DEA = 150°/2 = 75°$. Hence, $\angle BAE = 90° - \angle DAE = 15°$.

Now, note that $DA = DF$ and $\angle ADF = \angle ADE + \angle EDF = 60°$. Therefore, triangle ADF is equilateral, so $AF = AD$ and $\angle FAD = 60°$. Then $\angle EAF = \angle EAD - \angle FAD = 75° - 60° = 15°$. Therefore, $\angle BAF = \angle BAE + \angle EAF = 15° + 15° = 30°$. Since $AB = AF = DA = DE$, by SAS, triangles ADE and BAF are congruent.

Hence, $BF = AE$. But $AE = EF$, so $BF = EF$. Similarly, we can show $BE = EF$. Therefore, triangle BEF is equilateral. (We could also have shown $BE = EF$ by noticing that $\triangle AEF \cong \triangle AEB$ by SAS.)

3.46 Let $x = \angle BAC$. Since $EF = FA$, we have $\angle FEA = x$, so $\angle EFA = 180° - \angle FAE - \angle FEA = 180° - 2x$. Therefore, $\angle EFC = 180° - \angle EFA = 2x$.

Since $EF = EC$, we have $\angle ECF = 2x$, so $\angle FEC = 180° - \angle EFC - \angle ECF = 180° - 4x$. Then $\angle CEB = 180° - \angle CEF - \angle FEA = 3x$. Since $CB = CE$, $\angle CBE = 3x$. By a similar argument, $\angle BCG = 3x$. Then

$$\angle BAC + \angle ABC + \angle ACB = x + 3x + 3x = 180°, \text{ so } x = \boxed{25\tfrac{5°}{7}}.$$

3.47 Since $\overline{ED} \parallel \overline{YZ}$, we have $\angle DZE = \angle EZY = \angle ZED$, so $\triangle DEZ$ is isosceles with $DE = DZ$. Similarly, our parallel segments give us $\angle YED = \angle AYE = \angle CYE$. Therefore, $\triangle CEY$ is isosceles with $CE = CY$.

Putting $DE = DZ$ and $CE = CY$ together with $CD = DE - CE$ gives us $CD = DZ - CY$, as desired.

3.48 Suppose we are given the lengths of sides $\overline{XY}$ and $\overline{YZ}$ and the measure of $\angle X$. In the text we have already seen one case in which there's only one possible triangle $\triangle XYZ$ that fits these conditions – when the circle with center Y and radius YZ is tangent to the ray originating at X that forms the given $\angle X$ with side $\overline{XY}$. As you'll learn later in the text, this will occur when $\angle Z$ is a right angle.

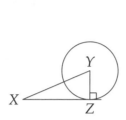 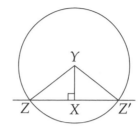

Another case in which there is one possible triangle is when $\angle X$ is 90°. When this occurs, as in the second diagram in the figure above, the circle drawn with center Y and radius YZ hits the line through X perpendicular to $\overline{XY}$ in two points, Z and Z'. Triangles $\triangle XYZ$ and $\triangle XYZ'$ are mirror images of each other and are therefore congruent. Hence, if the A of SSA is a 90° angle, then SSA *is valid*. You'll learn this in a later chapter as HL Congruence.

Finally, SSA Congruence 'works' when all four sides in question are equal in length. In this case, each triangle is isosceles, and each has another base angle equal to the angles we already know are equal. We can then see that the triangles are congruent by AAS Congruence.

3.49 Let $\overline{AY}$ and $\overline{XZ}$ meet at B, and $\overline{ZD}$ and $\overline{AY}$ meet at C. If we can prove that $\triangle BCZ \cong \triangle YCZ$, then we can conclude that $\overline{ZD} \perp \overline{AY}$. Since $AX = XY$, we have $\angle XYA = \angle A$. We have $\angle ZBC = \angle ABX = 90° - \angle A$ (from right triangle $\triangle BXA$) and $\angle ZYC = \angle ZYX - \angle AYX = 90° - \angle AYX = 90° - \angle A$. Therefore, $\angle ZBC = \angle ZYC$. 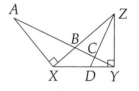 Together with $\angle YZC = \angle BZC$ (given) and $ZC = ZC$, we have $\triangle ZBC \cong \triangle ZYC$ by AAS Congruence. Therefore, $\angle BCZ$ and $\angle YCZ$ are congruent and add to 180°, so they must both be right angles, which means $\overline{ZD} \perp \overline{AY}$.

3.50 Since $RB = AB$, $\angle RBC = \angle RBA + \angle ABC = 60° + \angle ABC = \angle PBC + \angle ABC = \angle PBA$, and $BC = PB$, we have $\triangle RBC \cong \triangle ABP$ by SAS Congruence. Therefore, $RC = AP$. Similarly, we have $AB = RA$, $\angle BAQ = \angle BAC + \angle CAQ = \angle BAC + \angle RAB = \angle RAC$, and $AC = AQ$, so $\triangle RAC \cong \triangle BAQ$ by SAS, so $RC = BQ$. Therefore, $AP = BQ = CR$, as desired.

3.51 In the diagram, $AB = 13$, $BC = BD = 10$, and $\angle BAC = 40°$. Therefore, $\triangle ABC$ and $\triangle ABD$ are the two possible triangles that have sides of length 13 and 10 and an angle of 40° opposite the side of length 10. Let $\angle ACB = x$. Therefore, $\angle BDC = x$ and $\angle ADB = 180° - x$. From triangle $\triangle ADB$, we have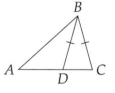

$$\angle ABD = 180° - \angle BAD - \angle ADB = x - 40°.$$

From triangle $\triangle ABC$, we have

$$\angle ABC = 180° - \angle BAC - \angle ACB = 140° - x.$$

These two angles are the possible $\angle B$ we seek, and their sum is $\boxed{100°}$.

CHAPTER 4

Perimeter and Area

Exercises for Section 4.1

4.1.1 Since $\angle Y = \angle Z$, we have $XZ = XY$, so $XZ = 5\sqrt{2}$. Therefore, the perimeter of $\triangle XYZ$ is $XY + YZ + XZ = \boxed{10 + 10\sqrt{2}}$.

4.1.2 Since D is the midpoint of $\overline{AC}$, equilateral triangle ADE has side length $AB/2 = 2$. Since G is the midpoint of $\overline{AE}$, equilateral triangle EFG has side length $AE/2 = 1$. Then the perimeter of figure $ABCDEFG$ is $AB + BC + CD + DE + EF + FG + GA = 4 + 4 + 2 + 2 + 1 + 1 + 1 = \boxed{15}$.

4.1.3

(a) Let triangles ABC and DEF be congruent. Then $AB = DE$, $AC = DF$, and $BC = EF$, so $AB+AC+BC = DE + DF + EF$. We conclude that congruent triangles must have equal perimeters.

(b) Two triangles with the same perimeter need not be congruent. For example, consider a triangle with side lengths 3, 4, and 5, and a second triangle with side lengths 4, 4, and 4. They both have a perimeter of 12, yet are clearly not congruent.

(c) No. For example, consider equilateral triangle ABC with side length equal to 2. Let M be on $\overline{AC}$ such that M is very close to A. Then, consider placing a point N on $\overrightarrow{AB}$ that we can slide along ray $\overrightarrow{AB}$. No matter where we put N, $\angle NAM = \angle BAC = 60°$. When N is close to A, the perimeter of $\triangle NAM$ is very small. As we move N farther and farther away from A, the perimeter of $\triangle MAN$ increases. In fact, we can make this perimeter as large as we want. At some point as we slide N from very close to A (where the perimeter of $\triangle MAN$ is less than 6) to a point very far from N (where the perimeter of $\triangle MAN$ is more than 6), we will hit a point at which the perimeter of $\triangle MAN$ is exactly 6. Such a triangle would not be congruent to our equilateral $\triangle ABC$. (Note that this is not a perfectly rigorous proof, which would require some tools we haven't yet discovered.)

4.1.4 We can simply add up the sides:

$$AB + BC + CD + DE + EF + FG + GH + HA = 6 + 3 + 5 + 4 + 3 + 2 + 4 + 9 = \boxed{36}.$$

Another approach is to notice that the total length along the 'tops' is the same as that of the bottom, 9, so the top and bottom together contribute 18. The 'left' and 'right' each contribute 6, but there's a little extra to add from part of the 'U' portion $CDEF$. We must add FE and its equivalent along $\overline{CD}$. This gives us an additional $2EF = 6$, for a total of $2(9) + 2(6) + 2(3) = 36$, as before.

4.1.5 Let the shortest side have length x. Then, the other two sides have lengths $1.5x$ and $2x$. The perimeter of the triangle is 45, so $45 = x + 1.5x + 2x$, or $45 = 4.5x$. Solving this equation, we find $x = 10$, which means the shortest side has length $\boxed{10}$.

Exercises for Section 4.2

4.2.1

(a) The area of right triangle $\triangle ABC$ is $(AB)(BC)/2 = (4)(7)/2 = \boxed{14}$.

(b) The altitude to base $\overline{DF}$ from E has length 4.8, so the area of $\triangle DEF$ is $(4.8)(10)/2 = \boxed{24}$.

(c) The area of rectangle $GHIJ$ is $(14)(8) = \boxed{112}$.

4.2.2 Since all of the sides of a square are equal, each side has length $36/4 = 9$. The area of the square is $9^2 = \boxed{81}$.

4.2.3

(a) $\overline{KO}$ is the altitude to side $\overline{LM}$, so $[KLM] = (KO)(LM)/2 = (6)(7)/2 = \boxed{21}$.

(b) $\overline{LN}$ is the altitude to side $\overline{KM}$, so $[KLM] = (LN)(KM)/2$. From the first part, we have $[KLM] = 21$, so $LN = (2[KLM])/KM = (42)/12 = \boxed{7/2}$.

4.2.4 Each wall has area $(24)(8) = 192$ square feet, so we need enough paint to cover $2(192) = 384$ square feet. Since each gallon covers 80 square feet, we need $384/80 = \boxed{4.8}$ gallons.

4.2.5 Let the side of the square poster be x, in inches. Then the area of the square poster is x^2, and the area of the rectangular poster is $(x + 2)(x - 2) = x^2 - 4$, so their positive difference is $\boxed{4 \text{ in}^2}$.

4.2.6 Let x and y be the side length of the garden and the width of the path, respectively, in meters. The perimeter of the garden is $4x = 64$, so $x = 16$.

Solution 1: The slickest approach is to notice that since the garden is a square with side length 16, its area is $16^2 = 256$. Therefore, the area of the garden and the path together is $256 + 228 = 484$. Since the outside boundary of the path forms a square, the total area bounded by the path and the garden is the square of the length of one outer side of the path. Therefore, each outer side of the path has length $\sqrt{484} = 22$. So, the path requires $4(22) = \boxed{88}$ meters of fencing.

Solution 2: In terms of x and y, the side length of the square containing the garden and the path is $x + 2y = 2y + 16$. The area of the path is then $(2y + 16)^2 - 16^2 = 4y^2 + 64y + 256 - 256 = 4y^2 + 64y$. We are given that this equals 228, so $4y^2 + 64y = 228$. Rearranging this gives $y^2 + 16y - 57 = 0$, so $(y + 19)(y - 3) = 0$. Since y is positive, $y = 3$.

The amount of fencing needed to surround the path is the perimeter of the big square, which is $4(2y + 16) = \boxed{88}$ meters.

Exercises for Section 4.3

4.3.1

(a) We have

$$[PQR] = \frac{(PR)(QV)}{2} = \boxed{18}$$

$$[PTQ] = \frac{(PT)(QV)}{2} = \boxed{12}$$

$$[QTR] = \frac{(TR)(QV)}{2} = \boxed{6}$$

(b) From the given information, we have $TR/PT = \boxed{1/2}$, and from the first part, we have $[QTR]/[PTQ] = 6/12 = \boxed{1/2}$. After reading this section, you should not be surprised that these two are equal.

4.3.2

(a) Since $\triangle ACD$ and $\triangle BCD$ share base $\overline{CD}$, the ratio of their areas is the ratio of their altitudes to $\overline{CD}$. Since $AE = 2 + 3 = 5$ and $BE = 3$, we have $[BCD]/[ACD] = BE/AE = \boxed{3/5}$.

(b) $ADBC$ is what is leftover after we cut $\triangle BCD$ out of $\triangle ACD$. In the first part, we found that $[BCD]$ is $3/5$ of $[ACD]$. Therefore, the 'leftover' $[ADBC]$ is the remaining $2/5$ of $[ACD]$, so $[ADBC]/[ADC] = \boxed{2/5}$.

(c) We can simply take the ratio of the answers to our first two parts to find:

$$\frac{[BCD]}{[ADBC]} = \frac{[BCD]/[ACD]}{[ADBC]/[ACD]} = \frac{3/5}{2/5} = \boxed{\frac{3}{2}}.$$

You shouldn't be too surprised to see that $[BCD]/[ADBC] = BE/AB$. See if you can use the Same Base/Same Altitude principle to prove this will be true no matter what AB and BE are.

4.3.3 Jean's triangle has a base of 8 feet and a height of 10 feet, so it has area $(8)(10)/2 = 40$ ft^2. George's triangle has a base of 10 feet and a height of 10 feet, so it has area $(10)(10)/2 = 50$ ft^2. Therefore, Jean is correct: George's triangle will use more paint than hers. We could have seen this without computing the areas by noting that the heights of the two triangles are the same, but George's triangle has a longer base than Jean's.

4.3.4 *Solution 1:* Let the triangle have base b and height h, so it has area $bh/2$. If the base is doubled to $2b$ and the height is tripled to $3h$, then the area of the new triangle is $(2b)(3h)/2 = 3bh$. Therefore, the area increases by a factor of

$$\frac{3bh}{\frac{1}{2}bh} = \boxed{6}.$$

Solution 2: Doubling the base multiplies the area by two and tripling the height multiplies the area by 3. Doing both of these will therefore multiply the area by $2 \cdot 3 = \boxed{6}$.

4.3.5 First, note that $[WXYZ] = (WX)(XY)$, while $[WPX] = (WX)(PX)/2 = (WX)(XY)/4$ (since the height from P to $\overline{WX}$ equals $XY/2$). Therefore, $[WPX] = [WXYZ]/4$. We are also given $[PQX] = [WXYZ]/6$. We therefore have $[PQX]/[WPX] = ([WXYZ]/6)/([WXYZ]/4) = 2/3$. Since $[PQX]$ and $[WPX]$ share an altitude from P, we have $[PQX]/[WPX] = QX/WX$. Therefore, $QX/WX = 2/3$. Since $WX = 8$, we have $QX = \boxed{16/3}$.

Review Problems

4.13

(a) $[ABC] = (AB)(AC)/2 = \boxed{54}$.

(b) $[DEF] = (6)(11)/2 = \boxed{33}$.

(c) $[GHIJ] = (GH)(HI) = \boxed{96}$.

(d) $[KLMN] = (KL)(KN) = (KN)^2 = \boxed{196}$.

(e) $[OQS] = (QS)(OP)/2 = \boxed{12}$.

4.14 Since the area of a triangle is (base)(height)/2, we have height = 2(Area)/(base) = 2(42)/7 = $\boxed{12}$. We cannot find the length of either of the other two sides of the triangle, nor can we find the sum of the lengths of these sides. Therefore, it is not possible to find the perimeter of this triangle without more information.

4.15 Let the side length of the square be s. Then, the area is s^2. Since the area of the square is 75, we have $s^2 = 75$. Solving $s^2 = 75$ gives $s = \sqrt{75} = 5\sqrt{3}$ as the side length of the square. There are four such sides, so the perimeter is $4(5\sqrt{3}) = \boxed{20\sqrt{3}}$.

4.16 Since $[ABC] = 24$ and $[ABC] = (AB)(BC)/2$, we have $BC = 2[ABC]/(AB) = 2(24)/6 = 8$. Therefore, the perimeter of $\triangle ABC$ is $6 + 8 + 10 = \boxed{24}$. Similarly, since $[XYZ] = (XY)(YZ)/2$, we have $YZ = 2[XYZ]/(XY) = 2(30)/5 = 12$, so the perimeter of $\triangle XYZ$ is $5 + 12 + 13 = \boxed{30}$.

4.17 Let the width of the rectangle be w, so the length is $4w - 2$. In terms of w, the perimeter then is $2(w) + 2(4w - 2) = 10w - 4$. Since the perimeter of the rectangle is given as 51, we have $10w - 4 = 51$, so $w = 5.5$. Therefore, the length is $4w - 2 = 20$, so the area is $(5.5)(20) = \boxed{110}$.

4.18 Since I want to leave a 2 foot space between the edge of the carpet and the wall, the region we want to carpet is only 20 feet by 12 feet (since we subtract 2 from all sides of the room, we subtract 4 from both dimensions). Therefore, the area of the rug is $(20)(12) = 240$ ft^2. Since the carpet costs \$2.50 per square foot, the room will cost $(240)(\$2.50) = \boxed{\$600}$ to carpet.

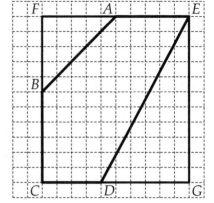

4.19   We can find the area of the figure by viewing it as a rectangle with a couple right triangles cut out. The rectangle is $FEGC$ as shown. We see that $FE = 10$ and $FC = 11$, so the area of $FEGC$ is $(11)(10) = 110$. We then find $[AFB] = (AF)(FB)/2 = 12.5$ and $[EDG] = (EG)(DG)/2 = (6)(11)/2 = 33$. Therefore, the area of $ABCDE$ is $110 - 12.5 - 33 = \boxed{64.5}$.

4.20   Let x and y be the dimensions of the rectangle, in centimeters. Then the perimeter of the rectangle is $2(x + y) = 28$, so $x + y = 14$, and the area of the rectangle is $xy = 48$. Substituting $y = 14 - x$ into $xy = 48$, we get $x(14 - x) = 48$, so $x^2 - 14x + 48 = (x - 6)(x - 8) = 0$. The solutions are $x = 6$ and $x = 8$, which lead to $y = 8$ and $y = 6$, respectively. Therefore, the rectangle is $\boxed{6 \text{ cm by } 8 \text{ cm}}$. (Notice that we could also have used trial-and-error to solve the equations $x + y = 14$ and $xy = 48$.)

4.21   *Solution 1:* Let the base and height of the original triangle be b_1 and h, respectively. Then the height of the new triangle is $4h$. Let the base of the new triangle be b_2. Then

$$\frac{1}{2}b_1 h = \frac{1}{2}b_2 \cdot 4h,$$

so $b_2 = b_1/4$. Hence, the base must be $\boxed{\text{divided by } 4}$ to leave the area unchanged.

Solution 2: Multiplying the height by 4 multiplies the area by 4. We must then divide the base by 4, which divides the area by 4, in order to leave the area unchanged.

4.22   Let the common altitude be h. Since $[ABC] = (h)(BC)/2 = (h)(AC)/2$, we have $BC = AC$.

4.23

(a)   Since $[ABCD] = (CD)(BC)$ and $[BCD] = (CD)(BC)/2$, we have $[BCD] = [ABCD]/2 = 36/2 = \boxed{18}$.

(b) $\triangle BCD$ and $\triangle BED$ have the same altitude from B. Therefore, the ratio of their areas is the ratio of the bases to which this altitude is drawn: $[BED]/[BCD] = DE/DC$. Since $DE = 2EC$ and $EC + DE = DC$, we have $DE = 2DC/3$, so $DE/DC = 2/3$. Therefore, $[BED]/[BCD] = DE/DC = 2/3$, so $[BED] = (2/3)[BCD] = \boxed{12}$.

4.24   The perimeter of the star is the total perimeter of all the outer triangles minus the perimeter of $BDFHJ$, since the latter is included in the sum of the triangles, but not in the star. Therefore, the perimeter of the star is $(5)(23) - 51 = \boxed{64}$.

4.25   Let E be the point on $\overline{AB}$ where the two triangles meet. Then the area of the shaded region is

$$[ADE] + [BCE] = \frac{1}{2}AE \cdot AD + \frac{1}{2}BE \cdot BC = \frac{1}{2}AE \cdot 8 + \frac{1}{2}BE \cdot 8 = 4AE + 4BE = 4AB = \boxed{48 \text{ cm}^2}.$$

4.26 $\triangle WOZ$ and $\triangle WYZ$ share an altitude from Z, so $WO/WY = [WOZ]/[WZY] = 12/30 = 2/5$. We also have $[WXY] = [WXYZ] - [WZY] = 20$. Since $[WOX]/[WXY] = WO/WY = 2/5$, we have $[WOX] = (2/5)[WXY] = \boxed{8}$.

Challenge Problems

4.27 We split *ABCD* into a rectangle and a right triangle by drawing $\overline{CE}$ as shown. The area of *ABCD* is the sum of the areas of the pieces we thus form: $[ABCD] = [ABCE] + [CED] = (AB)(BC) + (CE)(ED)/2 = 30 + 9 = \boxed{39}$.

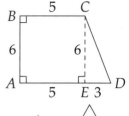

4.28 The gardener requires a post every 6 feet, so he needs $96/6 = \boxed{16}$ posts.

4.29 The key to this problem is realizing that for each segment, we divide it into three smaller segments and end up with four copies of the smaller segment. Hence, the perimeter increases by a factor of 4/3 with each step. Therefore, the second figure has perimeter $(4/3)(12) = \boxed{16}$ and the third has perimeter $(4/3)(16) = \boxed{64/3}$.

4.30 After each step, the perimeter of the shape is 4/3 the perimeter from the previous step. Therefore, the perimeter grows and grows without bound – the perimeter is infinite. However, we can contain the whole figure on the page (play with it and see). How can this be?!? (If you aren't convinced, look up Koch's snowflake on the internet.)

4.31 Label the vertices *A*, *B*, *C*, *D*, *E*, *F*, *G*, *H*, *I*, and *J*, as shown.

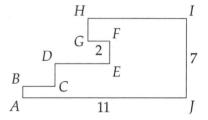

The sum of the vertical lengths *AB*, *CD*, *EF*, and *GH* must be equal to $IJ = 7$. When we sum the horizontal lengths *BC*, *DE*, *FG*, and *HI*, the total includes the length *AJ* plus the length *FG* two more times, so $BC + DE + FG + HI = AJ + 2 \cdot 2 = 15$.

Therefore, the perimeter is $7 + 7 + 11 + 15 = \boxed{40}$.

4.32 We find areas in steps. Since $BX/CX = 1/2$, we have $BX/BC = 1/3$. Therefore, $[ABX] = [ABC]/3 = 144/3 = 48$, since $[ABX]$ and $[ABC]$ share an altitude from *A*. Since $BY/AY = 1/3$, we have $BY/BA = 1/4$. Since $[BYX]/[ABX] = BY/AB = 1/4$, we have $[BYX] = 48/4 = \boxed{12}$.

4.33 Let x be the side of the smallest square. Then the side of the second smallest square is $2x$, the second largest, $4x$, and the largest, $8x$. Then the perimeter of the figure, starting at the lower left corner and going clockwise, is $x + x + x + 2x + 2x + 4x + 4x + 8x + 8x + 8x + 4x + 2x + x = 46x = 115$. Therefore, $x = 115/46 = 2.5$. Then the area of the figure is $x^2 + (2x)^2 + (4x)^2 + (8x)^2 = 85x^2 = \boxed{531.25}$.

4.34 Let *K* denote the area of triangle *ABC*. Let h_a and h_c denote the height from *A* to $\overline{BC}$ and from *C* to $\overline{AB}$, respectively. Then

$$K = \frac{1}{2}BC \cdot h_a = \frac{1}{2}AB \cdot h_c.$$

(a) We are given that

$$\frac{1}{2}AB \cdot h_a = 12 \quad \text{and} \quad \frac{1}{2}BC \cdot h_c = 27.$$

Therefore,

$$\frac{1}{2}AB \cdot h_a \cdot \frac{1}{2}BC \cdot h_c = 12 \cdot 27 = 324.$$

But

$$\frac{1}{2}AB \cdot h_a \cdot \frac{1}{2}BC \cdot h_c = \frac{1}{2}AB \cdot h_c \cdot \frac{1}{2}BC \cdot h_a = K^2,$$

so $K^2 = 324$, and $K = \sqrt{324} = \boxed{18}$.

(b) By the same calculations as in part (a), $K^2 = 120 \cdot 150 = 18000$, so $K = \sqrt{18000} = \boxed{60\sqrt{5}}$.

4.35 The two triangles share the same base in $\overline{PW}$, so all we have to do is to show that the altitude from X in $\triangle XPW$ has the same length as the altitude from Z in $\triangle ZPW$. These two altitudes are shown: $\overline{XB}$ and $\overline{ZA}$, respectively. These are also altitudes of triangles $\triangle XWY$ and $\triangle ZYW$, respectively. Since $\triangle WYZ$ and $\triangle WXY$ have the same area (each is half $[WXYZ]$ because the triangles are congruent by SAS or ASA), we have $(WY)(ZA)/2 = [ZWY] = [WXY] = (WY)(XB)/2$. Therefore, $ZA = XB$, so $[WPX] = [WPZ]$.

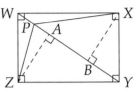

4.36 The area of triangle ABC is given by

$$\frac{1}{2}AB \cdot h_c = h_c,$$

where h_c is the distance from C to line $\overleftrightarrow{AB}$. The point on the circle farthest away from line AB is the point $(3,3)$. Hence, the maximum possible area of triangle ABC is $\boxed{3}$.

4.37 Triangle $\triangle AXF$ and $\triangle BXF$ share an altitude from X, so $[AXF]/[BXF] = AF/FB$. Similarly, $[AFC]/[BFC] = AF/FB$. Therefore, we have:

$$\frac{[AXC]}{[BXC]} = \frac{[ACF] - [AXF]}{[BCF] - [BXF]}$$
$$= \frac{(AF/FB)[BCF] - (AF/FB)[BXF]}{[BCF] - [BXF]}$$
$$= (AF/FB)\left(\frac{[BCF] - [BXF]}{[BCF] - [BXF]}\right) = \frac{AF}{FB}.$$

4.38 Since $[PQRS] = (PS)(SR)$, $[PSB] = (SB)(PS)/2$, and $[PBS]/[PQRS] = 1/4$, we have

$$\frac{(SB)(PS)/2}{(PS)(SR)} = \frac{1}{4},$$

from which we find that $SB = SR/2$. Therefore, B is the midpoint of $\overline{RS}$, so $BR = RS/2$. Similarly, we find that $[PAQ] = (PQ)(AQ)/2$, $[PQRS] = (PQ)(QR)$, and $[PQA]/[PQRS] = 1/5$, so

$$\frac{(PQ)(AQ)/2}{(PQ)(QR)} = \frac{1}{5}.$$

Therefore, $AQ = \frac{2}{5}RQ$, so $AR = \frac{3}{5}RQ$.

Combining these, we have $[ABR] = (AR)(BR)/2 = \frac{3}{20}(RS)(RQ)$. Since $[PQRS] = (RS)(RQ)$, we have $[ABR]/[PQRS] = 3/20$.

Finally, since $[ABP] = [PQRS] - [PBS] - [RAB] - [PQA]$, we have

$$\frac{[ABP]}{[PQRS]} = 1 - \frac{[PBS]}{[PQRS]} - \frac{[RAB]}{[PQRS]} - \frac{[PQA]}{[PQRS]} = \boxed{\frac{2}{5}}.$$

4.39 Since $[CBE] = (CB)(EC)/2$, $[ADE] = (AD)(DE)/2$, and $[ABE] = (AB)(AD)/2$ (since the altitude from E to $\overline{AB}$ is the same as the distance from D to $\overline{AB}$), our given equation becomes

$$\frac{(CB)(EC)}{2} - \frac{(AD)(DE)}{2} = \frac{(AB)(AD)}{2} - \frac{(CB)(EC)}{2}.$$

Since $AD = BC$, we have

$$\frac{(AD)(EC)}{2} - \frac{(AD)(DE)}{2} = \frac{(AB)(AD)}{2} - \frac{(AD)(EC)}{2},$$

or $EC - DE = AB - EC$. This gives us $2EC - DE = AB$. Since $AB = CD = DE + EC$, we have $2EC - DE = DE + EC$, or $EC = 2DE$. Therefore, $DE = CD/3 = AB/3$. Finally, we have

$$\frac{[ABE]}{[ADE]} = \frac{(AB)(AD)/2}{(DE)(AD)/2} = \frac{AB}{DE} = \boxed{3}.$$

4.40 Cut the cake into seven pieces as shown.

Note that the each piece has frosting four units wide on the side, so all we need to show now is that each piece has the same amount of frosting on the top (which also shows that the size of each piece is the same).

To see why this is the case, imagine instead that we cut the cake into 28 pieces, as shown in the second diagram. Each of these 28 pieces is a triangle with base 1 and height $7/2$, so each has area

$$\frac{1}{2} \cdot 1 \cdot \frac{7}{2} = \frac{7}{4}.$$

Therefore, each of the original seven pieces has area $4 \cdot 7/4 = 7$, so each has an equal amount of cake and frosting.

CHAPTER **5**

_____ **Similar Triangles**

Exercises for Section 5.1

5.1.1 If triangles ABC and YXZ are similar, then

$$\frac{AB}{YX} = \frac{AC}{YZ} = \frac{BC}{XZ}.$$

It follows that $AB/BC = YX/XZ$ and $AC \cdot YX = AB \cdot YZ$. Hence, statements (a), (b), and (d) are true. Statements (c) and (e) are not necessarily true.

5.1.2 If triangles ABC and ADB are similar, then $AB/AD = AC/AB$, so $AB^2 = AC \cdot AD$. We are given that $AC = 4$ and $AD = 9$, so $AB^2 = 4 \cdot 9 = 36$. Therefore, $AB = \boxed{6}$.

Exercises for Section 5.2

5.2.1

(a) Since $\overline{AB} \parallel \overline{DE}$, we have $\angle A = \angle D$ and $\angle B = \angle E$, so $\triangle ABC \sim \triangle DEC$ by AA Similarity. Therefore, $AC/CD = BC/CE = AB/DE = 7/21 = 1/3$, so $AC = CD/3 = \boxed{16/3}$ and $BC = CE/3 = \boxed{4}$.

(b) Since $\triangle FGI \sim \triangle FHJ$ by AA (because $\overline{GI} \parallel \overline{HJ}$ gives us $\angle FGI = \angle FHJ$ and $\angle FIG = \angle FJH$), we have $GI/HJ = FG/FH = 4/13$. Therefore, $HJ = GI(13/4) = \boxed{91/4}$.

(c) Since $\angle LOM = 90° - \angle NOM = \angle MNO$ and $\angle LMO = \angle OMN$, we have $\triangle LMO \sim \triangle OMN$ by AA Similarity. Therefore, $ON/LO = MN/OM = OM/LM = 1.2/1.6 = 3/4$, so $ON = OL(3/4) = \boxed{1.5}$ and $MN = OM(3/4) = \boxed{0.9}$.

(d) Since $\angle PQR = \angle PST$ and $\angle P$ equals itself, we have $\triangle PQR \sim \triangle PST$ by AA Similarity. So, $PQ/PS = PR/PT = 9/20$. Therefore, $PS = PQ(20/9) = 160/9$, so $RS = PS - PR = 160/9 - 9 = \boxed{79/9}$.

5.2.2 $\boxed{\text{Yes.}}$ Let the vertex angle of an isosceles triangle have measure x and each of the base angles have measure y. Then $x + 2y = 180°$, or $y = 90° - x/2$. Hence, if the vertex angles of each of two isosceles triangles have the same measure, x, then all four base angles of the two triangles have the same measure, $90° - x/2$. Therefore, the triangles have the same angle measures and must be similar.

5.2.3

(a) Since $WXYZ$ is a square, we have $\angle WZY = \angle ZYX = 90°$. Therefore, $\angle WZY + \angle ZYX = 180°$, so $\overline{WZ} \parallel \overline{XY}$.

(b) Since $\overline{WZ} \parallel \overline{XY}$, we have $\angle AZM = \angle MYB$ and $\angle MAZ = \angle MBY$. Furthermore, we are given $ZM = MY$, so $\triangle AZM \cong \triangle BYM$ by AAS Congruence. Therefore, $AZ = BY$.

(c) From the triangle congruence in the previous part, we have $AM = BM$, so $\triangle AMX \cong \triangle BMX$ by SAS Congruence. Therefore, $XA = XB$.

(d) Since $WXYZ$ is a square, all of its angles are right angles. Specifically, $\angle Z = \angle XYM = 90°$. Furthermore, since $\angle ZMY$ is a straight angle, we have $\angle AMZ + 90° + \angle XMY = 180°$, so $\angle AMZ = 90° - \angle XMY$. From right triangle $\triangle XMY$, we have $\angle YXM = 180° - \angle XYM - \angle XMY = 90° - \angle XMY$ also, so $\angle YXM = \angle AMZ$. Therefore, $\triangle AZM \sim \triangle MYX$ by AA Similarity. Hence, $AZ/ZM = MY/XY$. Since $WXYZ$ is a square, we have $XY = ZY$, so $ZM = MY = XY/2$. Therefore, we have

$$AZ = \frac{(ZM)(MY)}{XY} = \frac{(XY/2)(XY/2)}{XY} = \frac{XY}{4}.$$

5.2.4

(a) Since $AB = AC$, $\angle ABC = \angle ACB$. From $\triangle ABC$ we have $\angle ABC + \angle ACB = 180° - \angle BAC = 180° - 36° = 144°$, so $\angle ABC = \angle ACB = 144°/2 = 72°$. Since $\angle ABD = \angle CBD$ and $\angle ABD + \angle CBD = \angle ABC = 72°$, we have $\angle ABD = \angle CBD = 72°/2 = 36°$. Therefore, $\angle A = \angle DBC$ and $\angle ACB = \angle BCD$, so $\triangle ACB \sim \triangle BCD$ by AA Similarity.

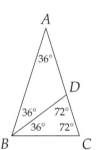

(b) Let $x = AB$. Then $AC = x$. From $\triangle ACB \sim \triangle BCD$ we have $\angle BDC = \angle ABC$. We also have $\angle ABC = \angle ACB$ and $\angle ACB = \angle BCD$, so $\angle BDC = \angle BCD$. Therefore, $\triangle BCD$ is isosceles, and $BD = BC = 1$. Also, $\angle ABD = \angle BAD$, so $AD = BD = 1$. Therefore, $CD = AC - AD = x - 1$. The similar triangles from part (a) give us $AB/BC = BC/CD$, so we have

$$\frac{x}{1} = \frac{1}{x-1},$$

or $x(x-1) = 1$. Rearranging gives $x^2 - x - 1 = 0$. By the quadratic formula, we have

$$x = \frac{1 \pm \sqrt{5}}{2}.$$

But $\frac{1 - \sqrt{5}}{2} < 0$ and x must be positive, so $AB = x = \boxed{(1 + \sqrt{5})/2}$.

5.2.5 Since $\overline{AB} \parallel \overline{CD}$, we have $\angle A = \angle DCE$ and $\angle ABE = \angle CDE$, so $\triangle ABE \sim \triangle CDE$ by AA Similarity. Therefore,

$$\frac{4}{x} = \frac{5}{5 + AC}.$$

Cross-multiplying gives $20 + 4AC = 5x$, so $AC = \frac{5x}{4} - 5$. Similarly, $\overline{FG} \parallel \overline{AB}$, so $\triangle FGH \sim \triangle ABH$, which gives

$$\frac{7}{7 + y + 5 + AC} = \frac{5}{x}.$$

Substituting for AC gives:

$$\frac{7}{7 + y + 5x/4} = \frac{5}{x}.$$

Cross-multiplying gives $7x = 35 + 5y + 25x/4$, so $3x/4 = 35 + 5y$, or $x = \boxed{140/3 + 20y/3}$.

Exercises for Section 5.3

5.3.1   Since $AB/AD = AC/AE$ and $\angle BAC = \angle DAE$, we have $\triangle ABC \sim \triangle ADE$ by SAS Similarity. Therefore $BC/DE = AB/AD = 3/5$, so $DE = (BC)(5/3) = \boxed{10}$.

5.3.2   Since M is the midpoint of $\overline{FG}$, we have $FM = MG$. Since F is the midpoint of $\overline{JM}$, we have $FJ = FM$. Therefore, $JM = 2FM = 2MG$. Similarly, $IM = 2MH$. So, $IM/MH = JM/MG = 2$ and $\angle GMH = \angle JMI$, so $\triangle IMJ \sim \triangle HMG$ by SAS Similarity. Therefore $\angle IJM = \angle HGM$, so $\overline{IJ} \parallel \overline{GH}$.

5.3.3   Since $WZ^2 = (WX)(WY)$, we have $WZ/WX = WY/WZ$. Also, $\angle ZWX = \angle YWZ$, so we have $\triangle WZX \sim \triangle WYZ$. This similarity gives the desired $\angle WZX = \angle WYZ$.

5.3.4

(a) $\angle QBC = \angle QBP = 90° - \angle QPB = 90° - \angle RPC = \angle PCR = \angle QCB$.

(b)   Because $\angle QBC = \angle QCB$, we have $QB = QC$. Since $QB = QC$ and $QA = QR$, we have $QB/QC = QA/QR$, so $\triangle QBC \sim \triangle QAR$ by SAS Similarity. Therefore, $\angle QCB = \angle QRA$ and $\overline{PB} \parallel \overline{RA}$ as desired.

Exercises for Section 5.4

5.4.1   We are given:

$$\frac{\text{Base length first triangle}}{\text{Leg length first triangle}} = \frac{\text{Base length second triangle}}{\text{Leg length second triangle}},$$

and a simple rearrangement turns this into

$$\frac{\text{Base length first triangle}}{\text{Base length second triangle}} = \frac{\text{Leg length first triangle}}{\text{Leg length second triangle}}.$$

Clearly, this equality holds with the other legs of the triangles, so the two triangles are similar by SSS Similarity. Therefore, the corresponding angles of the two triangles are the same. Specifically, the vertex angles of the two triangles are equal.

Exercises for Section 5.5

5.5.1   Since $\overline{XY} \parallel \overline{QR}$, we have $\angle PXY = \angle PQR$ and $\angle PYX = \angle PRQ$, so $\triangle PXY \sim \triangle PQR$ by AA Similarity. Therefore, $PY/PR = XY/QR = 1/3$, so $PY/(PY + 8) = 1/3$. Cross-multiplying gives $3PY = PY + 8$, so $PY = \boxed{4}$.

5.5.2

(a) The parallel lines in the diagram give us $\angle ECD = \angle FDB$ and $\angle EDB = \angle FBD$, so $\triangle FBD \sim \triangle EDC$. Since $[EDC] = 25[FBD]$ and the two triangles are similar, each side of $\triangle EDC$ is $\sqrt{25} = 5$ times each corresponding side of $\triangle FBD$. Therefore, $CD/DB = \boxed{5}$.

(b) Since $\overline{AB} \parallel \overline{ED}$, we have $\angle A = \angle CED$ and $\angle B = \angle EDC$, so $\triangle CAB \sim \triangle CED$. Since $CD/DB = 5$ from the first part, we have $CD/CB = CD/(CD + DB) = CD/(CD + CD/5) = 5/6$. Therefore, the ratio of corresponding sides in $\triangle CED$ and $\triangle CAB$ is $5/6$, and the ratio of their areas is $(5/6)^2 = \boxed{25/36}$.

(c) $\triangle EDC$ takes up $25/36$ of $\triangle ABC$. $[BDF] = [EDC]/25 = [ABC]/36$, so it takes up another $1/36$ of $\triangle ABC$, leaving $10/36 = 5/18$ of the area of $\triangle ABC$ for $AEDF$. The parallel lines give us $\angle EFD = \angle AEF$ and $\angle AFE = \angle FED$, which combined with $EF = EF$ gives $\triangle AEF \cong \triangle DFE$ by ASA. Therefore, $[AEF] = [AEDF]/2$. Since $[AEDF]$ is $5/18$ of $[ABC]$, we have $[AEF] = (5/36)[ABC]$, or $[AEF]/[ABC] = \boxed{5/36}$.

5.5.3

(a) Since $\overline{ZA} \parallel \overline{WX}$, we have $\angle AZC = \angle CXW$ and $\angle CWX = \angle CAZ$, so $\triangle CAZ \sim \triangle CWX$ by AA Similarity. Therefore, $ZC/XC = AC/CW$.

(b) Following essentially the same logic as in the previous part, we have $\triangle DXB \sim \triangle DZW$, so $XD/ZD = DB/WD$.

(c) Since $ZC = XD$, we also have $ZD = ZC + CD = XD + CD = XC$, so $ZC/XC = XD/ZD$. Combining this with both of the first two parts gives $DB/WD = XD/ZD = ZC/XC = AC/WC$. Adding 1 to both sides of $DB/WD = AC/WC$ gives $(DB + WD)/WD = (AC + WC)/WC$, or $BW/WD = AW/WC$. Therefore, we have $\triangle CWD \sim \triangle AWB$ by SAS Similarity. From this, we have $\angle WCD = \angle WAB$, so $\overline{CD} \parallel \overline{AB}$.

5.5.4

(a) Since $PR = PQ$, we have $\angle R = \angle PQR = \angle WQZ$. Since $\overline{ZX} \parallel \overline{QY}$, we have $\angle ZXR = \angle QYR = 90°$. So, $\angle ZXR = \angle ZWQ$, and we have $\triangle QWZ \sim \triangle RXZ$ by AA Similarity.

(b) First, we note that $RQ = RZ - QZ$, which looks a lot like the expression we want to prove. Since $\overline{XZ} \parallel \overline{YQ}$, we have $\angle RYQ = \angle RXZ$ and $\angle RQY = \angle RZX$, so $\triangle RYQ \sim \triangle RXZ$. This similarity gives us $RZ/ZX = RQ/YQ$, so $RQ = (RZ/ZX)(YQ)$. From $\triangle QWZ \sim \triangle RXZ$ in the last part, we have $RZ/ZX = QZ/ZW$, so $QZ = (RZ/ZX)(ZW)$. Substituting these into $RQ = RZ - QZ$ gives

$$\frac{(RZ)(YQ)}{ZX} = RZ - \frac{(RZ)(ZW)}{ZX}.$$

Multiplying this equation by ZX/RZ gives the desired $YQ = ZX - ZW$.

5.5.5 We extend $\overrightarrow{RY}$ and $\overrightarrow{RX}$ to meet $\overleftrightarrow{PQ}$ at C and D, respectively, as shown. Since $\angle RPA = \angle APQ$, $PX = PX$, and $\angle RXP = \angle DXP$, we have $\triangle RXP \cong \triangle DXP$ by ASA Congruence. Similarly, we have $\triangle CYQ \cong \triangle RYQ$. Therefore, $RX = XD$ and $RY = YC$, so X and Y are midpoints of $\overline{RD}$ and $\overline{RC}$, respectively. So, we have $RY/RC = RX/RD = 1/2$, which gives us $\triangle RYX \sim \triangle RCD$ by SAS Similarity. Therefore, we have $\angle RYX = \angle RCD$, so $\overline{XY} \parallel \overline{DC}$. Since $\overline{DC}$ is on the same line as $\overleftrightarrow{PQ}$, we have $\overline{XY} \parallel \overline{PQ}$.

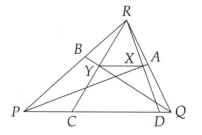

Exercises for Section 5.6

5.6.1 We follow essentially the same procedure as described in the text. Let our given segment of length 1 be $\overline{AZ}$. We then construct $\overline{AB}$ of any length (preferably small!), then copy it 4 times along $\overrightarrow{AB}$ as shown to get point E such that $AE = 5AB$. We then draw $\overline{EZ}$ and construct a line through B parallel to $\overline{EZ}$ (as described earlier in the text). Where this line meets $\overline{AZ}$ we call point C. Since $\overline{BC} \parallel \overline{EZ}$, we have $\triangle ABC \sim \triangle AEZ$. Since $AB/AE = 1/5$, we have $AC/AZ = 1/5$ from our similar triangles. Since $AZ = 1$, we have $AC = 1/5$, as desired.

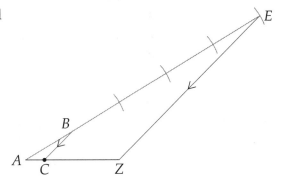

To construct a segment of length $2\frac{2}{3}$, we start by again letting our segment with length 1 be $\overline{AZ}$. We copy it again along $\overrightarrow{AZ}$ to find point Y such that $AY = 2$. We then construct a segment with length $1/3$ exactly as described in the text, then copy this length twice past point Y on $\overrightarrow{AY}$ to reach point X such that $AX = 2\frac{2}{3}$ as desired.

5.6.2 If $\triangle ABC \sim \triangle XYZ$ and $[XYZ] = 9[ABC]$, then the sides of $\triangle XYZ$ must be $\sqrt{9} = 3$ times as long as the sides of $\triangle ABC$. The easiest way to construct such a triangle is to start with $\triangle ABC$, then extend $\overline{AB}$ past B and $\overline{AC}$ past C. We locate point Y on $\overrightarrow{AB}$ such that $AY = 3AB$ by copying $\overline{AB}$ twice on $\overrightarrow{AB}$ past point B. Similarly,

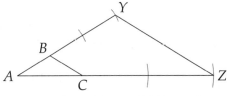

we find Z on $\overrightarrow{AC}$ such that $AC = 3AZ$. Since $AY/AB = 3$, $AZ/AC = 3$, and $\angle YAZ = \angle BAC$, we have $\triangle ABC \sim \triangle AYZ$ by SAS Similarity. The sides of $\triangle AYZ$ are three times as long as those of $\triangle ABC$, and the triangles are similar, so $\triangle AYZ$ has 9 times the area of $\triangle ABC$. Therefore, $\triangle AYZ$ is our desired triangle.

Review Problems

5.22

(a) Since $\overline{AB} \parallel \overline{CE}$, we have $\angle DAB = \angle DEC$ and $\angle DBA = \angle DCE$. Therefore, $\triangle DAB \sim \triangle DEC$ by AA Similarity.

(b) Since $\angle IHJ = \angle FHG$ (vertical angles) and $\angle GFH = \angle HIJ$, we have $\triangle HFG \sim \triangle HIJ$ by AA Similarity.

(c) There are no two triangles in this diagram that must be similar.

(d) Since $OP/OQ = OS/OR$ (both ratios equal 2/3) and $\angle QOP = \angle ROS$, we have $\triangle POQ \sim \triangle SOR$ by SAS Similarity.

(e) Since $\angle T = \angle W$ and $\angle U = \angle X$, we have $\triangle TUV \sim \triangle WXY$ by AA Similarity.

(f) Since $AB/FE = BC/DF = AC/DE$, we have $\triangle ABC \sim \triangle EFD$ by SSS Similarity.

5.23 Since $\angle A = \angle R$ and $\angle B = \angle P$, we have $\triangle ABC \sim \triangle RPQ$. Therefore, we have $AB/RP = BC/PQ = AC/RQ$. Substitution gives $y/12 = x/6 = 3/8$, from which we find $y = (12)(3/8) = \boxed{9/2}$ and $x = (6)(3/8) = \boxed{9/4}$.

5.24 Since $\overline{PQ} \parallel \overline{BC}$, we have $\angle APQ = \angle ABC$ and $\angle AQP = \angle ACB$. Therefore, $\triangle APQ \sim \triangle ABC$ by AA Similarity. Since $AB = 12$, $PB = 9$, and $AP = AB - PB$, we have $AP = 3$. Our similarity gives us $AQ/AC = AP/AB = 3/12 = 1/4$. So, $AQ = AC/4 = \boxed{9/2}$.

5.25 To maximize the perimeter of the second triangle, we should maximize the ratio of corresponding side lengths between the two triangles. If the side of length 4 cm in the first triangle corresponds to the side of length 36 cm in the second triangle, then this ratio is $36/4 = 9$. If it is the side of length 6 cm, then this ratio is $36/6 = 6$, and if it is the side of length 9 cm, then this ratio is $36/9 = 4$.

Hence, the maximum ratio is 9, and the maximum perimeter of the second triangle is $9 \cdot (4 + 6 + 9) = \boxed{171 \text{ cm}}$.

5.26 Since $\overline{DB} \parallel \overline{EC}$, we have $\triangle DAB \sim \triangle EAC$. Therefore, we must have $AD/AE = AB/AC$. However, using the lengths given in the diagram, we have $AD/AE = 2/(6.5) = 4/13$ and $AB/AC = 3.5/9.5 = 7/19$. These two ratios are not equal! Therefore, the diagram given in the problem is impossible.

5.27 Since $AB/AE = 6/20 = 3/10$ and $AC/AD = 9/30 = 3/10$, we have $AB/AE = AC/AD$. This, combined with $\angle BAC = \angle EAD$ gives us $\triangle ABC \sim \triangle AED$. Therefore, $ED/BC = DA/AC$, so $DE/13 = 30/9$. Finally, we have $DE = 13(30/9) = \boxed{130/3}$.

5.28 Since $\angle PQR = \angle TSR$ and $\angle PRQ = \angle TRS$, we have $\triangle PQR \sim \triangle TSR$. Therefore, we have $PR/RT = QR/RS = PQ/ST$. We also have $PR = RS$ and are given a number of side lengths. Making these substitutions gives
$$\frac{PR}{12} = \frac{6}{PR} = \frac{8}{10}.$$
From $PR/12 = 8/10$, we have $PR = 9.6$. From $6/PR = 8/10$, we have $PR = (6)(10/8) = 60/8 = 7.5$. But PR can't have two different values! Therefore, the diagram in the problem is impossible.

5.29 From $\triangle WYZ$ we have $\angle YZW = 180° - \angle WYZ - \angle YWZ = 90° - \angle YWZ$. Also, $\angle XWY = \angle XWZ - \angle YWZ = 90° - \angle YWZ$. Therefore, $\angle YZW = \angle XWY$. Together with $\angle ZYW = \angle WYX$, this gives us $\triangle WYX \sim \triangle ZYW$ by AA Similarity. Therefore, we have $WY/YZ = XY/WY$. Substitution gives $WY/6 = 4/WY$, so $WY^2 = 24$. Taking the square root of both sides gives $\boxed{WY = 2\sqrt{6}}$.

Similarly, $\angle YVZ = 90° - \angle YZV = \angle WZY$ and $\angle VYZ = \angle WYZ$, so $\triangle VYZ \sim \triangle ZYW$ by AA. Therefore, $VY/YZ = YZ/WY$, so $VY = YZ^2/WY = (36)/(2\sqrt{6}) = 18/\sqrt{6} = \boxed{3\sqrt{6}}$.

5.30

(a) Since $BM = CN$ and $AB = AC$, we have $AB - BM = AC - CN$. Therefore, $AM = AN$.

(b) From our first part we have $AM = AN$. Since we also have $AB = AC$, we have $AM/AB = AN/AC$. This, combined with $\angle BAC = \angle MAN$, gives us $\triangle MAN \sim \triangle BAC$ by SAS Similarity. Therefore, the angles of $\triangle MAN$ equal those of $\triangle BAC$, so the angles of $\triangle MAN$ are each 60° as well. Thus, $\triangle MAN$ is equilateral. We also could note that since $AM = AN$ and $\angle A = 60°$, we have $\angle ANM = \angle AMN = (180° - \angle A)/2 = 60°$.

5.31

(a) Since $\triangle ABC \sim \triangle YZX$ and the area of $\triangle YZX$ is 9 times the area of $\triangle ABC$, the side lengths of $\triangle YZX$ are $\sqrt{9} = 3$ times the corresponding side lengths of $\triangle ABC$. Therefore, $YZ = 3AB = \boxed{27}$, where the order of the vertices in the triangle similarity relationship tells us which side of $\triangle ABC$ corresponds to $\overline{YZ}$ in $\triangle YZX$.

(b) By the same reasoning as part (a), we have $XZ = 3BC = 36$. Letting our desired altitude have length h, we have $[YZX] = (h)(XZ)/2$. Since we are given $[YZX] = 360$, we have $(h)(XZ)/2 = 360$. Since $XZ = 36$, we have $h = \boxed{20}$. (We could also have found the corresponding height in $\triangle ABC$ and multiplied it by 3.)

5.32 Let $AE = x$, so $BE = 25 - x$. Since $\triangle AED \sim \triangle BCE$, we have $AE/AD = BC/BE$. Substitution gives $x/12 = 12/(25 - x)$. Cross-multiplying gives $x(25 - x) = 144$, so we have $x^2 - 25x + 144 = 0$. Therefore, we have $(x - 9)(x - 16) = 0$, so $x = 9$ or $x = 16$. When $x = 16$, we have $AE = 16$ and $BE = 25 - AE = 9$. We are given $AE < BE$, so we discard this possibility. For $x = 9$, we have $\boxed{AE = 9}$ and $BE = 16$.

5.33 Since $\overline{WY} \parallel \overline{XR}$, we have $PY/YR = PW/WX = 3/2$. Since $\overline{XY} \parallel \overline{QR}$, we have $PX/XQ = PY/YR = 3/2$. Therefore, $XQ = (PX)(2/3) = \boxed{20/3}$.

5.34 Since $\overline{AP} \parallel \overline{RC}$, we have $\angle P = \angle CRB$ and $\angle PAB = \angle RCB$, so $\triangle BRC \sim \triangle BPA$. Therefore, $BC/BA = CR/PA$. Similarly, $\overline{BQ} \parallel \overline{CR}$ gives us $\triangle QBA \sim \triangle RCA$, so $AC/BA = CR/BQ$. Adding these two equations gives us

$$\frac{CR}{AP} + \frac{CR}{BQ} = \frac{BC}{BA} + \frac{AC}{BA} = \frac{BC + AC}{BA} = \frac{BA}{BA} = 1.$$

Dividing both sides of this by CR gives $\frac{1}{AP} + \frac{1}{BQ} = \frac{1}{CR}$, as desired.

5.35 Label the vertices A, B, and C, so that $BC = 12$, $AB = 20$, and the right angle is at C. Let D be the foot of the perpendicular from C to AB.

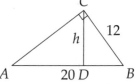

Then $\angle ACB = \angle ADC = \angle CDB = 90°$. Also, $\angle ACD = 90° - \angle CAD = \angle ABC$, and $\angle CBD = \angle ABC$. Therefore, triangles ABC, ACD, and CBD are similar.

Let $AD = x$, so $DB = 20 - x$. From $\triangle CDB \sim \triangle ACB$, we have $DB/BC = BC/AB$, so $20 - x = 12^2/20$, from which we find $x = 64/5$. Therefore, $AD = 64/5$ and $DB = 20 - x = 36/5$. From $\triangle ACD \sim \triangle CBD$, we have $CD/BD = AD/CD$, so $CD^2 = (AD)(DB)$. Therefore, $h = CD = \sqrt{(AD)(DB)} = \boxed{48/5}$.

5.36 Since $\overline{DE} \parallel \overline{BC}$, $\angle ADE = \angle ABC$ and $\angle AED = \angle ACB$. Then by AA similarity, triangles ABC and ADE are similar. Therefore,

$$\frac{AB}{AD} = \frac{AC}{AE}.$$

Since $AB = AD + DB$ and $AC = AE + EC$, we have

$$\frac{AD + DB}{AD} = \frac{AE + EC}{AE},$$

so

$$1 + \frac{DB}{AD} = 1 + \frac{EC}{AE}.$$

Subtracting 1 from each side gives $DB/AD = EC/AE$, and a little rearranging gives the desired $AD/AE = DB/EC$.

Challenge Problems

5.37 We have $AD/AE = BD/EC$. We would like to show that $AB/AC = AD/AE$, so we can then conclude that $\triangle ABC \sim \triangle ADE$ by SAS Similarity. A little algebra, and noting that we are given $AD = (AE)(BD)/EC$, does it for us:

$$\frac{AB}{AC} = \frac{AD + DB}{AE + EC} = \frac{\frac{(AE)(BD)}{EC} + BD}{AE + EC} = \frac{BD\left(\frac{AE}{EC} + 1\right)}{AE + EC}$$

$$= \frac{BD\left(\frac{AE+EC}{EC}\right)}{AE + EC} = \frac{(BD)(AE + EC)}{(EC)(AE + EC)} = \frac{BD}{EC}.$$

Since $BD/EC = AD/AE$, we have $AB/AC = BD/EC = AD/AE$. We also have $\angle BAC = \angle DAE$, so $\triangle ABC \sim \triangle ADE$ by SAS Similarity. Our similar triangles give us $\angle ABC = \angle ADE$, so $\overline{BC} \parallel \overline{DE}$.

5.38 The third angle (i.e. the other base angle) of each of the isosceles triangles equals $180°$ minus the sum of the other two angles. Therefore, the two 'other' base angles of each of the triangles are the same. Since both triangles have two base angles that have this measure, the two triangles have two angle measures in common. Therefore, the triangles are similar by AA Similarity.

5.39 The area of $CDEFG$ can be calculated by taking the area of triangle ABC and subtracting the areas of triangles ADE and BFG. To find the areas of our little right triangles, we need FG, FB, AE, and DE. We can use similar triangles to find FG. We have $\angle GFB = \angle CHB = 90°$ and $\angle GBF = \angle CBH$, so triangles GFB and CHB are similar. Then

$$\frac{FG}{CH} = \frac{FB}{HB},$$

so

$$FG = \frac{CH \cdot FB}{HB} = \frac{24 \cdot 6}{18} = \boxed{8}.$$

We have $DE = FG = 8$, and we know $FB = 6$. Since $AC = CB$, we have $\angle A = \angle B$. Together with $\angle AED = \angle BFG$ and $DE = GF$, this gives $\triangle DEA \cong \triangle GFB$ by AAS Congruence. Therefore, $AE = BF = 6$.

Hence,

$$[CDEFG] = [ABC] - [ADE] - [BFG]$$

$$= \frac{1}{2}AB \cdot CH - \frac{1}{2}AE \cdot DE - \frac{1}{2}BF \cdot GF$$

$$= \frac{1}{2}(36 \cdot 24) - \frac{1}{2}(6 \cdot 8) - \frac{1}{2}(6 \cdot 8) = \boxed{384}.$$

5.40 First, we rewrite our given equation as $AD/AE = AB/AC$. This, combined with $\angle DAE = \angle CAB$ gives us $\triangle DAE \sim \triangle BAC$ by SAS Similarity. Therefore, $\angle ADE = \angle ABC$, so $\angle CBE + \angle CDE = \angle CBA + \angle CDE = \angle ADE + \angle CDE = 180°$, because $\angle ADE$ and $\angle CDE$ together make a straight angle.

We can also rewrite our given equation as $AD/AB = AE/AC$. This, combined with $\angle BAD = \angle CAE$ gives us $\triangle ABD \sim \triangle ACE$ by SAS Similarity. Therefore, we have $\angle ADB = \angle CEA$, so $\angle ADB + \angle BEC = \angle CEA + \angle BEC = 180°$, as desired.

5.41 Because $\overline{FE} \parallel \overline{BC}$, we have $\angle DFE = \angle FDB$. Because $\overline{BF} \parallel \overline{DE}$, we have $\angle BFD = \angle FDE$. Combined with $DF = DF$, we have $\triangle BDF \cong \triangle EFD$ by ASA Congruence. Therefore, $BF = DE$. Similarly, we can show $\triangle AFE \cong \triangle DEF$, so $AF = DE$. Since $AF = DE$ and $BF = DE$, we have $AF = BF$, so F is the midpoint of $\overline{AB}$.

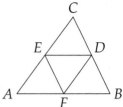

We could finish the problem with more pairs of congruent triangles, or we can use what we know about similarity. We'll show the latter approach. Since $\overline{FE} \parallel \overline{CD}$, we have $\triangle AFE \sim \triangle ABC$ by AA ($\angle AFE = \angle B$ and $\angle AEF = \angle C$), so $AE/AC = AF/AB = 1/2$. Therefore, E is the midpoint of $\overline{AC}$. Similarly, since $\overline{DE} \parallel \overline{AB}$, we have $\triangle CDE \sim \triangle CBA$, so $CD/CB = CE/AC = 1/2$, and D is the midpoint of $\overline{BC}$.

5.42 As described in the text, since $\overline{PS} \parallel \overline{QT}$, we have $QP/QR = TU/TR$. Similarly, since $\overline{PQ} \parallel \overline{ST}$, we have $SU/SP = TU/TR$. Combining these equations gives the desired $SU/SP = QP/QR$.

5.43 We are given that C is the midpoint of $\overline{YZ}$. Therefore, triangles XYC and XZC have equal bases $YC = CZ$ and the same altitude, so $[XYC] = [XZC]$. But $[XYC] + [XZC] = [XYZ] = 8$, so $[XYC] = [XZC] = 4$.

Next, let M be the intersection of $\overline{AB}$ and $\overline{CX}$. Since A is the midpoint of $\overline{XY}$, $XA = XY/2$. Since B is the midpoint of $\overline{XZ}$, $XB = XZ/2$. Therefore, triangles AXB and YXZ are similar. Then $\angle XAB = \angle XYZ$, so $\overline{AB}$ and $\overline{YZ}$ are parallel. Furthermore, $\angle AXM = \angle YXC$, so triangles XAM and XYC are similar. The ratio between their sides is $XA/XY = 1/2$, so the ratio between their areas is

$$\frac{[XAM]}{[XYC]} = \left(\frac{1}{2}\right)^2 = \frac{1}{4}.$$

Hence, $[XAM] = [XYC]/4 = 1$. Then the area of the shaded region is $[AMCY] = [XYC] - [XAM] = 4 - 1 = \boxed{3}$.

5.44 Let ABC be an arbitrary triangle, and let D, E, and F be the midpoints of $\overline{BC}$, $\overline{CA}$, and $\overline{AB}$, respectively. Then $AF = AB/2$, $AE = AC/2$, and $\angle FAE = \angle BAC$, so by SAS, triangles FAE and BAC are similar with ratio $1/2$. Hence, $FE = BC/2$.

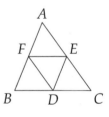

Similarly, $DF = AC/2$ and $DE = AB/2$. Therefore,

$$DE + DF + EF = \frac{1}{2}(AB + AC + BC).$$

In other words, the perimeter of triangle DEF is half of the perimeter of triangle ABC.

In the original problem, let P_n denote the perimeter of the nth triangle. Then $P_2 = P_1/2$, $P_3 = P_2/2$, $P_4 = P_3/2$, and so on, until $P_{10} = P_9/2$. We can combine these and write $P_n = \left(\frac{1}{2}\right)^{n-1} P_1$. Therefore,

$$\frac{P_{10}}{P_3} = \frac{(\frac{1}{2})^9 P_1}{(\frac{1}{2})^2 P_1} = \left(\frac{1}{2}\right)^7 = \boxed{\frac{1}{128}}.$$

5.45 Since $\overline{AB}$ and $\overline{CD}$ are parallel, $\angle EFG = \angle EAB$ and $\angle EGF = \angle EBA$. Therefore, triangles EFG and EAB are similar, so

$$\frac{AE}{EF} = \frac{AB}{FG} = \frac{5}{2}.$$

But $AE = EF + FA$, so

$$\frac{EF + FA}{EF} = 1 + \frac{FA}{EF} = \frac{5}{2},$$

so $FA/EF = 3/2$.

Now, let H be the foot of the perpendicular from E to $\overline{FG}$. Since $\overline{EH}$ and $\overline{AD}$ are both perpendicular to $\overline{CD}$, $\angle EHF = \angle ADF = 90°$, and $\angle EFH = \angle AFD$. Therefore, triangles EFH and AFD are similar, so we get

$$\frac{EH}{AD} = \frac{EF}{FA} = \frac{2}{3}.$$

Then $EH = \frac{2}{3}AD = 2$. Hence, the base of triangle ABE is $AB = 5$, the height of triangle ABE is $EH + AD = 2 + 3 = 5$, so its area is $(AB)(5)/2 = \boxed{25/2}$.

We also could have solved this problem by finding $[AFGB]$ and noting that $([ABE] - [AFGB])/[ABE] = (EF/AE)^2$.

5.46 $\overline{CC_1}$ is an altitude from the right angle of $\triangle ABC$. As we learned in the text, this gives us similar triangles. Specifically, from right angle ACB and right triangle $\triangle BCC_1$, we have $\angle ACC_1 = 90° - \angle BCC_1 = \angle CBC_1$. Combined with $\angle ACB = \angle AC_1C$, this gives us $\triangle AC_1C \sim \triangle ACB$. Therefore, we have $BC/AB = CC_1/AC$, so $CC_1 = (AC)(BC/AB) = (BC)(AC/AB) = (4/5)(BC)$.

In exactly the same way, $\overline{C_1C_2}$ is an altitude from the right angle of $\triangle AC_1C$. And in the same way as before, we can show that $\triangle AC_1C_2 \sim \triangle ACC_1$. This gives us $C_1C_2/AC_1 = CC_1/AC$, so $C_1C_2 = (CC_1)(AC_1/AC)$. Our previous similarity gives us $AC_1/AC = AC/AB = 4/5$, so we have $C_1C_2 = (4/5)(CC_1)$.

We can continue in this manner, showing at every step that $C_nC_{n+1} = (4/5)C_{n-1}C_n$. In other words, each term in our series is $4/5$ of the one before it. Since our first term is $CC_1 = (4/5)(BC) = 12/5$, our sum is

$$S = \frac{12}{5} + \left(\frac{12}{5}\right)\left(\frac{4}{5}\right) + \left(\frac{12}{5}\right)\left(\frac{4}{5}\right)^2 + \left(\frac{12}{5}\right)\left(\frac{4}{5}\right)^3 + \left(\frac{12}{5}\right)\left(\frac{4}{5}\right)^4 + \cdots$$

This is a geometric series with first term $12/5$ and common ratio $4/5$, so its sum is $\frac{12/5}{1-4/5} = \boxed{12}$. If you don't see why that is the sum, try multiplying our expression for S above by $4/5$ to get:

$$4S/5 = \quad \left(\frac{12}{5}\right)\left(\frac{4}{5}\right) + \left(\frac{12}{5}\right)\left(\frac{4}{5}\right)^2 + \left(\frac{12}{5}\right)\left(\frac{4}{5}\right)^3 + \left(\frac{12}{5}\right)\left(\frac{4}{5}\right)^4 + \cdots,$$

then subtract this equation from the original expression for S.

5.47 The large equilateral triangle is similar to each of the small equilateral triangles, and the ratio between their sides is n. Then the ratio between the area of the large equilateral triangle and the small equilateral triangle is n^2, so the large equilateral triangle contains n^2 small equilateral triangles.

The first row contains 1 small equilateral triangle, the second row contains 3 small equilateral triangles, and so on, until the nth row contains $2n - 1$ small equilateral triangles. So counting row by row, the large equilateral triangle contains $1 + 3 + 5 + \cdots + (2n - 1)$ small equilateral triangles.

We've found two different ways to count all the little equilateral triangles. Since we are counting the same thing in both cases, our counts must be equal! Therefore, we have shown that for all positive integers n,

$$1 + 3 + 5 + \cdots + (2n - 1) = n^2.$$

In other words, the sum of the first n odd positive integers is n^2.

5.48 Since $\overline{PX}$ is parallel to $\overline{AD}$, $\angle BAD = \angle BPX$ and $\angle BDA = \angle BXP$. Therefore, triangles BAD and BPX are similar, with

$$\frac{PX}{AD} = \frac{BP}{BA} = \frac{1}{2}.$$

Also, since $\overline{RY}$ is parallel to $\overline{CB}$, $\angle DRY = \angle DCB$ and $\angle DYR = \angle DBC$. Therefore, triangles DRY and DCB are similar, with

$$\frac{RY}{CB} = \frac{DR}{DC}.$$

We can see that $RC = BP$ by noting that $\triangle RCB \cong \triangle BPR$ by ASA Congruence ($\angle BRC = \angle RBP$ and $\angle CBR = \angle PRB$ due to our parallel lines). Therefore, $DR = DC - RC = 5$, so $RY/BC = DR/DC = 5/8$. Dividing $PX = (1/2)(AD)$ by $RY = (5/8)(BC)$, and noting that $AD = BC$, gives

$$\frac{PX}{RY} = \frac{\frac{1}{2}AD}{\frac{5}{8}BC} = \boxed{\frac{4}{5}}.$$

5.49 Since E is the midpoint of $\overline{CD}$, $EC = CD/2 = AB/2$. Since $AF = FG = GB$, we have $AF = AB/3$.

Now, $\overline{AF}$ is parallel to $\overline{CE}$, so $\angle HAF = \angle HCE$ and $\angle HFA = \angle HEC$. Therefore, triangles HAF and HCE are similar, and we have

$$\frac{HF}{HE} = \frac{AF}{CE} = \frac{AB/3}{AB/2} = \frac{2}{3}.$$

Draw segment $\overline{AE}$. Then triangles AHF and AHE have bases $\overline{HF}$ and $\overline{HE}$, respectively, and equal altitudes to these bases, so

$$\frac{[AHF]}{[AHE]} = \frac{HF}{HE} = \frac{2}{3}.$$

But

$$[AHF] + [AHE] = [AEF] = \frac{1}{2}AF \cdot AD = \frac{1}{2} \cdot \frac{1}{3}AB \cdot AD = \frac{1}{6}[ABCD] = \frac{70}{6} = \frac{35}{3}.$$

Then, since $[AHE] = \frac{3}{2}[AHF]$, we have

$$[AHF] + \frac{3}{2}[AHF] = \frac{5}{2}[AHF] = \frac{35}{3},$$

so

$$[AHF] = \frac{2}{5} \cdot \frac{35}{3} = \boxed{\frac{14}{3}}.$$

(Alternatively, we could note that $[AHF]/[AEF] = 2/5$ from noting that $[AHF]/[AEF] = HF/EF$.)

5.50 From our construction of a perpendicular bisector, we know how to construct a right angle. We also know how to copy an angle. Therefore, we know how to construct a square. Specifically, we can construct square $BCYX$ outside the triangle on side $\overline{BC}$, as shown. We do so by constructing a line through B perpendicular to $\overline{BC}$, then drawing an arc with radius BC. Where this arc hits our line gives us point X. Similarly, we can find point Y. Let D be the intersection of $\overline{AX}$ and $\overline{BC}$, and let E be the intersection of $\overline{AY}$ and $\overline{BC}$. We then draw lines through D and E perpendicular to $\overline{BC}$; these lines hit $\overline{AB}$ and $\overline{AC}$ at G and F as shown.

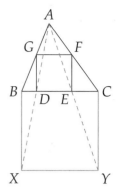

To prove $DEFG$ is a square, we note that $\overline{DG} \parallel \overline{EF} \parallel \overline{BX} \parallel \overline{CY}$ since all are perpendicular to the same line. Therefore, $\triangle AGD \sim \triangle ABX$ and $\triangle AFE \sim \triangle ACY$, so $GD = (AD/AX)(BX)$ and $FE = (AE/AY)(CY)$. Since $\overline{BC} \parallel \overline{XY}$ (because $\angle CBX + \angle BXY = 180°$), $\triangle ADE \sim \triangle AXY$. Therefore, $AD/AX = AE/AY = DE/XY$. This, combined with $BX = CY = XY$, gives us

$$GD = \left(\frac{AD}{AX}\right)(BX) = \left(\frac{AD}{AX}\right)(CY) = \left(\frac{AE}{AY}\right)(CY) = FE = \left(\frac{AE}{AY}\right)(CY) = \left(\frac{AE}{AY}\right)(XY) = DE.$$

Finally, we note that

$$\frac{AG}{AB} = \frac{AD}{AX} = \frac{AE}{AY} = \frac{AF}{AC},$$

so $\triangle AGF \sim \triangle ABC$. Therefore, $\angle AGF = \angle ABC$ and $\overline{GF} \parallel \overline{BC}$, so $\overline{GF}$ is perpendicular to $\overline{DG}$ and $\overline{EF}$. Moreover, $GF/BC = AG/AB = AD/AX$, so

$$GF = (BC)\left(\frac{AG}{AB}\right) = (BC)\left(\frac{AD}{AX}\right) = (BX)\left(\frac{AD}{AX}\right) = GD,$$

which completes our proof that all the angles of $DEFG$ equal 90° and all its sides have the same length.

Alternatively, we could construct a square $WXYZ$ with $\overline{WX}$ on $\overline{BC}$ and Z on $\overline{AB}$ as shown. (Start by choosing W, draw a line through it perpendicular to $\overline{BC}$ to find Z, then build the rest of the square.) We then draw $\overrightarrow{BY}$; where it meets $\overline{AC}$ is point F. We then use F to construct the square by drawing a line through F parallel to $\overline{BC}$ to find G, then lines through G and F perpendicular to $\overline{FG}$ to find D and E. We can prove $DEFG$ is a square first by noting all its angles are right angles by construction. Next, we see that $\triangle BYX \sim \triangle BFE$ and $\triangle BZY \sim \triangle BGF$. Therefore, $GF/ZY = BF/BY = FE/YX$. Since $ZY = YX$, we have $GF = EF$. EF and GD are each the distance between parallel segments $\overline{GF}$ and $\overline{DE}$, so $GD = EF$. Similarly, $ED = FG$. Therefore, all the sides of $DEFG$ are equal, as are all its angles, so it is a square.

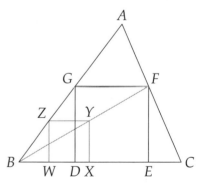

5.51 *Solution 1:* Since $\overline{SP} \parallel \overline{AB}$ and $\overline{TQ} \parallel \overline{AC}$, we have $\angle XPQ = \angle ABC$ and $\angle PXQ = \angle BTQ = \angle BAC$. Therefore, triangles ABC and XPQ are similar. Hence,

$$\frac{[ABC]}{[XPQ]} = \left(\frac{BC}{PQ}\right)^2 = \frac{BC^2}{PQ^2}.$$

Also, triangles XBC and XPQ have bases $\overline{PQ}$ and $\overline{BC}$, respectively, and equal altitudes to these bases, so

$$\frac{[XBC]}{[XPQ]} = \frac{BC}{PQ}.$$

Therefore,

$$\frac{[XBC]}{[ABC]} = \frac{[XBC]/[XPQ]}{[ABC]/[XPQ]} = \frac{BC/PQ}{BC^2/PQ^2} = \frac{PQ}{BC}.$$

By similar reasoning,

$$\frac{[XCA]}{[ABC]} = \frac{RS}{CA} \quad \text{and} \quad \frac{[XAB]}{[ABC]} = \frac{TU}{AB}.$$

Therefore,

$$\frac{PQ}{BC} + \frac{RS}{CA} + \frac{TU}{AB} = \frac{[XBC]}{[ABC]} + \frac{[XCA]}{[ABC]} + \frac{[XAB]}{[ABC]} = \frac{[ABC]}{[ABC]} = 1.$$

Solution 2: (If you haven't learned about parallelograms yet, return to this solution after learning about them.) Since $\overline{SP} \parallel \overline{AB}$ and $\overline{TQ} \parallel \overline{AC}$, we have $\angle XPQ = \angle ABC$ and $\angle PXQ = \angle BTQ = \angle BAC$. Therefore, triangles ABC and XPQ are similar. By the same argument, triangles XPQ, SXR, TUX, and ABC are all similar to each other. From $\triangle SXR \sim \triangle ABC$ we have $RS/CA = XR/BC$. From $\triangle TUX \sim \triangle ABC$, we have $TU/AB = UX/BC$.

Since both pairs of opposite sides of $XRCQ$ are parallel, $XRCQ$ is a parallelogram. Therefore, $XR = QC$. Similarly, $XUBP$ is a parallelogram, so $UX = BP$. Therefore,

$$\frac{PQ}{BC} + \frac{RS}{CA} + \frac{TU}{AB} = \frac{PQ}{BC} + \frac{XR}{BC} + \frac{UX}{BC} = \frac{PQ}{BC} + \frac{QC}{BC} + \frac{BP}{BC} = \frac{BC}{BC} = 1.$$

CHAPTER 6

Right Triangles

Exercises for Section 6.1

6.1.1 In each part, we directly apply the Pythagorean Theorem.

(a) $AB = \sqrt{BC^2 - AC^2} = \boxed{8}$.

(b) $DF = \sqrt{EF^2 + DE^2} = \boxed{\sqrt{130}}$.

(c) $GH = \sqrt{IH^2 - GI^2} = \sqrt{108} = \boxed{6\sqrt{3}}$.

(d) $KL = \sqrt{JK^2 + JL^2} = \boxed{26}$.

(e) $MN = \sqrt{ON^2 - MO^2} = \sqrt{12} = \boxed{2\sqrt{3}}$.

(f) $QR = \sqrt{PQ^2 - PR^2} = \boxed{8}$.

6.1.2 Let ABC be a triangle, with a right angle at C. Then $\angle A + \angle B + \angle C = \angle A + \angle B + 90° = 180°$, so $\angle A + \angle B = 90°$. Thus, $\angle A$ and $\angle B$ are complementary.

6.1.3 Let x be the length of the shorter leg, in centimeters. Then the length of the longer leg is $3x + 3$, and the hypotenuse is $(3x + 3) + 1 = 3x + 4$. By the Pythagorean Theorem, $x^2 + (3x + 3)^2 = (3x + 4)^2$. Rearranging gives $x^2 - 6x - 7 = 0$, which factors as $(x - 7)(x + 1) = 0$. Therefore, $x = 7$ because x must be positive.

The sides of the triangle are then 7, 24, and 25, so the area of the triangle is $(7)(24)/2 = \boxed{84}$.

6.1.4 Let ABC be a right triangle, with hypotenuse $\overline{BC}$. Then by the Pythagorean Theorem, $AB^2 + AC^2 = BC^2$. Since $AC > 0$, $BC^2 = AB^2 + AC^2 > AB^2$, so $BC > AB$. Since $AB > 0$, $BC^2 = AB^2 + AC^2 > AC^2$, so $BC > AC$. Therefore, the hypotenuse, $\overline{BC}$, is longer than the other two sides, $\overline{AB}$ and $\overline{AC}$.

6.1.5 If the sum of the squares of the smaller two numbers equals the square of the largest number in each triple, then that triple can be the side lengths of a right triangle.

(a) $6^2 + 8^2 = 10^2$. Therefore, these can be the sides of a right triangle.

(b) $4^2 + 5^2 = 41 \neq 6^2$. These cannot be the sides of a right triangle.

(c) $9^2 + (3\sqrt{3})^2 = 81 + 27 = 108 = (6\sqrt{3})^2$. These can be the sides of a right triangle.

(d) $(5/8)^2 + (3/2)^2 = 25/64 + 9/4 = 25/64 + 144/64 = 169/64 = (13/8)^2$. These can be the sides of a right triangle.

(e) $(2\sqrt{2})^2 + (3\sqrt{2})^2 = 8 + 18 = 26 \neq 5^2$. These cannot be the sides of a right triangle.

(f) $(1.2)^2 + (3.5)^2 = 1.44 + 12.25 = 13.69 = (3.7)^2$. These can be the side lengths of a right triangle.

6.1.6 Since $(\sqrt{6})^2 + (\sqrt{7})^2 = (\sqrt{13})^2$, our triangle is a right triangle. Therefore, its area is half the product of its legs, or $\boxed{\sqrt{42}/2}$.

6.1.7 Our triangle is shown at right, and we have drawn an altitude to the base of our isosceles triangle. Since $\triangle ABC$ is isosceles, we have $BM = MC$, so $BM = 3$. Since $\triangle AMB$ is a right triangle, we have $AM = \sqrt{AB^2 - BM^2} = \sqrt{64 - 9} = \sqrt{55}$. Therefore, the area of the triangle is $(BC)(AM)/2 = \boxed{3\sqrt{55}}$.

6.1.8

(a) Since $\overleftrightarrow{AB} \parallel \overleftrightarrow{CD}$, we have $\angle XAB + \angle AXY = 180°$. Since $\angle AXY = 90°$, we have $\angle XAB = 90°$.

(b) Since $\overleftrightarrow{AB} \parallel \overleftrightarrow{CD}$, we have $\angle BXY = \angle ABX$. Since we also have $\angle BAX = \angle BYX$ and $BX = BX$, we have $\triangle AXB \cong \triangle YBX$ by AAS Congruence. Therefore, $AX = BY$, so A and B are equidistant from $\overleftrightarrow{CD}$.

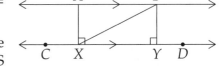

6.1.9 When the top of the ladder rests against the wall and the other end is touching the ground, the ladder, wall, and ground together form a right triangle of which the ladder is the hypotenuse. When the top of the ladder touches the wall 24 feet above the ground, this triangle has one leg of 24 feet and hypotenuse of 25 feet. Therefore, its other leg is $\sqrt{625 - 576} = 7$ feet.

After it slides, the hypotenuse is still 25 feet, but now one leg is 20 feet instead of 24 feet. Therefore, the other leg has length $\sqrt{625 - 400} = 15$ feet. Thus, Nathan has moved the base of the ladder a distance of $15 - 7 = \boxed{8 \text{ feet}}$.

6.1.10 First consider the triangle with side lengths 13, 13, and 10, which is isosceles. The altitude of this triangle to the side with length 10 divides the triangle into two right triangles, each with hypotenuse 13 and a leg with length 5. By the Pythagorean Theorem, the length of the altitude is $\sqrt{13^2 - 5^2} = \sqrt{169 - 25} = \sqrt{144} = 12$. Therefore, the area of this triangle is $(10)(12)/2 = 60$.

We can split our triangle along this altitude into two triangles with sides 5, 12, and 13, then glue these two together along the sides of length 5, thus making a triangle with altitude 5, base length 24 and other two sides of length 13. Therefore, $x = \boxed{24}$.

If we don't think of this clever dissection of our initial triangle, we can use algebra. In the triangle with side lengths 13, 13, and x, we use the same steps as before to find that the length of the altitude to the side with length x is $\sqrt{13^2 - (x/2)^2}$, so the area of this triangle is

$$\frac{1}{2}x \cdot \sqrt{13^2 - \frac{x^2}{4}} = \frac{1}{2}x \cdot \frac{1}{2}\sqrt{676 - x^2} = \frac{x\sqrt{676 - x^2}}{4} = 60,$$

so $x\sqrt{676-x^2} = 240$. Squaring, we get $x^2(676-x^2) = 57600$, and rearranging gives $x^4 - 676x^2 + 57600 = 0$. This factors as $(x^2 - 100)(x^2 - 576) = 0$, so $x = \pm\sqrt{100} = \pm 10$ or $x = \pm\sqrt{576} = \pm 24$. We are told that x is not equal to 10, and x cannot be negative, so the solution is $x = \boxed{24}$.

Exercises for Section 6.2

6.2.1

(a) Since one leg of the right triangle is half the hypotenuse, the triangle is a 30-60-90 triangle with the 30° angle opposite the shorter leg. Therefore, $\angle B = \boxed{30°}$ and $\angle C = \boxed{60°}$. Since $\triangle ABC$ is a 30-60-90 triangle and $\overline{AB}$ is opposite the 60° angle, we have $AB = AC\sqrt{3} = \boxed{10\sqrt{3}}$. We also could have used the Pythagorean Theorem to find $AB = \sqrt{BC^2 - AC^2} = \boxed{10\sqrt{3}}$.

(b) $\angle F = 180° - \angle D - \angle E = \boxed{30°}$. $\triangle DEF$ is a 30-60-90 triangle. Leg $\overline{DE}$ is opposite the 30° angle, so $DE = DF/2 = \boxed{1}$. $\overline{EF}$ is opposite the 60° angle, so $EF = (DE)\sqrt{3} = \boxed{\sqrt{3}}$.

(c) Since $IH = GH$, $\angle G = \angle I$. Since $\angle H = 90°$, we have $\angle G + \angle I = 90°$, so $\angle G = \angle I = \boxed{45°}$. Since $\triangle GHI$ is a 45-45-90 triangle, we have $GI = GH\sqrt{2} = \sqrt{12} = \boxed{2\sqrt{3}}$.

(d) $\triangle JKL$ is a 45-45-90 triangle, so $JL = KL = \boxed{6}$ and $JK = LK\sqrt{2} = \boxed{6\sqrt{2}}$. Since $\angle JLM = 90°$ and leg $\overline{JL}$ has half the length of hypotenuse $\overline{JM}$ in triangle $\triangle JLM$, we know that $\triangle JML$ is a 30-60-90 triangle. Therefore, $LM = JL\sqrt{3} = \boxed{6\sqrt{3}}$, $\angle JML = \boxed{30°}$, and $\angle LJM = \boxed{60°}$.

6.2.2 *Solution 1:* A triangle with sides of length 1, $\sqrt{3}$, and 2 is a right triangle because $(1)^2 + (\sqrt{3})^2 = 4 = 2^2$. A triangle whose sides are in the ratio $1 : \sqrt{3} : 2$ is similar to this triangle by SSS Similarity. Therefore, its angles must be the same as the angles of a triangle with sides of length 1, $\sqrt{3}$, and 2. Specifically, this means one of them must be a right angle.

Solution 2: A triangle with sides in the ratio $1 : \sqrt{3} : 2$ must have sides of lengths x, $x\sqrt{3}$, and $2x$ for some real number x. Since $(x)^2 + (x\sqrt{3})^2 = 4x^2 = (2x)^2$, these side lengths satisfy the Pythagorean Theorem. Therefore, a triangle with these sides as lengths must be a right triangle.

6.2.3

(a) Since $AE = AH = 1$ and $\angle HAE = 90°$, we have $\angle AHE = \angle AEH = 45°$. Similarly, $\angle BEF = \angle BFE = 45°$, so $\angle HEF = 180° - \angle AEH - \angle BEF = 180° - 45° - 45° = 90°$. It can be shown in the same way that $\angle EFG = \angle FGH = \angle GHE = 90°$, so $EFGH$ is a rectangle.

(b) We have $HE = \sqrt{2}AE = \sqrt{2}$, and $EF = \sqrt{2}BE = 2\sqrt{2}$. Therefore, the area of $EFGH$ is $HE \cdot EF = \sqrt{2} \cdot 2\sqrt{2} = \boxed{4}$.

6.2.4 We draw an altitude from Q to $\overline{PR}$ as shown. Since $PQ = QR$, $\angle P = \angle R$. Combining this with $\angle QMP = \angle QMR$ and $PQ = QR$, we have $\triangle QMP \cong \triangle QMR$ by AAS Congruence. Therefore, $\angle PQM = \angle RQM = 60°$, so $\triangle PMQ$ and $\triangle RMQ$ are 30-60-90 triangles. Since $\angle P = 180° - \angle PMQ - \angle PQM = 30°$, we know

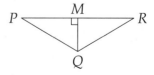

that $QM = PQ/2 = 4$, since QM must be 1/2 the length of the hypotenuse. Since $\overline{PM}$ is opposite the $60°$ angle in right triangle $\triangle PQM$, we have $PM = QM\sqrt{3} = 4\sqrt{3}$. Since $\triangle PQM \cong \triangle RQM$, we have $RM = PM = 4\sqrt{3}$. Therefore, $PR = PM + RM = 8\sqrt{3}$, so $[PQR] = (QM)(PR)/2 = \boxed{16\sqrt{3}}$.

6.2.5 We draw altitude $\overline{BH}$ from B to $\overline{AC}$ as shown. Since $\triangle ABH$ is a 30-60-90 triangle, we have $BH = AB/2 = 10$. Since $\triangle HBC$ is a 45-45-90 triangle, we have $BC = BH\sqrt{2} = \boxed{10\sqrt{2}}$.

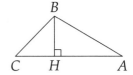

Exercises for Section 6.3

6.3.1

(a) Since $\{3,4,5\}$ is a Pythagorean triple and $300:400:500 = 3:4:5$, a triangle with side lengths 300, 400, and 500 is a right triangle.

(b) Since $\{3,4,5\}$ is a Pythagorean triple and $36:48:60 = 3:4:5$, a triangle with side lengths 36, 48, and 60 is a right triangle.

(c) Since $(\sqrt{5})^2 + (\sqrt{12})^2 = 17 \neq (\sqrt{13})^2$, a triangle with sides $\sqrt{5}$, $\sqrt{12}$, and $\sqrt{13}$ is not a right triangle.

(d) Since $\{8,15,17\}$ is a Pythagorean triple and $20:37.5:42.5 = 8:15:17$, a triangle with side lengths 20, 37.5, 42.5 is a right triangle.

(e) Since $1.44^2 + 1.96^2 = 5.9152 \neq (2.4)^2$, a triangle with side lengths 1.44, 1.96, and 2.4 is not a right triangle.

(f) Since $\{5,12,13\}$ is a Pythagorean triple and $15:36:39 = 5:12:13$, a triangle with side lengths 15, 36, and 39 is a right triangle.

6.3.2 For each part, let our legs have lengths a and b, and our hypotenuse have length c. From the Pythagorean Theorem, we must have $a^2 + b^2 = c^2$.

(a) $\boxed{\text{No.}}$ Suppose a and b are both odd. Then $a^2 + b^2$ must be even. Since $a^2 + b^2 = c^2$, we know that c^2 is even, too. Therefore, if c is an integer, it must be even. Hence, it is impossible for all three side lengths to be odd integers.

(b) $\boxed{\text{No.}}$ We have two cases to investigate.

First, we must check if it is possible for both legs to be even integers and the hypotenuse odd. If a and b are even, then $a^2 + b^2 = c^2$ tells us that c^2 must be even, too. Therefore, if c is an integer, it must be even. So, it is impossible for a and b to be even and c be odd.

Second, we check what happens if one of a or b is even and the other odd. In this case, $a^2 + b^2$ is odd, so if c is an integer, it must be odd. Therefore, it is impossible to have an odd leg, an even leg, and an even hypotenuse.

(c) $\boxed{\text{Yes.}}$ A triangle with sides 3/5, 4/5, and 1 is a right triangle since $(3/5)^2 + (4/5)^2 = 1^2$.

(d) $\boxed{\text{No.}}$ If both legs are integers, then $a^2 + b^2 = c^2$ is an integer. The square of a simple fraction (i.e., a reduced fraction with an integer other than 1 as the denominator) cannot be an integer, so since c^2 is an integer, c cannot be a fraction.

If one leg is an integer and the other is a simple fraction, then $a^2 + b^2$ is a simple fraction. Therefore, c^2 is a fraction. Since c^2 is a fraction, it is impossible for c to be an integer.

Therefore, it is impossible to have two sides of a right triangle with integer lengths and the other with a simple fraction as its length.

6.3.3 When she is finished, Susie has ridden south 10 miles and east 24 miles. We can draw a right triangle with the 10 miles south as one leg, the 24 miles east as another leg, and her distance from her starting point as the hypotenuse. Since $\{5, 12, 13\}$ is a Pythagorean triple, so is $\{10, 24, 26\}$ (which is double the aforementioned triple). Therefore, Susie is $\boxed{26 \text{ miles}}$ from her starting point.

6.3.4 Since $\{a, b, c\}$ is a Pythagorean triple, there is a right triangle ABC with sides a, b, and c. Since $na/a = nb/b = nc/c$, a triangle with sides na, nb, nc is similar to $\triangle ABC$ by SSS Similarity. The angles of our new triangle are therefore equal to those of $\triangle ABC$, so one of its angles must be a right angle. Therefore, $\{na, nb, nc\}$ is a Pythagorean triple.

6.3.5

(a) Since the legs of our right triangle have lengths 18 and b and the hypotenuse has length c, we have $18^2 + b^2 = c^2$. Therefore, we have $c^2 - b^2 = 18^2$.

(b) Factoring the left-hand side of $c^2 - b^2 = 18^2$ from part (a) gives $(c - b)(c + b) = 18^2$. Therefore, our problem is now finding all positive integer pairs (b, c) that satisfy the equation

$$(c - b)(c + b) = 18^2 = 324 = 2^2 \cdot 3^4.$$

Each pair of integers that multiply to 324 will give us a solution. However, if c and b are integers, then $c - b$ and $c + b$ must be both odd or both even. If they're both odd, then they can't multiply to 324, so they have to both be even. This limits our possibilities. Further, $c + b > c - b$, and now we can go through our cases.

(a) $c + b = 2 \cdot 3^4 = 162$, $c - b = 2$. Adding these equations gives $2c = 164$, so $c = 82$ and substitution gives $b = 80$. A quick check shows that $\{18, 80, 82\}$ is a Pythagorean triple.

(b) $c + b = 2 \cdot 3^3 = 54$, $c - b = 2 \cdot 3 = 6$. Adding these gives us $2c = 60$, so $c = 30$ and substitution gives $b = 24$. Therefore, this triple is $\{18, 24, 30\}$, which is a 3-4-5 triangle.

There are no more cases to consider (all others give us either an odd $c + b$ or $c - b$, or give us $c + b \le c - b$), so this shows that these are the only two Pythagorean triples with 18 as a leg.

6.3.6 Let the legs have lengths a and b. Since the hypotenuse has length 97, we have $a^2 + b^2 = 97^2$. We rearrange this and use the difference of squares factorization to give $a^2 = 97^2 - b^2 = (97 - b)(97 + b)$. This at least gives us a guide to use a little trial-and-error to find a and b. Since $(97 - b)(97 + b)$ must be a perfect square, we try to find values of b that make both $97 - b$ and $97 + b$ perfect squares. We do so by solving for b in $97 - b = 1^2$, $97 - b = 2^2$, etc., then checking if $97 + b$ is a perfect square as well. Since $97 - b = 5^2$ gives us $b = 72$ and $97 + 72 = 169 = 13^2$, we have our desired a and b. Testing we see that $\boxed{\{65, 72, 97\}}$ is indeed a Pythagorean triple.

Exercises for Section 6.4

6.4.1

(a) Since $AB/PQ = AC/QR$ and $\angle A = \angle Q$, we have $\triangle PQR \sim \triangle BAC$ by SAS Similarity (we could also say LL Similarity).

(b) These triangles are not similar. (We see this by noting that $\triangle DEF$ is a 30-60-90 triangle, but $\triangle WXY$ is not.)

(c) Since $AB/FG = BC/EG$ and $\angle A = \angle F = 90°$, we have $\triangle CAB \sim \triangle EFG$ by HL Similarity.

(d) From the Pythagorean Theorem (or our knowledge of Pythagorean triples), we find $MO = 25$ and $QR = 96$. From here, we can use SAS, SSS, or HL Similarity to prove that $\triangle MNO \sim \triangle PQR$.

6.4.2 Let O be the center of the circle and M be the midpoint of $\overline{AB}$. A radius of a circle that is perpendicular to a chord of the circle bisects the chord. Therefore, the altitude from O to $\overline{AB}$ meets $\overline{AB}$ at its midpoint, M. Since $\triangle OMA$ is a right triangle with hypotenuse $\overline{OA}$ and we are given that $OM = 3$ and $OA = 9$, we have $AM = \sqrt{OA^2 - OM^2} = 6\sqrt{2}$. Since M is the midpoint of $\overline{AB}$, we have $AB = 2AM = \boxed{12\sqrt{2}}$.

6.4.3

(a) From the Pythagorean Theorem we have $AC^2 = BC^2 - AB^2$ and $MO^2 = NO^2 - MN^2$.

(b) From $AB/BC = MN/NO$, we have $AB = (BC)(MN/NO)$. Using our expressions from the previous part, we have

$$\frac{AC^2}{MO^2} = \frac{BC^2 - AB^2}{NO^2 - MN^2}.$$

(c) Substituting our expression for AB into our expression for AC^2/MO^2, we have

$$\frac{AC^2}{MO^2} = \frac{BC^2 - \left(\frac{(BC)(MN)}{NO}\right)^2}{NO^2 - MN^2} = \frac{BC^2 - BC^2\left(\frac{MN^2}{NO^2}\right)}{NO^2 - MN^2} = \frac{BC^2\left(1 - \frac{MN^2}{NO^2}\right)}{NO^2 - MN^2}$$

$$= \frac{BC^2\left(\frac{NO^2 - MN^2}{NO^2}\right)}{NO^2 - MN^2} = \frac{BC^2(NO^2 - MN^2)}{NO^2(NO^2 - MN^2)} = \frac{BC^2}{NO^2}.$$

(d) Since $AC^2/MO^2 = BC^2/NO^2$, we have $AC/MO = BC/NO$ (since all lengths are positive). We can rearrange our given $AB/BC = MN/NO$ to have $AB/MN = BC/NO$. We thus have $AB/MN = BC/NO = AC/MO$, so by SSS, we have $\triangle ABC \sim \triangle MNO$.

6.4.4 We have $\angle ABC = 90°$, so by the Pythagorean Theorem, we have $AC = \sqrt{AB^2 + BC^2} = \sqrt{4^2 + 3^2} = \sqrt{25} = 5$. We also have $\angle CDA = 90° - \angle DAC = \angle CAB$, so triangles ABC and DCA are similar. Therefore, $CD/AB = AC/BC$, so

$$CD = AB \cdot \frac{AC}{BC} = 4 \cdot \frac{5}{3} = \boxed{\frac{20}{3}}.$$

6.4.5

(a) Since $\angle NMD = 180° - \angle NMB - \angle BMA = 180° - 90° - \angle BMA = 90° - \angle BMA = \angle ABM$ and $\angle MDN = \angle A$, we have $\triangle MDN \sim \triangle BAM$.

(b) Since $\triangle MDN \sim \triangle BAM$, we have $MN/BM = MD/AB = 1/2$ (since M is the midpoint of $\overline{AD}$ and $AD = AB$). Since we also know $AM/AB = 1/2$, we have $MN/BM = AM/AB$. Combined with $\angle BMN = \angle A = 90°$, this gives $\triangle BAM \sim \triangle BMN$ by SAS Similarity (or LL Similarity). Therefore, $\angle ABM = \angle MBN$.

Exercises for Section 6.5

6.5.1

(a) Our semiperimeter is $(8 + 8 + 8)/2 = 12$, so our area is $\sqrt{12(12-8)(12-8)(12-8)} = \boxed{16\sqrt{3}}$.

(b) Our semiperimeter is $(4 + 5 + 6)/2 = 15/2$, so the area is

$$\sqrt{(15/2)(15/2 - 4)(15/2 - 5)(15/2 - 6)} = \sqrt{(15/2)(7/2)(5/2)(3/2)} = \boxed{\frac{15}{4}\sqrt{7}}.$$

(c) Our semiperimeter is $(12 + 35 + 37)/2 = 42$, so the area is

$$\sqrt{(42)(42 - 12)(42 - 35)(42 - 37)} = \sqrt{(42)(30)(7)(5)} = \boxed{210}.$$

(d) Our semiperimeter is $(6\sqrt{2} + 7\sqrt{2} + 9\sqrt{2})/2 = 11\sqrt{2}$, so the area is

$$\sqrt{(11\sqrt{2})(11\sqrt{2} - 6\sqrt{2})(11\sqrt{2} - 7\sqrt{2})(11\sqrt{2} - 9\sqrt{2})} = \sqrt{(11\sqrt{2})(5\sqrt{2})(4\sqrt{2})(2\sqrt{2})} = \boxed{4\sqrt{110}}.$$

6.5.2 We know HI, so if we can find the area of $\triangle GHI$, we can find the length of the altitude from G to $\overline{HI}$. Since the semiperimeter of $\triangle GHI$ is $(5 + 7 + 8)/2 = 10$, the area of $\triangle GHI$ is $\sqrt{10(10 - 5)(10 - 7)(10 - 8)} = \sqrt{10(5)(3)(2)} = 10\sqrt{3}$. Letting our desired height be h, we can also write $[GHI] = (h)(HI)/2$. Therefore, we have $h = (10\sqrt{3})(2)/HI = \boxed{20\sqrt{3}/7}$.

Exercises for Section 6.6

6.6.1 *45-45-90 Triangle:* We start with a segment $\overline{AB}$, then construct a line n through A perpendicular to $\overline{AB}$ as described in the text. We then draw a circle with center A and radius AB. Call one of the points where the circle meets n point C. Then we have $\angle BAC = 90°$ and $AB = AC$. Therefore, drawing $\overline{BC}$ completes our 45-45-90 triangle. The construction is shown atop the next page.

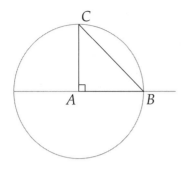

Figure 6.1: Constructing a 45-45-90 Triangle

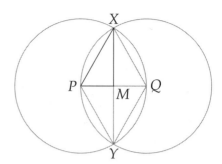

Figure 6.2: Constructing a 30-60-90 Triangle

30-60-90 Triangle: We start with segment $\overline{PQ}$ and proceed with our construction of an equilateral triangle (as described in Chapter 3), constructing circles with centers P and Q and radius PQ. Let the two points where these circles meet be X and Y, and the point where $\overline{XY}$ meets $\overline{PQ}$ be M. Since $\angle XPY = 120°$ and $PX = PY$, we have $\angle PXY = \angle PYX = 30°$. Since $\angle PXM = \angle PXY = 30°$ and $\angle XPM = 60°$, we have $\angle XMP = 180° - \angle PXY - \angle XPM = 90°$, so $\triangle XPM$ is a 30-60-90 triangle.

6.6.2 *Solution 1:* We can use our construction from the first Exercise in this section to start with a segment of length 1 and create a 45-45-90 triangle with legs of length 1. The hypotenuse will then have length $\sqrt{2}$. We can then construct a right triangle with legs of length 1 and $\sqrt{2}$. The hypotenuse of this triangle will have length $\sqrt{3}$.

Solution 2: In the previous problem we constructed a 30-60-90 triangle. If we did that construction with $PQ = 2$, we would have $PM = PX/2 = PQ/2 = 1$ and $XM = PM\sqrt{3} = \sqrt{3}$. Therefore, we can start with our segment of length 1, copy it onto the end of itself to get a segment of length 2. Then, we proceed with the construction described in the solution to the previous Exercise to construct a 30-60-90 triangle with a leg of length $\sqrt{3}$. (This solution is inspired by asking ourselves 'Where have we seen $\sqrt{3}$ before?')

Review Problems

6.26

(a) $BC = \sqrt{AC^2 + AB^2} = \sqrt{49 + 9} = \boxed{\sqrt{58}}$.

(b) $IG = \sqrt{HI^2 - HG^2} = \sqrt{441 - 338} = \boxed{\sqrt{103}}$.

(c) $JL = \sqrt{KJ^2 + KL^2} = \sqrt{17 + 32} = \boxed{7}$.

6.27

(a) Since the ratio of the length of leg $\overline{PQ}$ to the length of hypotenuse $\overline{PR}$ is $3.6/6 = 3/5$, we know $\triangle PQR$ is a 3-4-5 right triangle. Therefore, $QR = (4/3)PQ = \boxed{4.8}$.

(b) Since $SU/ST = 8/15$ and $\overline{SU}$ and $\overline{ST}$ are legs of right triangle STU, we know the triangle is a 8-15-17 right triangle. Therefore, $TU = (17/8)SU = \boxed{51}$.

(c) Since $VW/VX = 5/13$ and $\overline{VW}$ is a leg while $\overline{VX}$ is the hypotenuse, we know that $\triangle VWX$ is a

5-12-13 triangle. Therefore, $WX = (12/13)VX = \boxed{12\sqrt{2}}$.

6.28

(a) Since $MN = NO$, we know that $\triangle MNO$ is a 45-45-90 triangle. Therefore, $MN = NO = MO/\sqrt{2} = \boxed{4\sqrt{2}}$.

(b) Since $\triangle PQR$ is a 30-60-90 right triangle, we have $PR = 2PQ = \boxed{12\sqrt{3}}$ and $QR = PQ\sqrt{3} = \boxed{18}$.

6.29

(a) Since $11^2 + 16^2 = 377 \neq 19^2$, these cannot be the side lengths of a right triangle.

(b) Writing these with a common denominator, we have $3/15, 4/15, 5/15$. These numbers are in the ratio $3:4:5$, and therefore a triangle with these side lengths is a 3-4-5 right triangle.

(c) Since $(\sqrt{73})^2 + (2\sqrt{2})^2 = 81 = 9^2$, a triangle with these side lengths is a right triangle.

(d) Since $(\sqrt{0.5})^2 + (\sqrt{1.2})^2 = 1.7 \neq (\sqrt{1.3})^2$, these cannot be the side lengths of a right triangle.

(e) Writing these as $7(0.11), 24(0.11)$, and $25(0.11)$, we see that these numbers are in the ratio $7:24:25$. Therefore, a triangle with these side lengths is a 7-24-25 right triangle.

6.30 Let $\triangle XYZ$ have $XY = 3$, $\angle X = 90°$, and $\angle Y = 60°$. This is a 30-60-90 right triangle with $\angle Z = 30°$, so $XZ = XY\sqrt{3} = 3\sqrt{3}$ and $YZ = 2XY = 6$. Therefore, we have $AB = XY$, $BC = YZ$ and $\angle B = \angle Y$, so $\triangle ABC \cong \triangle XYZ$ by SAS Congruence. From this triangle congruence we have $\angle A = \angle X = 90°$, so $\triangle ABC$ must be a right triangle. Similarly, we can show that if we ever have a triangle in which one side is twice as long as the other, and the angle between the sides is $60°$, then the triangle must be a 30-60-90 triangle. (We can also solve this problem using SAS Similarity to show $\triangle ABC$ is similar to a 30-60-90 triangle.)

6.31

(a) The large square contains four triangles, plus a square of side length a and a square of side length b, so the total area of the large square is $4K + a^2 + b^2$.

(b) The large square again contains four triangles, plus a square of side length c, so the total area of the large square is $4K + c^2$.

(c) Since both large squares have area $(a + b)^2$, the two expressions for the area must be equal. So, $4K + a^2 + b^2 = 4K + c^2$, which implies that $a^2 + b^2 = c^2$, and the Pythagorean Theorem is proven.

6.32 Since $AB = BC = 1$ and $\angle ABC = 90°$, $\angle BAC = \angle ACB = 45°$. Since $\overline{BC} \parallel \overline{AD}$, we have $\angle BAD = 180° - \angle B = 90°$. Therefore, $\angle CAD = \angle BAD - \angle BAC = 45°$, so $\angle ADC = 45°$. We also have $AC = \sqrt{2} \cdot AB = \sqrt{2}$, so $AD = \sqrt{2} \cdot AC = 2$. Since $\overline{AE} \parallel \overline{CD}$, $\angle EAD = \angle ADC = 45°$. Therefore, $\triangle AED$ is a 45-45-90 triangle, so finally $AE = \sqrt{2} \cdot AD = \boxed{2\sqrt{2}}$.

6.33 Let the triangle be $\triangle ABC$ with $AB = BC$. We are given $AC = 10$. Let $\overline{BH}$ be the altitude to $\overline{AC}$. As we have seen several times, $\overline{BH}$ divides $\overline{AC}$ into two equal segments. Since $[ABC] = 60$, we know that $(BH)(AC)/2 = 60$, so $BH = 120/AC = 12$. We also have $AH = CH = 10/2 = 5$. Therefore, from right triangle ABH, we have $AB = 13$, so $BC = 13$ as well. Therefore, the perimeter of our triangle is $13 + 13 + 10 = \boxed{36}$.

6.34

(a) Since the side lengths are in the ratio $3:4:5$, we know our triangle is a right triangle with legs of length 6 and 8. Therefore, its area is $(6)(8)/2 = \boxed{24}$.

(b) We use Heron's Formula. The semiperimeter is $(3 + 4 + 6)/2 = 13/2$, so the area is

$$\sqrt{(13/2)(13/2 - 3)(13/2 - 4)(13/2 - 6)} = \sqrt{(13/2)(7/2)(5/2)(1/2)} = \boxed{\sqrt{455}/4}.$$

(c) Since these side lengths satisfy the Pythagorean Theorem, the triangle must be a right triangle with legs $\sqrt{6}$ and $\sqrt{24}$. Therefore, its area is $(\sqrt{6})(\sqrt{24})/2 = \boxed{6}$.

(d) Writing these with a common denominator, the numbers are $7/7$, $24/7$, and $25/7$. Therefore, a triangle with these side lengths is a 7-24-25 right triangle. Thus, its area is $(1)(24/7)/2 = \boxed{12/7}$.

(e) We use Heron's Formula. The semiperimeter is $(5+6+7)/2 = 9$, so the area is $\sqrt{9(9-5)(9-6)(9-7)} = \boxed{6\sqrt{6}}$.

6.35

(a) We have $\angle EFB = \angle CAB = 90°$ and $\angle ABC = \angle FBE$. Therefore, triangles EFB and CAB are similar.

(b) Applying the Pythagorean Theorem to $\triangle ABC$ gives $BC = 20$. Since $\triangle EFB \sim \triangle CAB$, we have $EF/EB = AC/BC = 12/20 = 3/5$. We are given $EB = AB/2 = 8$, so we have $EF = (3/5)(EB) = \boxed{24/5}$. Similarly, we can show $\triangle CDG \sim \triangle CBA$, so $DG/CD = AB/BC$ and $DG = (AB/BC)(CD) = \boxed{24/5}$.

(c) *Solution 1:* Since $[AEFGD] = [ABC] - [EFB] - [CDG]$ and $[ABC] = (AB)(AC)/2 = 96$, if we find $[CDG]$ and $[EFB]$, we can compute $[AEFGD]$. From $\triangle EFB \sim \triangle CAB$, we have $FB/EF = AB/AC$. Therefore, $FB = (AB/AC)(EF) = 32/5$, so $[EFB] = (EF)(FB)/2 = (32)(12)/25$. Similarly, from $\triangle CDG \sim \triangle CBA$, we have $CG/DG = AC/AB$, so $CG = (AC/AB)(DG) = 18/5$. Therefore, $[CDG] = (CG)(DG)/2 = (18)(12)/25$. Finally, we have

$$[AEFGD] = [ABC] - [EFB] - [CDG] = 96 - \frac{(32)(12)}{25} - \frac{(18)(12)}{25} = 96 - \frac{(50)(12)}{25} = \boxed{72}.$$

Solution 2: Let H be the foot of the perpendicular from A to $\overline{BC}$. Then $\angle AHB = \angle EFB = 90°$ and $\angle ABF = \angle EBF$, so triangles AHB and EFB are similar, and

$$\frac{[EFB]}{[AHB]} = \left(\frac{EB}{AB}\right)^2 = \frac{1}{4}.$$

Similarly, triangles DGC and AHC are similar, with

$$\frac{[DGC]}{[AHC]} = \left(\frac{DC}{AC}\right)^2 = \frac{1}{4}.$$

Hence,

$$[EFB] + [DGC] = \frac{1}{4}[AHB] + \frac{1}{4}[AHC] = \frac{1}{4}[ABC].$$

Then finally

$$[AEFGD] = [ABC] - [EFB] - [DGC] = [ABC] - \frac{1}{4}[ABC] = \frac{3}{4}[ABC] = \frac{3}{4} \cdot \frac{1}{2}(AB \cdot AC) = \frac{3}{8}(16 \cdot 12) = \boxed{72}.$$

6.36 Let the equilateral triangle have side length s. Drawing an altitude of the triangle splits it into two 30-60-90 triangles, from which we find that the altitude has length $s\sqrt{3}/2$. Since we are given that this altitude has length 8, we have $s\sqrt{3}/2 = 8$, so $s = 16/\sqrt{3} = 16\sqrt{3}/3$. Therefore, the area of the equilateral triangle is

$$\frac{\sqrt{3}}{4}s^2 = \frac{\sqrt{3}}{4} \cdot \frac{16^2}{3} = \boxed{\frac{64\sqrt{3}}{3}}.$$

6.37 By the Pythagorean Theorem, we have $PM = \sqrt{MN^2 - PN^2} = \sqrt{13^2 - 12^2} = 5$ and $PO = \sqrt{NO^2 - PN^2} = \sqrt{37^2 - 12^2} = \sqrt{1369 - 144} = \sqrt{1225} = 35$. (We could also have used Ptyhagorean triples to figure out PM and PO.) Therefore, the area of triangle MNO is

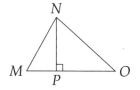

$$\frac{1}{2}MO \cdot PN = \frac{1}{2}(MP + PO) \cdot PN = \frac{1}{2}(5 + 35) \cdot 12 = \boxed{240}.$$

6.38 Since $\angle ACD = 90° - \angle DCB = \angle B$ and $\angle ADC = \angle CDB$, we have $\triangle ACD \sim \triangle CBD$. Therefore, $CD/DB = AD/CD$, so $CD^2 = (AD)(DB) = xy$ and $CD = \sqrt{xy}$.

6.39 Since $AB = AC$, $\angle ABF = \angle ACF$. Therefore, by AAS, triangles ABF and ACF are congruent. It follows that $BF = CF$.

We have $\angle FDB = \angle FEC = 90°$, $BF = CF$, and $\angle B = \angle C$, so by AAS, triangles FDB and FEC are congruent. Hence, $FD = FE$. (We might also solve this problem by showing $[ABF] = [ACF]$, so since $AB = AC$, the altitudes from F to $\overline{AB}$ and $\overline{AC}$ are equal.)

6.40 *Solution 1:* Since $\angle ACB = \angle DCE$ and $\angle ABC = \angle CDE$, we have $\triangle ABC \sim \triangle EDC$ by AA Similarity. Letting $BC = x$, we have $CD = BD - BC = 6 - x$. Our similarity gives us $BC/BD = AB/DE$, so $x/(6-x) = 3/5$. Cross-multiplying gives $5x = 3(6 - x)$, and solving this equation gives us $x = 9/4$. Since $BC/AB = 3/4$, we know that $\triangle ABC$ is a 3-4-5 right triangle. Therefore, $AC = AB(5/4) = 15/4$. Since $\triangle ABC \sim \triangle EDC$, we know that $\triangle EDC$ is also a 3-4-5 triangle. Therefore, $CE = (5/4)(DE) = 25/4$, so $AE = AC + CE = \boxed{10}$.

Solution 2: Take point F such that $BDEF$ is a rectangle. Then $EF = BD = 6$ and $BF = DE = 5$, so $AF = AB + BF = 3 + 5 = 8$. By the Pythagorean Theorem, $AE = \sqrt{AF^2 + EF^2} = \sqrt{8^2 + 6^2} = \sqrt{64 + 36} = \sqrt{100} = \boxed{10}$.

6.41 We seek a Pythagorean triple with 22 as one of the legs. There are several approaches we could take.

Solution 1: We use our knowledge of Pythagorean triples. Specifically, we know that $\{11, 60, 61\}$ is a Pythagorean triple, so multiplying these by two must give another triple: $\{22, 120, 122\}$.

Solution 2: Use Problem 6.16. We halve 22, then square the result, to get $(22/2)^2 = 121$. We subtract 1 to get the other leg, and add 1 to get the hypotenuse. As we proved in Problem 6.16, this will give us a Pythagorean triple starting from any even number. Starting with 22 gives us $\{22, 120, 122\}$.

Solution 3: Take the approach of Problem 6.3.5. We let the other leg be b and the hypotenuse c, so we have $22^2 + b^2 = c^2$. Rearranging this gives $c^2 - b^2 = 22^2$. We factor this to get $(c - b)(c + b) = 22^2$. As discussed in our solution to Problem 6.3.5, both $c - b$ and $c + b$ must be even. They can't be equal, since b is nonzero. Therefore, the only option that gives positive integers for c and b is $c + b = 2(121) = 242$ and $c - b = 2$. This gives $b = 120$ and $c = 122$ for the Pythagorean triple $\{22, 120, 122\}$.

The desired perimeter then is $22 + 120 + 122 = \boxed{264}$.

6.42 Let $AX = x$. Since $XA = XY$, we have $XY = x$. Since $\triangle XYZ$ is a 30-60-90 triangle, we have $XZ = XY\sqrt{3}$, so $x + 6 - \sqrt{12} = x\sqrt{3}$. Solving for x, we have:

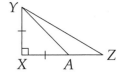

$$x = \frac{6 - \sqrt{12}}{\sqrt{3} - 1} = \frac{6 - 2\sqrt{3}}{\sqrt{3} - 1} \cdot \frac{\sqrt{3} + 1}{\sqrt{3} + 1} = 2\sqrt{3}.$$

Therefore, $XY = 2\sqrt{3}$ and $XZ = XA + AZ = 6$, so $[XYZ] = (XY)(XZ)/2 = \boxed{6\sqrt{3}}$.

6.43 The sides are $15 = 5 \cdot 3$, $20 = 5 \cdot 4$, and $25 = 5 \cdot 5$, so the triangle is a 3-4-5 right triangle. Therefore, the area of the triangle is $(15)(20)/2 = 150$. In any triangle, the shortest altitude is perpendicular to the longest side, which in this case is the hypotenuse 25. Letting this altitude have length h, we have $25h/2 = 150$, so $h = 2(150)/25 = \boxed{12}$.

Challenge Problems

6.44 Let the legs of the triangle be x and y. Our given information tells us that $xy/2 = 210$ and $x^2 + y^2 = 29^2$. Below are two solutions from here.

Solution 1: We can rearrange $xy/2 = 210$ to $xy = 420$ and try to guess integer solutions for x and y. We know that x and y are both less than 29, since the hypotenuse of length 29 must be the longest side of the triangle. This dramatically limits our options to try. We only have two to test: 15 and 28, and 20 and 21. Since $20^2 + 21^2 = 29^2$, we have our desired sides. (Notice also that a knowledge of Pythagorean triples would be very handy here!) Therefore, our perimeter is $20 + 21 + 29 = \boxed{70}$.

Solution 2: We can also use algebra to get the perimeter without even finding the sides. We notice that $(x + y)^2 = x^2 + 2xy + y^2 = 29^2 + 2(420) = 1681$, so $x + y = 41$. Therefore, the perimeter is $x + y + 29 = \boxed{70}$.

6.45 We can find OS several ways. If the diagram is valid, then each way should give the same answer. First, $[OQS] = (QS)(OP)/2 = 24$. We can also write $[OQS] = (OS)(QR)/2 = (3/2)(OS)$. Therefore, we have $(3/2)(OS) = 24$, so $OS = 16$. We can also use the Pythagorean Theorem to write $OS = \sqrt{OP^2 + PS^2} = \sqrt{16 + 225} = \sqrt{241}$. Since OS can't take on two values at the same time, the diagram must be invalid. (Make sure you see why the diagram is simply invalid, rather than this being a case where there are two possible values of OS!) Also, we could have used right triangles $\triangle ORQ$ and $\triangle RQS$ to find a third value for OS.

6.46 First, note that $\angle ADP = \angle PDB = \angle BDC = 90°/3 = 30°$. Since $\overline{AB} \parallel \overline{CD}$, $\angle PBD = \angle CDB = 30°$. Therefore, triangle BDP is isosceles with $PB = PD$. Triangle DBC is a 30-60-90 triangle. Since $BC = AD = 1$, we have $BD = 2BC = 2$.

Similarly, triangle DPA is a 30-60-90 triangle. Since $AD = 1$, $PD = (2/\sqrt{3}) \cdot AD = 2/\sqrt{3} = 2\sqrt{3}/3$.

Hence, the perimeter of triangle BDP is

$$PB + PD + BD = 2PD + BD = 2 \cdot \frac{2\sqrt{3}}{3} + 2 = \boxed{2 + \frac{4\sqrt{3}}{3}}.$$

6.47 Let a and b be the legs of the triangle. Then by the Pythagorean Theorem, $a^2 + b^2 = 8^2 = 64$. Since the area is 8, we have $ab/2 = 8$, so $ab = 16$. Then

$$(a + b)^2 = a^2 + 2ab + b^2 = (a^2 + b^2) + 2ab = 64 + 2 \cdot 16 = 96,$$

so $a + b = \sqrt{96} = 4\sqrt{6}$. Since the perimeter is $a + b + c$ and we know both $a + b$ and c, we don't even have to find a and b. Our perimeter is $a + b + c = (a + b) + c = \boxed{8 + 4\sqrt{6}}$.

If we wanted to find a and b, we could note that

$$(a - b)^2 = a^2 - 2ab + b^2 = (a^2 + b^2) - 2ab = 64 - 2 \cdot 16 = 32,$$

so $a - b = \sqrt{32} = 4\sqrt{2}$ (assuming $a \geq b$). Combining this with $a + b = 4\sqrt{6}$ gives $a = 2(\sqrt{6} + \sqrt{2})$ and $b = 2(\sqrt{6} - \sqrt{2})$.

6.48

(a) We have $AB = CD = 24$, $\angle ABE = \angle CDE = 90°$, and $\angle AEB = \angle CED$, so by AAS congruence, triangles ABE and CDE are congruent.

(b) By part (a), $AE = CE$, so triangle AEC is isosceles. Therefore, $\overline{EM} \perp \overline{AC}$, as shown many times in the text.

(c) Since $AB = 24 = 8 \cdot 3$, $BC = 32 = 8 \cdot 4$, and $\angle B = 90°$, triangle ABC is a 3-4-5 right triangle. Therefore, $AC = 8 \cdot 5 = 40$, so $MC = AC/2 = 20$. Since $\angle EMC = \angle ABC = 90°$ and $\angle ECM = \angle BCA$, triangles ABC and EMC are similar. Therefore, $EM/MC = AB/BC$, so

$$EM = \frac{MC \cdot AB}{BC} = \frac{20 \cdot 24}{32} = 15.$$

Hence, the area of the triangle is $(AC \cdot EM)/2 = (40 \cdot 15)/2 = \boxed{300}$.

6.49 Since $AB = 12 = 4 \cdot 3$, $BC = 16 = 4 \cdot 4$, and $AC = 20 = 4 \cdot 5$, we know that $\triangle ABC$ is a 3-4-5 right triangle with $\angle ABC = 90°$. Therefore, the area of the triangle is $(AB)(BC)/2 = 96$. Triangles ABD and ABC have equal heights from A to bases $\overline{AD}$ and $\overline{AC}$, respectively, so $[ABD]/[ABC] = AD/AC = 3/5$. Therefore, $[ABD] = (3/5)[ABC] = \boxed{288/5}$.

6.50 Let O be the center of the Ferris wheel, let A be the bottom of the wheel, and let B be the point on the wheel that is 10 feet from the bottom. Let C be the foot of the perpendicular from B to $\overline{OA}$, so $CA = 10$.

Then $OC = OA - CA = 10$, $OB = 20$, and $\angle OCB = 90°$. Hence, triangle BOC is a 30-60-90 triangle, so $\angle AOB = \angle COB = 60°$. Since there are 360° in a full revolution, and there is 1 minute, or 60 seconds, in a revolution, it takes $60 \cdot (60°/360°) = \boxed{10}$ seconds for the rider to go from point A to point B.

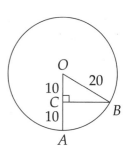

6.51

(a) We have $\angle DCF = \angle FCE = \angle ECB = 30°$, so triangles CEB and CFD are 30-60-90 triangles. Therefore, $BC = \sqrt{3}BE = \boxed{6\sqrt{3}}$.

(b) We have $AD = BC = 6\sqrt{3}$, so $DF = AD - AF = \boxed{6\sqrt{3} - 2}$.

(c) We have $CD = \sqrt{3}DF = 18 - 2\sqrt{3}$, so the area of rectangle $ABCD$ is $CD \cdot BC = (18 - 2\sqrt{3})(6\sqrt{3}) = \boxed{108\sqrt{3} - 36}$.

6.52 We can solve this either by using our knowledge of Pythagorean triples, or with algebra.

Solution 1: We know that $\{14, 48, 50\}$ and $\{30, 40, 50\}$ are both Pythagorean triples with hypotenuse 50. We notice that while the larger leg decreases 8, the shorter increases by $30 - 14 = 16$, which is exactly the situation described in the problem. Therefore, we conclude that the top of the ladder must be $\boxed{40\ \text{feet}}$ from the ground after the slide.

Solution 2: We can use algebra if we don't recognize the Pythagorean triples. Specifically, we let the ladder initially rest against the wall at a point x feet above the ground, and the base initially be y feet from the wall. After the slide, the top will be $x - 8$ feet from the ground and the base will be $y + 16$ feet from the wall. Applying the Pythagorean Theorem to both situations gives:

$$\begin{aligned} x^2 + y^2 &= 50^2 \\ (x - 8)^2 + (y + 16)^2 &= 50^2 \end{aligned}$$

Subtracting the second from the first yields $x = 2y + 20$ (after a fair amount of algebra). Substituting this into our first equation above gives $(2y + 20)^2 + y^2 = 50^2$, so $5y^2 + 80y + 400 = 2500$. Rearranging and dividing by 5 gives $y^2 + 16y - 420 = 0$. Factoring gives $(y - 14)(y + 30) = 0$. Since y must be positive, we have $y = 14$, from which we find $x = 48$, so the top of the ladder must be $48 - 8 = \boxed{40\ \text{feet}}$ from the ground after the slide.

6.53 Letting the legs be a and b, we have $a^2 + b^2 = 73^2$. Rearranging, we have $a^2 = 73^2 - b^2 = (73 - b)(73 + b)$. We try letting $73 - b$ equal perfect squares, but none of the resulting b's makes $73 + b$ a perfect square. Another possibility is that both $73 - b$ and $73 + b$ are the same constant multiple of two different perfect squares. The simplest such case is if this constant multiple is 2. So, we try letting $73 - b$ equal two times a perfect square, and check if $73 + b$ is also two times a perfect square. This works! $73 - b = 18$ gives us $b = 55$, from which we have $73 + b = 128$. Therefore, $a^2 = (18)(128)$, so $a = \sqrt{(18)(128)} = (3)(2)(8) = 48$. So, the desired Pythagorean triple is $\boxed{\{48, 55, 73\}}$.

6.54 Let x be the hypotenuse, and let y be the other leg, so $x^2 - y^2 = (x + y)(x - y) = 24^2 = 576 = (2^6)(3^2)$. We note first that it is impossible for $x + y$ and $x - y$ to both be odd, since their product is even. Further, we note that if one is odd and one even, then their sum, which equals $2x$, is odd. This means x would be a fraction. Since we want only integer solutions, we discard these cases. Finally, we must have $x + y > x - y$. So, we're ready to list the possibilities:

$x + y$	$x - y$	x	y
$2^5 \cdot 3^2$	2	145	143
$2^5 \cdot 3$	$2 \cdot 3$	51	45
2^5	$2 \cdot 3^2$	25	7
$2^4 \cdot 3^2$	2^2	74	70
$2^4 \cdot 3$	$2^2 \cdot 3$	30	18
$2^3 \cdot 3^2$	2^3	40	32
$2^2 \cdot 3^2$	2^4	26	10

Hence, there are $\boxed{7}$ distinct right triangles with a leg of length 24, namely those shown in the table above.

6.55

(a) If we have the area of $\triangle RST$, we can find the height to $\overline{ST}$. Using Heron's, or looking back at either Problem 6.21 or 6.23 in the text, we have $[RST] = 84$. Letting our desired height be h, we have $(h)(ST)/2 = 84$, so $h = 2(84)/14 = \boxed{12}$.

(b) Let the foot of the altitude from R be H. Since $RS = 13$ and $RH = 12$, we have $SH = 5$ from right triangle $\triangle RHS$. Since $SM = ST/2 = 7$, we have $HM = 2$. Therefore, triangle $\triangle RHM$ gives us

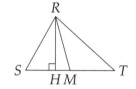

$$RM = \sqrt{RH^2 + RM^2} = \sqrt{144 + 4} = \boxed{2\sqrt{37}}.$$

6.56 In going from the first step to the second step, we add a triangle of side length $1/3$ onto each of the 3 sides of the original triangle. We also increase the number of sides from 3 to $3(4) = 12$, each of length $1/3$. In our next step, we add a triangle of side length $(1/3)/3 = 1/9$ to each of the 12 segments of length $1/3$, and we increase the number of sides from 12 to $12(4) = 48$. Each of these 48 sides has length $(1/3)/3 = 1/9$. And so on.

We therefore start with an area of $1^2\sqrt{3}/4$. We then add $3\left((1/3)^2\sqrt{3}/4\right)$, then $3 \cdot 4\left(((1/3)^2)^2\sqrt{3}/4\right)$, then $3 \cdot 4^2\left(((1/3)^3)^2\sqrt{3}/4\right)$, and so on. Similarly, we find that at each step after the first, we add an area equal to $4/9$ the area added the previous step because we add 4 times as many triangles, but each has side length $1/3$ of the triangles added in the previous step. Hence, our total area is:

$$
\begin{aligned}
\text{Area} &= \frac{1^2\sqrt{3}}{4} + \frac{3\left(\left(\frac{1}{3}\right)^2\sqrt{3}\right)}{4} + \frac{3 \cdot 4\left(\left(\frac{1}{3^2}\right)^2\sqrt{3}\right)}{4} + \frac{3 \cdot 4^2\left(\left(\frac{1}{3^3}\right)^2\sqrt{3}\right)}{4} + \cdots \\
&= \frac{1^2\sqrt{3}}{4} + \frac{3\left(\left(\frac{1}{3}\right)^2\sqrt{3}\right)}{4} + \frac{4}{9}\left(\frac{3\left(\left(\frac{1}{3}\right)^2\sqrt{3}\right)}{4}\right) + \left(\frac{4}{9}\right)^2\left(\frac{3\left(\left(\frac{1}{3}\right)^2\sqrt{3}\right)}{4}\right) + \cdots \\
&= \frac{1^2\sqrt{3}}{4} + \frac{\frac{3\left(\left(\frac{1}{3}\right)^2\sqrt{3}\right)}{4}}{1 - \frac{4}{9}} = \frac{1^2\sqrt{3}}{4} + \frac{9}{5} \cdot \frac{3\left(\left(\frac{1}{3}\right)^2\sqrt{3}\right)}{4} = \boxed{\frac{2\sqrt{3}}{5}}.
\end{aligned}
$$

(We used the formula for the sum of an infinite geometric series to simplify our expression. If you are unfamiliar with this formula, look back to the solution in this book to the problem in Chapter 4 with the same set-up. In that problem, we examined the perimeter of this figure, and found it to be infinite. Here, we find that the area is finite. Is this possible? Look up Koch's snowflake on the internet for an answer!)

6.57 Let $x = OX$ and $y = OY$. Then $OM = x/2$ and $ON = y/2$. Then by the Pythagorean Theorem applied to $\triangle XON$ and $\triangle MOY$, we have

$$19^2 = XN^2 = OX^2 + ON^2 = x^2 + \left(\frac{y}{2}\right)^2 = x^2 + \frac{y^2}{4},$$

$$22^2 = YM^2 = OY^2 + OM^2 = y^2 + \left(\frac{x}{2}\right)^2 = y^2 + \frac{x^2}{4}.$$

Adding the two equations, we get

$$\frac{5x^2 + 5y^2}{4} = 19^2 + 22^2 = 845,$$

so $x^2 + y^2 = 4/5 \cdot 845 = 676$. Therefore, $XY = \sqrt{x^2 + y^2} = \sqrt{676} = \boxed{26}$.

6.58 We present two solutions.

Solution 1: We have a 30° angle and we need to find a length, but we have no obvious tools to use to find the length. Therefore, we try introducing a 30-60-90 triangle by drawing the altitude from B to $\overline{AC}$. We let $AB = x$. Since $\triangle ABH$ is a 30-60-90 triangle, we have $BH = x/2$ and $AH = x\sqrt{3}/2$. Since $AC = AB = x$, we have $HC = AC - AH = x - x(\sqrt{3}/2)$. Now we can use the Pythagorean Theorem on $\triangle BCH$ to find x. We have $BH^2 + HC^2 = BC^2$, so we have

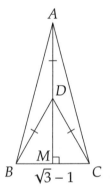

$$\left(\frac{x}{2}\right)^2 + \left(x - \frac{x\sqrt{3}}{2}\right)^2 = (\sqrt{3} - 1)^2.$$

Expanding both sides and doing a little algebra gives us $x^2(2 - \sqrt{3}) = 4 - 2\sqrt{3}$, from which we find $x^2 = 2$, so $x = \boxed{\sqrt{2}}$, since our length must be positive.

Solution 2: We use what we know about equilateral triangles. Since $AB = AC$, $\angle ABC = \angle ACB$. But $\angle ABC + \angle ACB = 180° - \angle BAC = 150°$, so $\angle ABC = \angle ACB = 150°/2 = 75°$.

Construct point D inside triangle ABC such that BCD is equilateral, so $\angle DCB = \angle DBC = 60°$ and $BD = DC = BC = \sqrt{3} - 1$. Then $\angle ABD = \angle ACD = \angle ABC - \angle DBC = 75° - 60° = 15°$.

Let M be the midpoint of $\overline{BC}$. Then $\overline{AM}$ is an altitude of triangle ABC, and by symmetry, D lies on $\overline{AM}$. Furthermore, $\angle BAM = \angle CAM = \angle BAC/2 = 15°$, so triangles ABD and ACD are isosceles, with $AD = BD = CD = BC = \sqrt{3} - 1$.

Then the height of equilateral triangle BCD is

$$DM = \frac{\sqrt{3}}{2} BC = \frac{\sqrt{3}}{2}(\sqrt{3} - 1) = \frac{3 - \sqrt{3}}{2},$$

so

$$AM = AD + DM = \sqrt{3} - 1 + \frac{3 - \sqrt{3}}{2} = \frac{1 + \sqrt{3}}{2}.$$

By the Pythagorean Theorem applied to $\triangle ABM$, we have

$$AB^2 = AM^2 + BM^2 = AM^2 + \left(\frac{BC}{2}\right)^2 = \frac{(1 + \sqrt{3})^2}{4} + \frac{(\sqrt{3} - 1)^2}{4} = \frac{1 + 2\sqrt{3} + 3 + 3 - 2\sqrt{3} + 1}{4} = \frac{8}{4} = 2,$$

so $AB = \boxed{\sqrt{2}}$.

6.59

(a) We draw altitude $\overline{AH}$ to form 30-60-90 triangle $\triangle ABH$. We thus have $AH = AB/2 = 3$ and $BH = AH\sqrt{3} = 3\sqrt{3}$. Then, $CH = BC - BH = 8 - 3\sqrt{3}$. We can then apply the Pythagorean Theorem to $\triangle AHC$ to find $AC = \sqrt{AH^2 + HC^2} = $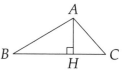

$$\sqrt{9 + (64 - 48\sqrt{3} + 27)} = \sqrt{100 - 48\sqrt{3}} = \boxed{2\sqrt{25 - 16\sqrt{3}}}.$$

(b) Again we draw altitude $\overline{AH}$, this time forming 45-45-90 $\triangle ABH$. We thus have $AH = AB/\sqrt{2} = 3\sqrt{2}$ and $BH = AH = 3\sqrt{2}$, so $CH = 8 - 3\sqrt{2}$. We then apply the Pythagorean Theorem as in the previous

part to find $AC = \sqrt{AH^2 + HC^2} = \sqrt{18 + (64 - 48\sqrt{2} + 18)} = \sqrt{100 - 48\sqrt{2}} = \boxed{2\sqrt{25 - 16\sqrt{2}}}.$

(c) This time we have to go outside triangle $\triangle ABC$ to build our useful right triangle, drawing altitude $\overline{AH}$ to the extension of $\overline{BC}$ as shown. (We are inspired to do this by noting that $180° - 135° = 45°$.) We have $AH = BH = AB/\sqrt{2} = 3\sqrt{2}$ as in the previous part. Right triangle AHC then gives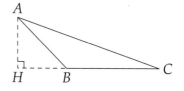

$$AC = \sqrt{AH^2 + HC^2} = \sqrt{18 + (64 + 48\sqrt{2} + 18)} = \boxed{2\sqrt{25 + 16\sqrt{2}}}.$$

CHAPTER 7

Special Parts of a Triangle

Exercises for Section 7.1

7.1.1

(a) ☐ False ☐. Below is an example in which $AB \neq CD$, but m is the perpendicular bisector of both $\overline{AB}$ and $\overline{CD}$.

(b) ☐ True ☐. Since m is perpendicular to both $\overline{AB}$ and $\overline{CD}$, m forms a right angle with each of $\overline{AB}$ and $\overline{CD}$. Looking at m as a transversal crossing $\overline{AB}$ and $\overline{CD}$, we see that the same side interior angles sum to 180°. Therefore, $\overline{AB} \parallel \overline{CD}$.

(c) ☐ True ☐. We have two cases to consider. Either A and C are on the same side of the perpendicular bisector, or they are on opposite sides. These possibilities are shown below.

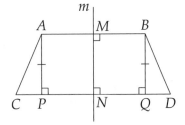

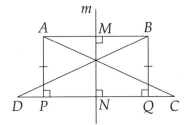

Figure 7.1: A and C on same side   Figure 7.2: A and C on opposite sides

Note that we assume that $AB \leq CD$ (for $AB > CD$, we can follow essentially the same proof). Let M and N be the midpoints of $\overline{AB}$ and $\overline{CD}$, respectively, so $AM = MB$ and $CN = ND$. Let P and Q be the feet of the perpendiculars from A and B to $\overline{CD}$, respectively.

Note that $APNM$ is a rectangle, so $AM = PN$. Similarly, $BQNM$ is also a rectangle, so $BM = QN$. Therefore, in the case at left above we have $CP = CN-PN = CN-AM = DN-BM = DN-QN = DQ$, and in the case at right above we have $CP = CN + NP = DN + AM = DN + MB = DN + NQ = DQ$. Furthermore, $AP = MN = BQ$. Therefore triangles APC and BQD are congruent in both cases by SAS (or LL) Congruence. We conclude that $AC = BD$.

7.1.2 Two intersecting lines form four angles of measure between $0°$ and $180°$. The set of all points equidistant from these two lines are the angle bisectors of these four angles. (Note that these four angle bisectors form two lines, which are perpendicular to each other. Can you prove this?)

7.1.3 The set of all points equidistant from two parallel lines is the line parallel to the two given lines and lying halfway in between.

Exercises for Section 7.2

7.2.1 The circumcenter of an acute triangle always lies $\boxed{\text{inside}}$ the triangle. The circumcenter of an obtuse triangle, however, always lies $\boxed{\text{outside}}$ the triangle. Examples of both are shown at right.

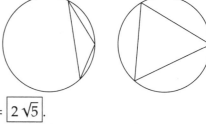

7.2.2 Since $\overline{OQ}$ is perpendicular to chord BC, $\overline{OQ}$ is the perpendicular bisector of chord $\overline{BC}$. This means $QC = BC/2 = 4$. Then by the Pythagorean Theorem, $OC^2 = OQ^2 + QC^2 = 2^2 + 4^2 = 20$, so $OC = \boxed{2\sqrt{5}}$.

7.2.3 Let triangle ABC be an equilateral triangle with side length 18, and let O be its circumcenter. Let P be the foot of perpendicular from O to $\overline{BC}$, so $PC = 9$. As we saw in Problem 7.8, triangle COP is a 30-60-90 triangle. Hence, radius $OC = (2/\sqrt{3})(PC) = (2/\sqrt{3})(9) = \boxed{6\sqrt{3}}$.

7.2.4 As shown in Problem 7.6, the circumradius of a right triangle is equal to half the hypotenuse. The hypotenuse of the triangle is 10, since $6^2 + 8^2 = 10^2$, so the circumradius is $10/2 = \boxed{5}$.

Exercises for Section 7.3

7.3.1 Since O is the incenter of triangle ABC, $\overline{OA}$ is the angle bisector of $\angle CAB$. Then $\angle OAB = \angle A/2 = \boxed{35°}$.

7.3.2 The incenter of any triangle is always $\boxed{\text{inside}}$ the triangle. We can see this by noting that each angle bisector is between two sides of the triangle. Only the interior of the triangle is between each of the three pairs of the sides of the triangle; therefore, the point at which all three angle bisectors meet must be inside the triangle.

7.3.3 $\boxed{\text{Yes}}$. For example, for an equilateral triangle, the incenter and circumcenter are the same point.

7.3.4 For all parts, we can find $\angle Q$ as follows. Let $q = \angle Q$ and $r = \angle PRQ$, so $\angle PRI = r/2$. Since $\angle RPI = (\angle QPR)/2$ and $\angle Q + \angle QPR + \angle PRQ = 180°$ we have $\angle RPI = \angle QPR/2 = (180° - \angle Q - \angle PRQ)/2 = (180° - q - r)/2$. In triangle PIR, $\angle PIR + \angle IRP + \angle RPI = 180°$, so we have $130° + r/2 + (180° - q - r)/2 = 180°$. Therefore, $\angle Q = q = \boxed{80°}$.

(a) In triangle PQR, $\angle PRQ = 180° - q - \angle RPQ = 180° - 80° - 30° = \boxed{70°}$, and $\angle PRT = \angle PRQ/2 = \boxed{35°}$.

(b) In triangle PQR, $\angle PRQ = 180° - q - \angle RPQ = 180° - 80° - 50° = \boxed{50°}$, and $\angle PRT = \angle PRQ/2 = \boxed{25°}$.

(c) In triangle PQR, $\angle PRQ = 180° - q - \angle RPQ = 180° - 80° - 80° = \boxed{20°}$, and $\angle PRT = \angle PRQ/2 = \boxed{10°}$.

(d) $\angle Q$ is uniquely determined by $\angle PIR$ (i.e., all we need is $\angle PIR$ to find $\angle Q$). To see why, observe that

$$\angle PIR = 180° - \frac{\angle RPQ + \angle PRQ}{2} \quad \text{and} \quad \angle Q = 180° - (\angle RPQ + \angle PRQ).$$

So, $\angle RPQ + \angle PRQ = 2(180° - \angle PIR)$ and $\angle RPQ + \angle PRQ = 180° - \angle Q$. Therefore, $2(180° - \angle PIR) = 180° - \angle Q$, or $\boxed{\angle Q = 2\angle PIR - 180°}$.

7.3.5 Since $\overline{PZ}$ is an angle bisector, we have $QZ/QP = RZ/RP$ by the Angle Bisector Theorem. Therefore, $4/6 = RZ/9$, so $RZ = \boxed{6}$.

7.3.6 Let $x = BM$, so $MC = 8 - x$. Since $\angle BAM = \angle CAM$, $\overline{AM}$ is an angle bisector of $\triangle ABC$. Therefore, by the Angle Bisector Theorem, we have $BM/BA = CM/CA$. Substitution gives $x/10 = (8 - x)/12$, so $x = \boxed{40/11}$.

7.3.7

(a) As we discovered in the text, we have $A = rs$, where A is the area, r is the inradius, and s is the semiperimeter. In this case, we have a right triangle with legs 3 and 4, so $A = (3)(4)/2 = 6$ and $s = (3 + 4 + 5)/2 = 6$. Therefore, $r = A/s = \boxed{1}$.

(b) We have an equilateral triangle, and its area is $A = 6^2\sqrt{3}/4 = 9\sqrt{3}$. The semiperimeter is $s = (6 + 6 + 6)/2 = 9$. Therefore, $r = A/s = 9\sqrt{3}/9 = \boxed{\sqrt{3}}$.

(c) We have an isosceles triangle, and its height to the base is $\sqrt{7^2 - (10/2)^2} = \sqrt{24} = 2\sqrt{6}$, so its area is $(10)(2\sqrt{6})/2 = 10\sqrt{6}$. The semiperimeter is $s = (7 + 7 + 10)/2 = 12$. Therefore, $r = A/s = 10\sqrt{6}/12 = \boxed{5\sqrt{6}/6}$.

(d) To find the area of this triangle, we will use Heron's formula, which states the area of a triangle with sides a, b, and c and semiperimeter s is $\sqrt{s(s - a)(s - b)(s - c)}$. In this case, $a = 5$, $b = 6$, and $c = 7$, so $s = (5 + 6 + 7)/2 = 9$. Hence, the area is

$$A = \sqrt{9(9 - 5)(9 - 6)(9 - 7)} = \sqrt{216} = 6\sqrt{6}.$$

Therefore, $r = A/s = 6\sqrt{6}/9 = \boxed{2\sqrt{6}/3}$.

Exercises for Section 7.4

7.4.1 Since the centroid of a triangle divides each median in a 2 : 1 ratio (with the longer piece connecting the centroid to a vertex), we have $AG/GD = BG/GE = 2/1$. Therefore, we have $GD = AG/2 = \boxed{9/2}$ and $BG = 2GE = \boxed{8}$.

7.4.2

(a) Our diagram is shown below.

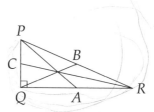

(b) Since triangle PQR has a right angle at Q, the midpoint of hypotenuse $\overline{PR}$ is the circumcenter of triangle PQR. Therefore, median $\overline{QB}$ is a circumradius of $\triangle PQR$ and its length equals half PR, so $QB = PR/2 = \boxed{13/2}$.

(c) Triangle PQA is a right triangle with legs 5 and $QR/2 = 6$, so the length of its hypotenuse, $\overline{AP}$, is $\sqrt{5^2 + 6^2} = \boxed{\sqrt{61}}$. Similarly, triangle RQC is a right triangle with legs 12 and $PQ = 5/2$, so the length of its hypotenuse, $\overline{RC}$, is $\sqrt{12^2 + (5/2)^2} = \boxed{\sqrt{601}/2}$.

7.4.3 We know that the medians $\overline{AD}$, $\overline{BE}$, and $\overline{CF}$ divide triangles ABC into six smaller triangles of equal area. The area of triangle ABC is $(8)(15)/2 = 60$, so each small triangle has area 10.

(a) As just computed, $[ABC] = \boxed{60}$.

(b) Triangle AFC consists of three small triangles, so its area is $\boxed{30}$.

(c) Triangle ACD consists of three small triangles, so its area is $\boxed{30}$.

(d) Triangle AEG is a small triangle, so its area is $\boxed{10}$.

(e) Quadrilateral $EGDC$ consists of two small triangles, so its area is $\boxed{20}$.

(f) Quadrilateral $AFGE$ consists of two small triangles, so its area is $\boxed{20}$.

(g) Triangle DEF is the medial triangle of triangle ABC, so by Problem 7.19, it has area $[ABC]/4 = 60/4 = \boxed{15}$.

(h) Note that $BF = BA/2$ and $BD = BC/2$. Therefore triangle BFD is similar to triangle ABC by SAS with ratio 1/2. Hence, the area of triangle BFD is $(1/2)^2 \cdot [ABC] = (1/4) \cdot 60 = \boxed{15}$. We could also note that $\triangle BFD \cong \triangle EDF$, so $[BFD] = [EDF] = \boxed{15}$.

(i) Triangle FAC is a right triangle with legs 15/2 and 8, so the length of its hypotenuse, $\overline{CF}$, is $\sqrt{8^2 + (15/2)^2} = \sqrt{481}/2$. The centroid G divides median $\overline{CF}$ such that $CG/CF = 2/3$, so the length of $\overline{CG}$ is

$$\frac{2}{3} \cdot \frac{\sqrt{481}}{2} = \boxed{\frac{\sqrt{481}}{3}}.$$

7.4.4 Let $\overline{XA}$ and $\overline{YB}$ meet at G. Because the centroid of a triangle cuts each median in a $2:1$ ratio, we have $XG = (2/3)XA$ and $YG = (2/3)YB$. Since $XA = YB$, we have $XG = YG$. Therefore, $\angle YXG = \angle XYG$, so $\angle AXY = \angle BYX$. This, combined with $XA = BY$ and $XY = XY$, gives $\triangle AXY \cong \triangle BYX$ by SAS Congruence. Therefore, $YA = XB$. Since $XZ = 2(XB)$ and $YZ = 2(YA)$, we have $XZ = YZ$, as desired.

Exercises for Section 7.5

7.5.1

(a) Since triangle PQB has a right angle at Q, $\angle PQB = 90° - \angle QPB = 90° - 32° = \boxed{58°}$.

(b) Since triangle CRP has a right angle at C, $\angle CRP = 90° - \angle CPR = 90° - \angle QPB = 90° - 32° = \boxed{58°}$.

(c) $\angle CQR$ is an exterior angle of triangle QPR, so $\angle CQR = \angle QPR + \angle QRP = 32° + 20° = \boxed{52°}$.

(d) $\angle CQB$ is an exterior angle of triangle QPB, so $\angle CQB = \angle QPB + \angle QBP = 32° + 90° = \boxed{122°}$.

7.5.2

(a) $[XYZ] = (XA)(YZ)/2 = \boxed{10}$.

(b) From right triangle $\triangle BYZ$, we have $BZ = \sqrt{YZ^2 - BY^2} = \boxed{4}$.

(c) Since $[XYZ] = (XZ)(YB)/2$, we have $(XZ)(YB)/2 = 10$. Therefore, $XZ = 20/YB = \boxed{20/3}$.

(d) $XB = XZ - BZ = \boxed{8/3}$.

(e) From right triangle $\triangle XBY$, we have $XY = \sqrt{XB^2 + BY^2} = \boxed{\sqrt{145}/3}$.

7.5.3 Let the altitude from B have length x and the altitude from C have length y. Since $x(AC)/2 = [ABC] = y(AB)/2$ and $AB = AC$, we have $x = y$, so the altitudes from B and C must have the same length.

7.5.4 By the previous problem, altitudes to equal bases of a triangle are equal in length. Since all three bases are equal in an equilateral triangle, all three altitudes are equal as well.

7.5.5 We have $\angle BAD = \angle CAD$, $AD = AD$, and $\angle ADB = 90° = \angle ADC$. Hence, triangles ADB and ADC are congruent by ASA. Therefore, corresponding lengths BD and DC are equal, so D is the midpoint of $\overline{BC}$.

Exercises for Section 7.6

7.6.1

(a) As shown in the text, since $\overline{RQ}$ connects the midpoints of two sides of $\triangle XYZ$, its length is half the length of the third side of the triangle. We can explicitly prove this by noting that $XR/XY = XQ/XZ$ and $\angle RXQ = \angle YXZ$, so $\triangle RXQ \sim \triangle YXZ$. Therefore, $RQ/YZ = XR/XY = 1/2$, so $RQ = YZ/2$. Since $\overline{AP}$ is the median to hypotenuse $\overline{YZ}$ of $\triangle AYZ$, its length is half $\overline{YZ}$ as well. Therefore, we have $AP = RQ$.

(b) Following steps similar to those of the first part, we have $AQ = XZ/2$ because $\overline{AQ}$ is the median to hypotenuse $\overline{XZ}$ of right triangle $\triangle AXZ$. We also have $RP = XZ/2$ because $\overline{RP}$ connects the midpoints of $\overline{XY}$ and $\overline{YZ}$ of $\triangle XYZ$. Therefore, $AQ = RP$. Since $AP = RQ$, $AQ = RP$, and $PQ = PQ$, we have $\triangle PRQ \cong \triangle QAP$. Corresponding angles $\angle PRQ$ and $\angle QAP$ of these triangles are therefore equal.

7.6.2 Since $\overline{CF}$ is the median to hypotenuse $\overline{AB}$ of $\triangle ACB$, we have $CF = AB/2 = BF$. Therefore, $\angle BCF = \angle B$. Since $\angle ADC = 90°$, we have $\angle ACD = 90° - \angle A$. Since $\angle ACB = 90°$, we have $90° - \angle A = \angle B$, so $\angle ACD = \angle B$. Therefore, $\angle BCD = \angle BCA - \angle ACD = 90° - \angle B$. Because $\overline{CE}$ bisects $\angle ACB$, we have $\angle ACE = \angle BCE = 45°$. Since $\angle B < 90°$, either we must have $\angle B = 90° - \angle B = 45°$ (in which case we have $\angle DCE = \angle ECF = 0°$), or $45°$ is between $\angle B$ and $90° - \angle B$. In the latter case, point E must be between D and F, as shown, because $\angle BCF = \angle B$, $\angle BCE = 45°$, and $\angle BCD = 90° - \angle B$.

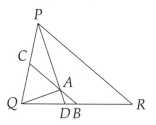

If $\angle B < 45°$, as shown, we have $\angle DCE = \angle ACE - \angle ACD = 45° - \angle B$ and $\angle ECF = \angle BCE - \angle BCF = 45° - \angle B$. Combining these gives the desired $\angle DCE = \angle ECF$. We can follow essentially the same steps to show $\angle DCE = \angle ECF$ if $90° > \angle B > 45°$.

7.6.3

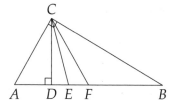

(a) Since $\overline{AC} \parallel \overline{PR}$, we have $\angle CAP = \angle RPA$. Since $\overline{PD}$ bisects $\angle QPR$, we have $\angle RPA = \angle CPA$. Therefore, $\angle CAP = \angle CPA$, so $AC = CP$. Since $\angle QAP = 90°$ and $\angle CAP = \angle CPA$, we have $\angle QAC = \angle QAP - \angle CAP = 90° - \angle CAP = 90° - \angle CPA$. From right triangle $\triangle QAP$, we also have $\angle AQP = 90° - \angle QPA = 90° - \angle CPA$. Therefore, $\angle QAC = \angle AQC$, so $AC = QC$. Since $AC = CP$ and $AC = QC$, we have $QC = CP$, so C is the midpoint of $\overline{PQ}$ (and $\overline{AC}$ is a median of $\triangle PAQ$).

(b) Since $\overline{BC} \parallel \overline{PR}$, we have $\angle QCB = \angle QPR$ and $\angle QBC = \angle R$, so $\triangle QCB \sim \triangle QPR$. Since $QC/QP = 1/2$ from part (a), our similarity gives us $QB/QR = QC/QP = 1/2$. Therefore, $QB = QR/2 = 9/2$. From the Angle Bisector Theorem, we have $PQ/QD = PR/DR$. Therefore, $8/QD = 10/(9 - QD)$. Cross-multiplying gives $10QD = 72 - 8QD$, so $QD = 4$. Finally, we have $DB = QB - QD = \boxed{1/2}$.

Exercises for Section 7.7

7.7.1 Given triangle ABC, we know that the circumcenter O lies on the perpendicular bisectors of the sides of the triangle. We construct the perpendicular bisectors of $\overline{AB}$ and $\overline{AC}$ as shown below. The intersection is circumcenter O. Now construct the circle with center O and radius OA. This gives us the desired circumcircle of triangle ABC.

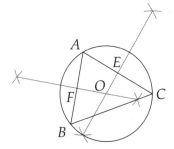

7.7.2 Draw a segment $\overline{AB}$. Now construct an arc using center A and radius AB. Similarly, construct an arc using center B and radius AB. The two arcs intersect at point C. By construction, radii $AC = AB$ and radii $BC = BA$. Hence, triangle ABC is an equilateral triangle. Draw segment $\overline{AC}$. We now have $\angle CAB = 60°$. We then construct the angle bisector $\overline{AD}$ of $\angle CAB$, and we get the desired $30°$ angle $\angle BAD$.

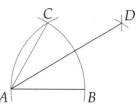

7.7.3 Draw a segment $\overline{AB}$. Construct the perpendicular bisector l of $\overline{AB}$, and let it intersect $\overline{AB}$ at M, the midpoint of $\overline{AB}$. Let P be a point on l other than M. Then $\angle PMA = 90°$, a right angle.

We know how to bisect an arbitrary angle, and bisecting $\angle PMA$ gives an angle of $90°/2 = 45°$.

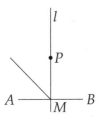

Figure 7.3: Diagram for Problem 7.7.3

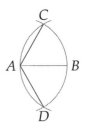

Figure 7.4: Diagram for Problem 7.7.4

7.7.4 Draw a segment $\overline{AB}$. Draw two arcs with radius AB, first using A as center then using B as center. The two arcs meet at two points, call them C and D as shown.

We have equal radii $AC = AB = AD$ and $BC = BA = BD$. Hence, both triangles ACB and ADB are equilateral triangles. This makes $\angle CAD = \angle CAB + \angle DAB = 60° + 60° = 120°$, as desired.

7.7.5 Given triangle ABC, we can find the orthocenter H by finding the intersection of two altitudes of the triangle.

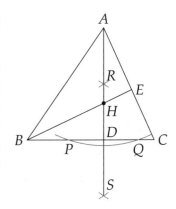

First, draw an arc centered at A with radius sufficiently large that the arc intersects line $\overleftrightarrow{BC}$ at two points P and Q as shown. Next, draw arcs of equal radius centered at P and Q. The two arcs intersect at two points R and S. Then $\overleftrightarrow{RS}$ is the perpendicular bisector of $\overline{PQ}$, so in particular, $\overleftrightarrow{RS}$ is perpendicular to $\overline{QR}$, which means that $\overleftrightarrow{RS}$ is perpendicular to $\overline{BC}$. Furthermore, $AP = AQ$ by construction, so A lies on the perpendicular bisector of $\overline{PQ}$, which means that A lies on $\overleftrightarrow{RS}$. Thus, $\overline{AD}$ is the altitude from A to $\overline{BC}$.

Similarly, construct the altitude from B to $\overline{AC}$. These two altitudes intersect at the orthocenter H.

7.7.6 By construction, points A and B lie on a circle centered at X, making $XA = XB$. We then constructed point Y, which lies on both the circle centered at B with radius XB and the circle centered at A with radius XA. Therefore, $YB = XB = XA = YA$. Looking at triangles YBX and YAX, we see that they are congruent by SSS, so corresponding angles are equal. In particular, $\angle BXY = \angle AXY$, so $\overline{XY}$ is the angle bisector of the given angle.

Review Problems

7.34

(a) *Area.* Since $\triangle PQR$ is a right triangle, we have $[PQR] = (PQ)(QR)/2 = \boxed{96}$.

(b) *Circumradius.* Since $\triangle PQR$ is a right triangle, its circumradius equals half its hypotenuse. Since the legs are in the ratio 3/4, $\triangle PQR$ is a 3-4-5 triangle, so $PR = PQ(5/3) = 20$. (We also could have used the Pythagorean Theorem.) The circumradius then is $PR/2 = \boxed{10}$.

(c) *Inradius.* The area of a triangle equals its inradius times its semiperimeter. Since the area of $\triangle PQR$ is 96 and the semiperimeter is $(12 + 16 + 20)/2 = 24$, the inradius of $\triangle PQR$ is $96/24 = \boxed{4}$.

7.35 Let M be the midpoint of $\overline{TU}$. Since $\overline{TU}$ is a chord of $\odot K$, the perpendicular bisector of $\overline{TU}$ goes through K. Since both M and K are on this perpendicular bisector, we have $\overline{MK} \perp \overline{TU}$. We are given $MK = 2$, and $TM = TU/2 = 7$, so the Pythagorean Theorem gives us $TK = \sqrt{TM^2 + KM^2} = \boxed{\sqrt{53}}$.

7.36

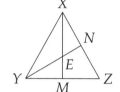

(a) Triangle XYM is a 30-60-90 triangle with $\angle XYM = 60°$, $\angle YMX = 90°$, $\angle MXY = 30°$.

(b) From 30-60-90 $\triangle XYM$, we have $XM = (\sqrt{3}/2) \cdot XY = (\sqrt{3}/2) \cdot 12 = \boxed{6\sqrt{3}}$.

(c) The area of equilateral triangle XYZ is given by

$$\frac{1}{2}YZ \cdot XM = \frac{1}{2} \cdot 12 \cdot 6\sqrt{3} = \boxed{36\sqrt{3}}.$$

(d) Recall that the centroid of a triangle divides each median into a ratio of $2 : 1$. Hence, $XE/EM = \boxed{2}$.

(e) The semiperimeter of triangle XYZ is $s = (12 + 12 + 12)/2 = 18$. Therefore, the inradius is $A/s = 36\sqrt{3}/18 = \boxed{2\sqrt{3}}$. We also might note that EM is the inradius, and $EM = XM/3 = \boxed{2\sqrt{3}}$.

(f) The inradius is EM, while the circumradius is XE. By part (d), $XE/EM = 2$, and by part (b), $XE + EM = 6\sqrt{3}$. Therefore, the circumradius is $XE = (2/3) \cdot 6\sqrt{3} = \boxed{4\sqrt{3}}$.

7.37 From right triangle $\triangle CFB$, we have $\angle FCB = 90° - \angle B$. From right triangle $\triangle HCD$, we have $\angle CHD = 90° - \angle HCD = 90° - (90° - \angle B) = \angle B$, as desired.

7.38 Let the triangle be $\triangle ABC$ and the lengths of the altitudes from A, B, and C, respectively, be h_A, h_B, and h_C. We have $[ABC] = (BC)(h_A)/2 = (AB)(h_C)/2 = (AC)(h_B)/2$. Since we are given $h_A = h_B = h_C$, we have $AB = BC = CA$, so the triangle must indeed be equilateral.

7.39 Since $\angle YXD = \angle ZXD$, we know that $\overline{XD}$ bisects $\angle YXZ$. From the Angle Bisector Theorem, we have $XY/YD = XZ/DZ = 2$. Therefore, $XY = 2YD = \boxed{6}$.

7.40 As described in the text, medians $\overline{AD}$, $\overline{BE}$, and $\overline{CF}$ divide triangles ABC into six smaller triangles of equal area. The area of triangle ABC is given as 48, so each small triangle has area $48/6 = 8$.

(a) $\triangle ADC$. Triangle ADC consists of three small triangles, so its area is $3 \cdot 8 = \boxed{24}$.

(b) $\triangle AGC$. Triangle AGC consists of two small triangles, so its area is $2 \cdot 8 = \boxed{16}$.

(c) $\triangle GFB$. Triangle GFB is a small triangle, so its area is $\boxed{8}$.

(d) $\triangle DEF$. Triangle DEF is the medial triangle of triangle ABC, so as described in the text, it has area $[ABC]/4 = 48/4 = \boxed{12}$.

(e) $\triangle AEF$. Note that $AE = AC/2$ and $AF = AB/2$. Therefore, by SAS, triangle AEF is similar to triangle ACB with ratio $1/2$. Hence, the area of triangle AEF is $(1/2)^2 \cdot [ABC] = (1/4) \cdot 48 = \boxed{12}$.

7.41

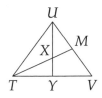

(a) We have $UT = UV$ and $\angle TUY = \angle VUY$. Hence, by SAS, triangles UYT and UYV are congruent. Then corresponding sides are equal, and in particular, $TY = YV$.

(b) The corresponding angles of triangle UYT and UYV are also equal, so $\angle UYT = \angle UYV$. The sum of these two angles makes a straight angle, or $180°$, so both angles are $90°$. Therefore, $\overline{UY} \perp \overline{TV}$.

(c) $TY = TV/2 = 36/2 = \boxed{18}$. Triangle UYT has a right angle at Y, so $UY = \sqrt{UT^2 - TY^2} = \sqrt{30^2 - 18^2} = \boxed{24}$.

(d) We have shown that $\overline{UY}$ is a median. We are also given that $\overline{TM}$ is a median, so their intersection X is the centroid. The centroid X divides median $\overline{UY}$ in the ratio $XU : XY = 2 : 1$, so $XY = UY/3 = \boxed{8}$, and $XU = (2/3)UY = \boxed{16}$.

(e) Triangle XYT has a right angle at Y, so $XT = \sqrt{TY^2 + XY^2} = \sqrt{18^2 + 8^2} = \boxed{2\sqrt{97}}$. Since X is the centroid, X divides median $\overline{TM}$ in the ratio $TX : XM = 2 : 1$. Therefore, $XM = TX/2 = \boxed{\sqrt{97}}$.

7.42 The two figures are similar to each other, with ratio $18/6 = 3$. The circumradii are then also in this same proportion, so the second circumradius must be $3 \cdot 2\sqrt{3} = \boxed{6\sqrt{3}}$.

7.43 In triangle ABC, $\angle ABC = 180° - \angle BAC - \angle ACB = 180° - 117° - 35° = 28°$. Since $\overline{BE}$ is an angle bisector, $\angle DBX = \angle ABC/2 = 28°/2 = 14°$. $\angle DXE$ is an exterior angle of triangle DXB, so $\angle DXE = \angle DBX + \angle XDB = 14° + 90° = \boxed{104°}$.

7.44 Since O is the circumcenter, it is equidistant from the vertices of the triangle. Therefore, $OP = OQ = OR$.

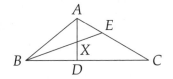

(a) $\angle RPO = \angle RPQ - \angle QPO = \boxed{22°}$. Since $OP = OQ$, we have $\angle OQP = \angle OPQ = \boxed{23°}$.

(b) *Solution 1:* From $\triangle OPQ$, we have $\angle POQ = 180° - \angle OPQ - \angle OQP = 134°$. Since $OR = OP$, we have $\angle ORP = \angle OPR = 22°$, so from $\triangle OPR$ we have $\angle POR = 180° - \angle ORP - \angle OPR = 136°$. Therefore, $\angle QOR = 360° - \angle POQ - \angle POR = 90°$. Since $OR = OQ$, we have $\angle ORQ = \angle OQR$. From $\triangle OQR$, we know these two angles sum to $180° - \angle PQR = 90°$. Therefore, each must equal $\boxed{45°}$.

Solution 2: As in the first solution, we have $\angle PRO = 22°$. Let $\angle ORQ = x$. Since $OR = OQ$, we have $\angle OQR = x$ as well. Since the sum of the angles in $\triangle PQR$ is $180°$, we have $2(23°) + 2(22°) + 2x = 180°$. Therefore, $x = \boxed{45°}$.

7.45 $\overline{BM}$ and $\overline{AN}$ are medians of $\triangle ABC$. Therefore, X is the centroid of $\triangle ABC$ and it divides $\overline{BM}$ and $\overline{AN}$ such that $BX/XM = AX/XN = 2/1$. Using our given information yields $XB = 2XM = \boxed{7}$ and $XN = AX/2 = \boxed{3.6}$.

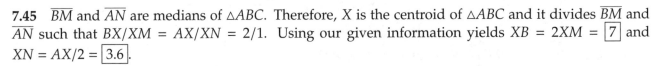

7.46

(a) Since medians $\overline{AX}$ and $\overline{BY}$ meet at O, the point O must be the centroid of the triangle. Therefore, $AO/AX = 2/3$, so $AO = 12(2/3) = \boxed{8}$. Since X is the midpoint of $\overline{BC}$, we have $BX = BC/2 = 5$. Since $OX = AX/3 = 4$, from right triangle $\triangle BOX$ we have $BO = 3$. Hence, $BY = (3/2)BO = \boxed{9/2}$.

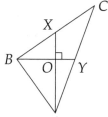

(b) From right triangle $\triangle ABO$ we have $AB = \sqrt{AO^2 + BO^2} = \sqrt{73}$. Median $\overline{CZ}$ must pass through the centroid, O, of $\triangle ABC$, so $CO + OZ = CZ$. Furthermore, we know that $CZ = 3OZ$ because O is the centroid. Since $\overline{OZ}$ is the median to hypotenuse $\overline{AB}$ of $\triangle AOB$, we have $OZ = AB/2 = \sqrt{73}/2$. Finally, we have $CZ = 3OZ = \boxed{3\sqrt{73}/2}$.

7.47 Using our area equalities, we have $[AFC] = [AFX] + [AEX] + [EXC] = [BFX] + [BXD] + [DXC] = [BFC]$. Since $\triangle AFC$ and $\triangle BFC$ have the same altitude from C, and they have the same area, we know that bases $\overline{AF}$ and $\overline{BF}$ have the same length. Therefore, $\overline{CF}$ is a median of the triangle. Similarly, we can show $\overline{BE}$ and $\overline{AD}$ are also medians, so X must be the centroid of $\triangle ABC$.

7.48

(a) $\angle EIC$ is an exterior angle of $\triangle BIC$, so $\angle EIC = \angle IBC + \angle ICB = (\angle ABC)/2 + (\angle ACB)/2$. Since $\angle ABC + \angle ACB = 180° - \angle A$ from $\triangle ABC$, we have $\angle EIC = (\angle ABC + \angle ACB)/2 = (180° - \angle A)/2 = 90° - \angle A/2$, as desired. (We also could have used the angles of $\triangle EBC$ and $\triangle EIC$ to achieve this result.)

(b) From the Angle Bisector Theorem, we have $AB/AE = BC/CE$. Therefore, $BC = (AB/AE)(CE) = \boxed{16/3}$.

(c) From the Angle Bisector Theorem, we have $BC/BF = AC/AF$. Since $AF = AB - BF = 8 - BF$, $AC = AE + CE = 10$, and $BC = 16/3$, we have $(16/3)/BF = 10/(8 - BF)$. Therefore, we have $30BF = 128 - 16BF$, so $46BF = 128$, from which we have $BF = \boxed{64/23}$.

7.49

(a) Let H be the point that is both the centroid and the orthocenter, and let $\triangle ABC$ be our triangle. $\overrightarrow{AH}$ must hit $\overline{BC}$ because we know that the centroid of a triangle is inside the triangle. Let $\overrightarrow{AH}$ meet $\overline{BC}$ at X. Since H is both the centroid and the orthocenter, $\overline{AX}$ is both a median and an altitude. Therefore, $\overline{AX} \perp \overline{BC}$ and $BX = XC$. Since $AX = AX$, $\angle AXB = \angle AXC = 90°$, and $BX = CX$, we have $\triangle AXB \cong \triangle AXC$ by SAS Congruence. Therefore, $AB = AC$ since they are corresponding sides of these triangles. Similarly, we can let $\overrightarrow{BH}$ meet $\overline{AC}$ at Y, and use $\overline{BY}$ (which is a median and an altitude) to prove $BA = BC$. Combining this with $AB = AC$ gives us $AB = AC = BC$, so $\triangle ABC$ must be equilateral if its centroid and orthocenter are the same point.

(b) Let our triangle be ABC and the point that is both the centroid and the incenter be I. Since the incenter of a triangle must be inside the triangle, I is inside the triangle. As in the first part, we let $\overrightarrow{AI}$ hit $\overline{BC}$ at point S. Since $\overline{AS}$ passes through the incenter, it is an angle bisector. Therefore, the Angle Bisector Theorem gives us $AB/AC = SB/SC$. Since $\overline{AS}$ passes through the centroid, it is a median. Therefore, S is the midpoint of $\overline{BC}$, so $SB = SC$. Therefore, we have $AB/AC = SB/SC = 1$, so $AB = AC$. Similarly, we extend $\overline{BI}$ past I to hit $\overline{AC}$ at T. We can then use $\overline{BT}$ to show that $AB = BC$. Therefore, $AB = AC = BC$, so $\triangle ABC$ is equilateral.

See if you can continue the work of these first two parts to prove that if any two of the incenter, centroid, orthocenter, and circumcenter coincide, then the triangle must be equilateral.

Challenge Problems

7.50 Let O be the circumcenter of the triangle. We are given that $\overline{GO}$ bisects $\angle IGH$; therefore, $\angle OGH = \angle OGI$. Since O is the circumcenter of $\triangle IGH$, it is equidistant from the vertices of $\triangle GHI$. Therefore, we have $GO = OH = OI$. From $OH = OG$, we have $\angle OHG = \angle OGH$. From $OI = GO$, we have $\angle OIG = \angle IGO$. Combining these angle equalities with the one from our angle bisector, we have $\angle OHG = \angle OGH = \angle OGI = \angle OIG$. Therefore, we have $\triangle GOH \cong \triangle GOI$ by AAS Congruence. Corresponding sides $\overline{GH}$ and $\overline{GI}$ of these triangles therefore have equal length.

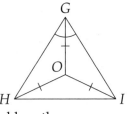

7.51 From right triangle $\triangle BAP$, we have $\angle BAP = 90° - \angle ABP = 90° - \angle ABC$. Since O is the circumcenter of $\triangle ABC$, we have $OA = OB = OC$. Therefore, $\angle CAO = \angle ACO$, $\angle ABO = \angle BAO$, and $\angle CBO = \angle BCO$. The sum of all six of these angles is the sum of the angles of $\triangle ABC$, which equals 180°. Therefore, we have $2(\angle CAO + \angle ABO + \angle CBO) = 180°$, or $\angle CAO + \angle ABO + \angle CBO = 90°$. Since $\angle ABO + \angle CBO = \angle ABC$, we have $\angle CAO + \angle ABC = 90°$. Therefore, $\angle CAO = 90° - \angle ABC$. Since we have $\angle BAP = 90° - \angle ABC$ and $\angle CAO = \angle CAQ$, the equation $\angle CAO = 90° - \angle ABC$ gives us the desired $\angle CAQ = \angle BAP$.

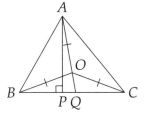

7.52

(a) Since $\triangle JKL$ is isosceles with $JK = JL$, we start by drawing altitude $\overline{JM}$ to the midpoint of $\overline{KL}$. Therefore, $KM = KL/2 = 20$ and from right triangle $\triangle JMK$, we have $JM = 15$. The area of $\triangle JKL$ is $(KL)(JM)/2 = \boxed{300}$.

(b) The area of a triangle equals the product of the triangle's inradius and its semiperimeter. Since the area of $\triangle JKL$ is 300 and its semiperimeter is $(25 + 25 + 40)/2 = 45$, the inradius is $300/45 = \boxed{20/3}$.

(c) Let the circumcenter be point O. Since $JK = JL$, median $\overline{JM}$ is perpendicular to side $\overline{KL}$. Therefore, $\overline{JM}$ is part of the perpendicular bisector of $\overline{KL}$ and thus must pass through O. Let $OK = x$, our circumradius. Since O is equidistant from K and J, we have $OJ = x$ as well. Therefore, $OM = JM - JO = 15 - x$. From right triangle $\triangle OKM$, we have $OK^2 = OM^2 + KM^2$. Therefore, $x^2 = (15 - x)^2 + 20^2$. Solving for x, we have $30x = 225 + 400$, so $x = \boxed{125/6}$.

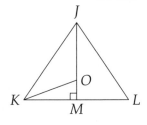

7.53

(a) Since $\overrightarrow{BI}$ passes through the incenter of $\triangle ABC$, it bisects $\angle ABC$. Therefore, $\angle ABI = \angle IBC$, so $\overline{BI}$ is an angle bisector of $\triangle ABA'$. Applying the Angle Bisector Theorem to bisector $\overline{BI}$ of $\triangle ABA'$, we have $A'I/IA = A'B/AB$, as desired.

(b) Since $\overline{AI}$ passes through the incenter, it bisects $\angle BAC$. The Angle Bisector Theorem then gives us $A'B/AB = A'C/AC$. Since we know $A'I/IA = A'B/AB$ from the first part, we only have to show $A'B/AB = BC/(AB + AC)$ to finish. We can rearrange $A'B/AB = A'C/AC$ to give $A'B/A'C = AB/AC$. We add 1 to both sides to get $A'B/A'C + 1 = AB/AC + 1$, so $(A'B + A'C)/A'C = (AB + AC)/AC$. Since $A'B + A'C = BC$, we have $BC/A'C = (AB + AC)/AC$, so $A'C/AC = BC/(AB + AC)$. Finally, this means we have

$$\frac{A'I}{IA} = \frac{A'B}{AB} = \frac{A'C}{AC} = \frac{BC}{AB + AC}.$$

7.54 Since $\overline{AH}$ is an altitude of $\triangle ABC$, $\triangle AHC$ is a right triangle. M is the midpoint of hypotenuse $\overline{AC}$ of this triangle, so $HM = MC = AM$. Since $MH = MC$, we have $\angle MHC = \angle MCH$. From $\triangle ABC$, we have $\angle MHC = \angle C = \boxed{30°}$.

7.55 We will show that it is impossible for two angle bisectors of a triangle to be perpendicular by showing that if two angle bisectors are perpendicular, then one of the angles of the triangle must equal $0°$. Let our triangle be $\triangle ABC$ and the incenter be I. Let the angle bisectors from A and C be the ones that are perpendicular, so that $\triangle AIC$ is a right triangle. Therefore, we have $\angle IAC + \angle ICA = 90°$. Since $\angle IAC = (\angle BAC)/2$ and $\angle ICA = (\angle BCA)/2$, we have $(\angle BAC)/2 + (\angle BCA)/2 = 90°$. Therefore, $\angle BAC + \angle BCA = 180°$. However, $\triangle ABC$ also gives us $\angle BAC + \angle BCA + \angle ABC = 180°$. Combining these last two equations gives $\angle ABC = 0°$, which is impossible. Therefore, it is not possible for two angle bisectors of a triangle to be perpendicular.

7.56 Let $a = BC$ and $b = CA$. In right triangle ANM, $NO = 12.5$ is the median to the hypotenuse $\overline{AM}$, so $OA = OM = 12.5$, and $MA = AO + OM = 25$. Since M is the midpoint of $\overline{AB}$, $MA = MB = 25$, so $AB = AM + MB = 50$. We are given that the perimeter is 112, so $a + b + 50 = 112$, or $a + b = 62$. By the Pythagorean Theorem on triangle ABC, $a^2 + b^2 = 50^2 = 2500$, so

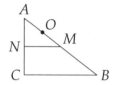

$$ab = \frac{(a + b)^2 - (a^2 + b^2)}{2} = \frac{62^2 - 50^2}{2} = 672.$$

Therefore, the area of $\triangle ABC$ is $ab/2 = 336$. $\triangle MNA \sim \triangle BCA$ by SAS Similarity with ratio 1/2, so $MN = a/2$ and $AN = b/2$. Hence, the area of $\triangle ANM$ is $(AN)(NM)/2 = ab/8 = 84$. Finally, we have $[MNCB] = [ABC] - [AMN] = \boxed{252}$.

7.57

(a) Since P is on the bisector of $\angle YBX$, it is equidistant from $\overrightarrow{BY}$ and $\overrightarrow{BX}$. These rays are contained by $\overleftrightarrow{AB}$ and $\overleftrightarrow{AC}$, respectively, so P is equidistant from $\overleftrightarrow{AB}$ and $\overleftrightarrow{AC}$.

(b) By the same reasoning as the previous part, Q is equidistant from $\overrightarrow{CX}$ and $\overrightarrow{CZ}$, so it is equidistant from $\overleftrightarrow{BC}$ and $\overleftrightarrow{AC}$.

(c) Since I_a is on the external bisector of $\angle B$, it is equidistant from $\overleftrightarrow{AB}$ and $\overleftrightarrow{AC}$ (part (a)). Therefore, $I_aX = I_aY$. Similarly, part (b) tells us that I_a is equidistant from $\overleftrightarrow{BC}$ and $\overleftrightarrow{AC}$, so $I_aX = I_aZ$. Therefore, $I_aX = I_aY = I_aZ$. Consequently, a circle with center I_a and radius I_aX will pass through X, Y, and Z. Moreover, it will be tangent to $\overleftrightarrow{AB}$, $\overleftrightarrow{BC}$, and $\overleftrightarrow{AC}$, since all points on these lines besides Y, X, and Z are farther from I_a than X is. (We can prove this in exactly the same way we proved that a triangle's incircle is tangent to all three sides of the triangle.)

(d) Since $I_aY = I_aZ$, we know that I_a is equidistant from $\overrightarrow{AB}$ and $\overrightarrow{AC}$. Therefore, it must be on the angle bisector of $\angle BAC$.

(e) We take a similar approach to the approach we used to show that $[ABC] = rs$, where r is the inradius and s the semiperimeter. We note that $[ABC] = [I_aBAC] - [I_aBC]$. Furthermore, we have

$[I_aBAC] = [I_aAB] + [I_aAC]$. Since $[I_aAB] = (AB)(I_aY)/2$, $[I_aAC] = (I_aZ)(AC)/2$, $[I_aBC] = (I_aX)(BC)/2$, and $I_aX = I_aY = I_aZ = r_a$, we have

$$
\begin{aligned}
[ABC] &= [I_aAB] + [I_aAC] - [I_aBC] \\
&= \frac{(r_a)(AB)}{2} + \frac{(r_a)(AC)}{2} - \frac{(r_a)(BC)}{2} \\
&= \frac{r_a}{2}(AB + AC - BC) \\
&= \frac{r_a}{2}(p - 2BC) \\
&= r_a(s - BC),
\end{aligned}
$$

where p is the perimeter of $\triangle ABC$ and s is its semiperimeter. Therefore, we have $r_a = [ABC]/(s-BC)$.

7.58

(a) We have $PM = PA/2$, $\angle MPN = \angle APB$, and $PN = PB/2$. Hence, triangles MPN and APB are similar with ratio $1/2$, so $MN = AB/2$, a constant.

(b) As P moves, the perimeter of triangle APB changes. For example, as P moves farther away from A and B, the perimeter increases.

(c) The area of triangle PAB is $(AB)(d)/2$, where d is the distance from P to $\overline{AB}$. But the length of the base $\overline{AB}$ is fixed, and since P moves parallel to $\overline{AB}$, so is d. Hence, the area of triangle PAB remains constant.

(d) Since triangles MPN and APB are similar with ratio $1/2$, $[MPN]/[APB] = (1/2)^2 = 1/4$, so $[MPN]$ always equals $[APB]/4$. Therefore, $[ABNM] = [APB] - [MPN] = 3[APB]/4$. Since $[APB]$ is constant (from part (c)), we know $[ABNM]$ is constant.

7.59

(a) Let r be the radius of the circle. Let S, T, and U be the midpoints of $\overline{AB}$, $\overline{CD}$, and $\overline{EF}$, respectively. Since $AB = CD = EF$, we get that $AS = SB = CT = TD = EU = UF$.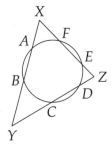

Triangle PSA has a right angle at S, so by the Pythagorean Theorem, $PS^2 = PA^2 - AS^2 = r^2 - AS^2$. Similarly, $PT^2 = PC^2 - CT^2 = r^2 - CT^2$, and $PU^2 = PE^2 - EU^2 = r^2 - EU^2$. Since $AS = CT = EU$, we have $PS = PT = PU$. In other words, P is equidistant from the sides of triangle ABC. Therefore, P is the incenter of triangle XYZ.

(b) Since P is the incenter of triangle XYZ, P lies on the angle bisectors of $\angle X$, $\angle Y$, and $\angle Z$. Therefore,

$$\angle XPY = 180° - \angle XYP - \angle YXP = 180° - \frac{1}{2}\angle XYZ - \frac{1}{2}\angle YXZ = 180° - \frac{1}{2}(45°) - \frac{1}{2}(60°) = \boxed{127.5°}.$$

7.60 Let the wall meet the ground at X and let the ladder be $\overline{AB}$ as shown. Let M be the midpoint of the ladder. When the ladder is completely against the wall, then M is $AB/2$ away from X. The same is true when the ladder has slid all the way to the ground. In between, we note that $\triangle ABX$ is a right triangle and $\overline{XM}$ is a median to the hypotenuse of this right triangle. Therefore, $XM = AB/2$ at all points throughout the ladder's slide, so M is always a constant distance from point X. Hence, the path the ladder traces out is the quarter-circle with center X and radius $AB/2$ that is shown in our diagram.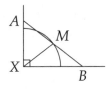

Our proof thus far only shows that every point that M passes through is on this quarter circle. We aren't quite finished; we must also show that M does in fact reach every point on the quarter circle. Therefore, we must show how we can start from any point N on the quarter circle and find a position of the ladder such that the midpoint of the ladder is at N. Let the length of our ladder be ℓ. Let point N be a point on the quarter circle with center X and radius equal to $\ell/2$. We draw a circle with center N and radius $\ell/2$. This circle hits the wall at C and the ground at D. (Make sure you see why this circle must hit the wall and the ground at points besides X.) Since this circle is the circumcircle of $\triangle CXD$, which is a right triangle, the center of this circle must be the midpoint of hypotenuse $\overline{CD}$. Furthermore, this hypotenuse is the diameter of the circle, so it has length ℓ. Since $CD = \ell$ and the midpoint of $\overline{CD}$ is N, we have found a location of the ladder such that its midpoint is at N.

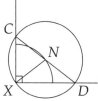

7.61 First, triangles ABM and ACB share the same altitude from A to $\overline{BC}$, and $BM = BC/2$, so $[ABM] = [ABC]/2 = 48/2 = 24$.

Now in triangle ABM, $\overline{BP}$ and $\overline{MN}$ are medians, so their intersection G is the centroid of triangle ABM. Since the three medians of a triangle divide the triangle into six small triangles of equal area, and triangle MGP is one of these six small triangles, we have

$$[MGP] = \frac{1}{6}[ABM] = \boxed{4}.$$

7.62 We use our Same Base/Same Altitude approach of Section 4.3 to show that $[ABG]/[ACG] = BD/CD$. Specifically, since $\triangle ABD$ and $\triangle ACD$ share an altitude from A, we have $BD/DC = [ABD]/[ACD]$. Similarly, we have $BD/DC = [BGD]/[GDC]$. Therefore, $[ABD]/[ACD] = [BGD]/[GDC]$. We can use this to prove

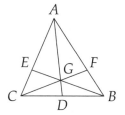

$$\frac{[ABG]}{[ACG]} = \frac{[ABD]}{[ACD]} = \frac{[BGD]}{[GDC]}$$

in a variety of ways. One way is to write $[ABD]/[ACD] = [BGD]/[GDC]$ as $[ABD]/[BGD] = [ACD]/[GDC]$. Therefore, we have

$$\frac{[ABG] + [BGD]}{[BGD]} = \frac{[ACG] + [GDC]}{[GDC]},$$

so $[ABG]/[BGD] = [ACG]/[GDC]$. Therefore, $[ABG]/[ACG] = [BGD]/[GDC]$, as desired. Hence, we conclude $[ABG]/[ACG] = BD/CD$. (This was also a Challenge Problem in Chapter 4. We will use this result in the rest of our solution.)

(a) Since $[ABG]/[ACG] = BD/CD$ and D is the midpoint of $\overline{BC}$, we have $BD = CD$, so $[ABG] = [ACG]$. Similarly, $[ABG]/[BCG] = AE/EC$ and $AE = EC$, so $[ABG] = [BCG]$. Therefore $[ABG] = [ACG] = [BCG]$.

(b) Since $AF/FB = [ACG]/[BGC]$ and $[ACG] = [BGC]$, we have $AF = FB$.

(c) We have shown that if we draw two medians from two vertices to two sides of a triangle, then draw a ray from the third vertex through the intersection point of our two medians, then this ray will hit the third side at its midpoint. Therefore, the line through the third vertex and this intersection point must contain our third median. In other words, we have proved that all of our medians are concurrent.

7.63 First, we note that $BC = 10$ from the Pythagorean Theorem applied to $\triangle ABC$. Next, since $\angle BAN = 90° - \angle CAN = 45° = \angle CAN$, we know that $\overline{AN}$ bisects $\angle BAC$. Therefore, we can apply the Angle Bisector Theorem to $\triangle ABC$ to find $AB/BN = AC/CN$. Since $CN = 10 - BN$, $AB = 6$, and $AC = 8$, we have $6/BN = 8/(10 - BN)$. Cross-multiplying gives $60 - 6BN = 8BN$, so $BN = 60/14 = 30/7$. We need AN, so we draw altitude $\overline{AH}$ to build right triangle $\triangle AHN$. We can find AH and BH by noting that $\triangle ABH \sim \triangle CBA$ by AA Similarity. Therefore, $BH/AB = AH/AC = AB/BC = 6/10 = 3/5$, so $BH = (3/5)(AB) = 18/5$ and $AH = (3/5)(AC) = 24/5$. Since $HN = BN - HB$, we have $NH = 30/7 - 18/5 = 24/35$. Finally, from right $\triangle AHN$ we have

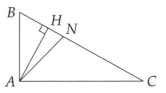

$$AN = \sqrt{AH^2 + HN^2} = \sqrt{\left(\frac{24}{5}\right)^2 + \left(\frac{24}{35}\right)^2} = \sqrt{\left(\frac{24}{5}\right)^2\left(1 + \left(\frac{1}{7}\right)^2\right)} = \frac{24}{5}\sqrt{1 + \frac{1}{49}} = \frac{24}{5}\sqrt{\frac{50}{49}} = \boxed{\frac{24\sqrt{2}}{7}}.$$

Notice how we did some clever algebraic manipulation to avoid squaring those nasty fractions. What we did there is essentially the same as noticing that the legs are in the ratio $1 : 7$, so the triangle is similar to a right triangle with legs 1 and 7. This simpler triangle has hypotenuse $\sqrt{1^2 + 7^2} = 5\sqrt{2}$, so the ratio of the sides of the triangle is $1 : 7 : 5\sqrt{2}$. Therefore, the hypotenuse is $5\sqrt{2}/1 = 5\sqrt{2}$ times as long as the shorter leg. So, we have $AN = (5\sqrt{2})(HN) = (24\sqrt{2})/7$, as before. (For a slick second solution, try drawing the square with diagonal $\overline{AN}$.)

7.64 Let O be the center of the circle. Since $\overleftrightarrow{XY}$ is the perpendicular bisector of chord $\overline{AC}$, it passes through O. Hence, $OY = 1$. Since $\triangle ABC$ is equilateral, each perpendicular bisector of a side of the triangle goes through the vertex opposite that side. Therefore, $\overline{BX}$ is a median of triangle ABC and O is the centroid of $\triangle ABC$. Therefore, O divides $\overline{BX}$ such that $OB/OX = 2/1$. Since $OB = 1$, $OX = OB/2 = 1/2$. Therefore, $XY = OY - OX = 1 - 1/2 = \boxed{1/2}$.

CHAPTER 8

Quadrilaterals

Exercises for Section 8.1

8.1.1 The sum of the angles in a quadrilateral is 360°, so we have $\angle A + \angle B + \angle C + \angle D = 360°$. Therefore, $100° + 50° + (\angle D + 30°) + \angle D = 360°$, so $\angle D = \boxed{90°}$.

8.1.2 The sum of the angles in a quadrilateral is 360°, so $(x) + (3x - 10°) + (27°) + (4x - 30°) = 360°$. Therefore, $x = 373°/8 = 46.625°$. The angles of the quadrilateral are then $x = 46.625°, 3x - 10° = 129.875°$, 27°, and $4x - 30° = 156.5°$. The largest angle is $\boxed{156.5°}$.

8.1.3

(a) $\boxed{\text{No}}$. For example, the angles could be 90°, 90°, 100°, and 80°.

(b) $\boxed{\text{Yes}}$. For example, the angles could be 100°, 100°, 100°, and 60°.

(c) Since all four angles sum to 360°, each angle must be $\boxed{90°}$.

8.1.4 As shown below, there are $\boxed{3}$ quadrilaterals.

8.1.5 Let *ABCD* be a concave quadrilateral as shown, where we call the vertex of the interior angle greater than 180° point *D*. In triangles *ADB* and *CDB*, we have

$$\angle BAD + \angle ABD + \angle BDA = 180°,$$
$$\angle BCD + \angle CBD + \angle BDC = 180°.$$

Adding these two equations gives

$$\angle BAD + (\angle ABD + \angle CBD) + \angle BCD + (\angle BDA + \angle BDC) = 360°.$$

Since $\angle ABD + \angle CBD = \angle ABC$ and $\angle BDA + \angle BDC = $ (reflex angle $\angle CDA$), we have the desired

$$\angle BAD + \angle ABC + \angle BCD + (\text{reflex angle } \angle CDA) = 360°.$$

Exercises for Section 8.2

8.2.1   Since the legs of the trapezoid are equal, the trapezoid is isosceles. Therefore, $\angle Q = \angle P = \boxed{83°}$. Since $\overline{PQ} \parallel \overline{RS}$, $\angle S = 180° - \angle P = \boxed{97°}$. Similarly, $\angle R$ is the supplement of $\angle Q$, which is also $\boxed{97°}$.

8.2.2   The area of the trapezoid is $(44 + 24)(18)/2 = \boxed{612}$.

8.2.3   Let the length of the shorter base be x, so the longer base is $x + 6$. We have $(x + x + 6)(8)/2 = 96$. Solving, we find that the shorter base has length $x = \boxed{9}$.

8.2.4   Since $\overline{BC} \parallel \overline{AD}$, we have $\angle ADC = 180° - \angle BCD = 180° - 126° = \boxed{54°}$. Similarly, $\angle ABC = 180° - \angle BAD = 180° - 72° = 108°$. Therefore, $\angle CBD = \angle ABC - \angle ABD = 108° - 88° = \boxed{20°}$.

8.2.5   Let $ABCD$ be an isosceles trapezoid with bases $\overline{AB}$ and $\overline{CD}$, so $\angle ADC = \angle BCD$. As we saw in the text, we have $AD = BC$. Therefore, triangles ADC and BCD are congruent by SAS (because we also have $CD = CD$). Corresponding lengths of these triangles are equal, so $AC = BD$.

8.2.6

(a) Since $\overline{AB} \parallel \overline{DC}$, we have $\angle XAB = \angle XDC$ and $\angle XBA = \angle XCD$, so $\triangle XAB \sim \triangle XDC$ by AA Similarity.

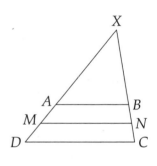

(b) Because $\triangle XAB \sim \triangle XDC$, we have $XD/XA = XC/XB$. Furthermore, we have $XD = XA + AD$ and $XC = XB + BC$, so our ratio equality is $(XA + AD)/XA = (XB + BC)/XB$, or $1 + AD/XA = 1 + BC/XB$. This gives the desired $AD/XA = BC/XB$. Since $AM = AD/2$ and $BN = BC/2$, we can divide both numerators of $AD/XA = BC/XB$ by 2 to get $AM/XA = BN/XB$.

(c) Adding 1 to both sides of $AM/XA = BN/XB$ gives

$$\frac{AM}{XA} + 1 = \frac{BN}{XB} + 1,$$

or $(AM + XA)/XA = (BN + XB)/XB$. Therefore, $XM/XA = XN/XB$.

(d) Since $XM/XA = XN/XB$ and $\angle AXB = \angle MXN$, we have $\triangle XAB \sim \triangle XMN$ by SAS Similarity. Therefore, $\angle XAB = \angle XMN$, so $\overline{AB} \parallel \overline{MN}$.

(e) Since $\overline{AB} \parallel \overline{MN} \parallel \overline{DC}$, we have $\triangle AXB \sim \triangle MXN \sim \triangle DXC$. Therefore, $AB/MN = XA/XM$ and $CD/MN = XD/XM$. We add these to get $(AB + CD)/MN = (XA + XD)/XM = (XM - AM + XM + DM)/XM = (2XM)/XM = 2$ (since $AM = DM$). Thus we have the desired $MN = (AB + CD)/2$.

8.2.7   Let the trapezoid be $ABCD$ with $\overline{AB} \parallel \overline{CD}$, $AB = 50$, and $CD = 75$. Let the diagonals meet at E. Since $\overline{AB} \parallel \overline{CD}$, we have $\angle ABD = \angle BDC$ and $\angle BAC = \angle ACD$. Therefore, $\triangle AEB \sim \triangle CED$, so $EC/AE = DE/EB = DC/AB = 75/50 = 3/2$. Letting $\overline{BD}$ be the short diagonal and $\overline{AC}$ the long one, we have $BE + ED = 35$ and $AE + EC = 120$. Since we also have $DE = (3/2)EB$ and $EC = (3/2)AE$, we can solve $BE + 3/2(BE) = 35$ to get $BE = 14$ and $AE + (3/2)AE = 120$ to get $AE = 48$. Since the sides of $\triangle AEB$ are in the ratio 7-24-25, we know that $\triangle AEB$ is a right triangle with right angle at

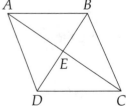

$\angle AEB$ (since $\overline{AB}$ is the longest side of the triangle). (We could also have used the Pythagorean Theorem to determine that $\triangle AEB$ is a right triangle.)

Therefore, we can dissect $ABCD$ into four right triangles, $\triangle AEB$, $\triangle BEC$, $\triangle CED$, and $\triangle DEA$. Since $AE = 48$ and $BE = 14$, we have $CE = AC - AE = 72$ and $DE = DB - BE = 21$. Therefore, we have:

$$[AEB] = \frac{(AE)(EB)}{2} = 336$$
$$[BEC] = \frac{(BE)(EC)}{2} = 504$$
$$[CED] = \frac{(CE)(ED)}{2} = 756$$
$$[DEA] = \frac{(DE)(EA)}{2} = 504.$$

The area of $ABCD$ is the sum of these, or $\boxed{2100}$. (Notice that the area of $ABCD$ equals half the product of the diagonals. Is this a coincidence?)

Exercises for Section 8.3

8.3.1

(a) Opposite sides in a parallelogram are equal. Therefore, $IU = TY = 6$ and $TI = YU = 8$. The perimeter of the parallelogram is then $6 + 8 + 6 + 8 = \boxed{28}$.

(b) $\boxed{\text{No}}$. The area of a parallelogram is not determined by the side lengths of the parallelogram. Below are two parallelograms with the same perimeter, but with very different areas.

8.3.2 Since $\overline{WX} \parallel \overline{YZ}$, we have $\angle W + \angle Z = 180°$. Similarly, $\angle Y + \angle Z = 180°$ because $\overline{XY} \parallel \overline{WZ}$. Therefore, $\angle W = \angle Y$.

8.3.3 Since $\overline{WO} \parallel \overline{RK}$, and $\overline{WR}$ is a transversal cutting the two parallel segments, we have $\angle OWR = \angle KRW$. We also have $WO = RK$ and $WR = WR$. Hence, triangles OWR and KRW are congruent by SAS. Therefore, $\angle ORW = \angle KWR$, and this implies $\overline{WK} \parallel \overline{RO}$. The opposite sides of $WORK$ are then parallel, so it must be a parallelogram.

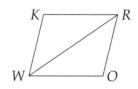

8.3.4 Let $ABCD$ be a parallelogram. Let E and F be the feet of the perpendiculars from A and B to $\overleftrightarrow{CD}$, respectively. Then triangles AED and BFC are right triangles, and $ABFE$ is a rectangle, so $AE = BF$. Since $ABCD$ is a parallelogram, we have $AD = BC$, so by HL congruence, triangles AED and BFC are congruent. Therefore,

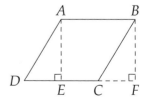

$$[ABCD] = [ABCE] + [AED] = [ABCE] + [BFC] = [ABFE].$$

In other words, the areas of parallelogram $ABCD$ and rectangle $ABFE$ are equal. The area of rectangle $ABFE$ is the base AB times the height BF. Therefore, the area of the parallelogram $ABCD$ is also the base AB times the height BF.

8.3.5 Point E is on line $\overleftrightarrow{EU}$ and point N is on line $\overleftrightarrow{NT}$, and the distance between lines $\overleftrightarrow{EU}$ and $\overleftrightarrow{NT}$ is given as 5. This means that the closest any point on one line can be to the other is 5 units. Therefore, the length of $\overline{EN}$ must be at least 5, and in particular, it cannot be 4.

8.3.6

(a) We have $AA' = AB + BA' = AB + CD$ and $DD' = DC + CD' = AB + CD$, so $AA' = DD'$. Furthermore, $\overline{AA'} \parallel \overline{DD'}$. As we saw in Exercise 8.3.3, $AA'D'D$ is a parallelogram because $AA' = DD'$ and $\overline{AA'} \parallel \overline{DD'}$.

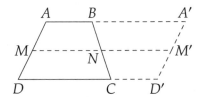

(b) Consider trapezoids $ABCD$ and $D'CBA'$. By construction, $D'C = AB$ and $BA' = CD$. Since $AA'D'D$ is a parallelogram, $A'D' = DA$. Therefore, corresponding sides of the two trapezoids are equal. Furthermore, our parallel segments give us $\angle DAB = \angle A'D'C$, $\angle ABC = \angle D'CB$, $\angle BCD = \angle CBA'$, and $\angle CDA = \angle BA'D'$. Thus, corresponding angles of the two trapezoids are also equal, so they are congruent.

Therefore, $[ABCD] = [D'CBA']$. But $[ABCD] + [D'CBA'] = [AA'D'D]$, so $[AA'D'D] = 2[ABCD]$. Finally, the area of the parallelogram is $[AA'D'D] = AA' \cdot h = (AB + CD) \cdot h$, so

$$[ABCD] = \frac{1}{2}[AA'D'D] = \frac{1}{2}(AB + CD) \cdot h.$$

(c) We have $AM = AD/2 = A'D'/2 = A'M'$, and $\overline{AM} \parallel \overline{A'M'}$. Hence, $AA'M'M$ is a parallelogram, so $MM' = AA' = AB + CD$. However, $\overline{MN}$ and $\overline{M'N}$ are corresponding parts of congruent trapezoids $ABCD$ and $D'CBA'$, so $MN = M'N$. Since $MM' = MN + NM'$, we have $MN = MM'/2 = (AB + CD)/2$.

8.3.7 Let the circumcenters of triangles ABE, BCE, CDE, and DAE be P, Q, R, and S, respectively. P lies on the perpendicular bisector of side $\overline{BE}$ of triangle ABE, and Q lies on the perpendicular bisector of side $\overline{BE}$ of triangle BCE. Hence, $\overline{PQ}$ is perpendicular to line $\overleftrightarrow{BD}$. Similarly, $\overleftrightarrow{RS}$ is also perpendicular to line $\overleftrightarrow{BD}$. Therefore, $\overline{PQ} \parallel \overline{RS}$. Similarly, both $\overleftrightarrow{QR}$ and $\overleftrightarrow{SP}$ are perpendicular to $\overleftrightarrow{AC}$, so they are parallel. Therefore, $PQRS$ is a parallelogram.

Exercises for Section 8.4

8.4.1 As shown earlier, the area of a rhombus is half the product of its diagonals. Hence, the area of $PQRS$ is $(6)(12)/2 = \boxed{36}$.

The diagonals bisect each other and are perpendicular, thus cutting the rhombus into four right triangles with legs $6/2 = 3$ and $12/2 = 6$. The hypotenuses of these triangles are the sides of the rhombus, and from the Pythagorean Theorem each has length $\sqrt{3^2 + 6^2} = 3\sqrt{5}$, so the perimeter of the rhombus is $4(3\sqrt{5}) = \boxed{12\sqrt{5}}$.

8.4.2

(a) The diagonals bisect each other, cutting the rhombus into four right triangles of hypotenuse $WX = 50$ and one leg $WY/2 = 48$. Then the other leg is $\sqrt{50^2 - 48^2} = \sqrt{196} = 14$, so diagonal $XZ = 2 \cdot 14 = \boxed{28}$.

(b) The area of the rhombus is $(28)(96)/2 = \boxed{1344}$.

(c) Since $WXYZ$ is a rhombus, it is a parallelogram. Therefore, its area equals the product of the length of one side and the distance from that side to the opposite side. Since the area of $WXYZ$ is 1344 and $WX = 50$, the distance (height) between $\overline{WX}$ and $\overline{YZ}$ is $1344/50 = \boxed{672/25}$.

8.4.3 Let E be the intersection of the diagonals. Then

$$[ABCD] = [AEB] + [BEC] + [CED] + [DEA]$$
$$= \frac{1}{2}AE \cdot EB + \frac{1}{2}BE \cdot EC + \frac{1}{2}CE \cdot ED + \frac{1}{2}DE \cdot EA$$
$$= \frac{1}{2}(AE \cdot EB + BE \cdot EC + CE \cdot ED + DE \cdot EA)$$
$$= \frac{1}{2}(AE + EC)(BE + ED)$$
$$= \frac{1}{2}AC \cdot BD.$$

8.4.4

(a) Triangles TUW and VUW are congruent by SSS. Therefore, $\angle TUW = \angle VUW$. Since the two angles combined make $60°$, each angle is $30°$.

(b) Diagonal $\overline{TV}$ cuts the rhombus into two equilateral triangles of side 10. Hence, $[TUVW] = 2[TUV] = 2(10^2\sqrt{3})/4 = \boxed{50\sqrt{3}}$.

Exercises for Section 8.5

8.5.1

(a) Opposite sides in a rectangle are equal. Therefore, the perimeter is $8 + 12 + 8 + 12 = \boxed{40}$.

(b) The segments $\overline{PS}$, $\overline{PO}$, and $\overline{OS}$ make a right triangle with $\overline{PS}$ as hypotenuse. Therefore,

$$PS = \sqrt{PO^2 + OS^2} = \sqrt{8^2 + 12^2} = \sqrt{208} = \boxed{4\sqrt{13}}.$$

(c) The area of rectangle $POST$ is simply the length times the width, which is $8 \cdot 12 = \boxed{96}$.

8.5.2 Let the width of the rectangle be x. Then the length of the rectangle is $2x - 1$, and the perimeter is $2[x + (2x - 1)]$. Setting this equal to 36 gives $x = 19/3$. The area of the rectangle is then $x(2x - 1) = (19/3)(35/3) = \boxed{665/9}$.

8.5.3 Triangle EYR is isosceles with $YE = YR$, so $\angle ERY = \angle YER$. Angle $\angle WYE = x$ is an exterior angle of triangle EYR, so $\angle ERY + \angle YER = \angle WYE = x$, from which we have $\angle ERY = \boxed{x/2}$. Then $\angle YRT = \angle ERT - \angle ERY = 90° - \angle ERY = \boxed{90° - x/2}$.

8.5.4 The other diagonal of the rectangle is a radius of the circle, which is 9 cm. The two diagonals in a rectangle are equal, so $RQ = \boxed{9 \text{ cm}}$.

8.5.5 The frame has length $36 + 2 + 2 = 40$ and width $24 + 2 + 2 = 28$, so the area of the frame is $40 \cdot 28 - 36 \cdot 24 = 256$ square inches. The cost of the frame is then $\$1.50 \cdot 256 = \boxed{\$384}$.

8.5.6 Let $ABCD$ be a rectangle, and let E, F, G, and H be midpoints of sides $\overline{AB}$, $\overline{BC}$, $\overline{CD}$, and $\overline{DA}$, respectively. In triangles AEH and BEF, we have $AH = AD/2 = BC/2 = BF$, $AE = AB/2 = BE$, and $\angle A = 90° = \angle B$. Hence, the two triangles are congruent by SAS. Therefore, $HE = EF$. Similarly, $EF = FG$, $FG = GH$, and $GH = HE$. Thus, all four sides of $EFGH$ are equal, so $EFGH$ is a $\boxed{\text{rhombus}}$.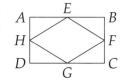

8.5.7 Triangle HEG and rectangle $IEGJ$ share the same altitude from H to $\overline{EG}$, so $[IEGJ] = 2[HEG]$. On the other hand, triangle HEG is half of rectangle $EFGH$, so $[EFGH] = 2[HEG]$. Therefore, $[IEGJ] = [EFGH] = \boxed{48}$.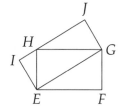

Exercises for Section 8.6

8.6.1 $EF = \sqrt{80} = \boxed{4\sqrt{5}}$, and $EG = EF\sqrt{2} = \boxed{4\sqrt{10}}$.

8.6.2 Since $BM/BA = BO/BD = 1/2$ and $\angle MBO = \angle ABD$, triangles BMO and BAD are similar with ratio $1/2$ by SAS Similarity, so $AD = 2MO = 8$. Therefore, the area of square $ABCD$ is $8^2 = \boxed{64}$.

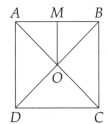

Figure 8.1: Diagram for Problem 8.6.2

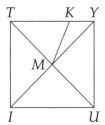

Figure 8.2: Diagram for Problem 8.6.3

8.6.3 First, $\angle MTK = \angle UTY = \boxed{45°}$. Since $TK = TM$, we have $\angle MKT = \angle TMK$. From $\triangle MTK$, we have $\angle MKT + \angle TMK = 180° - \angle MTK = 135°$, so $\angle TMK = 135°/2 = \boxed{67.5°}$.

8.6.4 Since every rectangle is a parallelogram, the diagonals of a rectangle bisect each other. Therefore, if the diagonals of a rectangle are perpendicular, they divide the rectangle into four triangles that are congruent by SAS Congruence. The hypotenuses of these triangles, which are the sides of the rectangle, are therefore equal in length. Thus, the sides of the rectangle are all equal in length, so the rectangle is a square.

8.6.5 Since $\overline{AC}$ is a diagonal of square $ABCD$, $AC = AB\sqrt{2}$. Therefore, $[ACFG]/[ABCD] = AC^2/AB^2 = (AC/AB)^2 = \boxed{2}$.

8.6.6

(a) Since triangle ABE is equilateral, $AE = AB = \boxed{4}$.

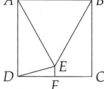

(b) $ABCD$ is a square of side length 4, so its area is $4^2 = \boxed{16}$.

(c) ABE is an equilateral triangle of side length 4, so its area is $4^2\sqrt{3}/4 = \boxed{4\sqrt{3}}$.

(d) $\angle DAE = \angle DAB - \angle EAB = 90° - 60° = \boxed{30°}$. Triangle DAE is isosceles with $DA = AB = AE$. Since $\angle DAE = 30°$, $\angle DEA = (180° - \angle DAE)/2 = (180° - 30°)/2 = \boxed{75°}$.

(e) The area inside square $ABCD$ but outside triangle ABE is simply $\boxed{16 - 4\sqrt{3}}$.

(f) Let F be the foot of perpendicular from E to $\overline{CD}$, so $FC = 2$. (Make sure you see why F must be the midpoint of $\overline{CD}$.) The altitude from E to $\overline{AB}$ in triangle AEB is $(\sqrt{3}/2)AE = 2\sqrt{3}$, so $EF = 4 - 2\sqrt{3}$. Then by the Pythagorean Theorem applied to right triangle EFC, we have

$$EC^2 = EF^2 + FC^2 = (4 - 2\sqrt{3})^2 + 2^2 = 32 - 16\sqrt{3}.$$

We then observe that $(2\sqrt{6} - 2\sqrt{2})^2 = 32 - 16\sqrt{3}$, so $EC = \boxed{2\sqrt{6} - 2\sqrt{2}}$.

Exercises for Section 8.7

8.7.1 $\boxed{\text{False}}$. The quadrilateral shown at right is an example of a quadrilateral that has perpendicular diagonals, but the quadrilateral shown is not a rhombus.

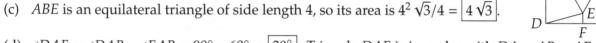

8.7.2 First, if XYZ is an equilateral triangle, then $XY = YZ$ and all angles are 60°.

Next we must show that $\triangle XYZ$ is equilateral if $XY = YZ$ and $\angle X = 60°$. Since $XY = YZ$, XYZ is an isosceles triangle with base angles $\angle X$ and $\angle Z$. So $\angle Z = \angle X = 60°$, which gives us $\angle Y = 180° - \angle Z - \angle X = 60°$. Since $\angle X = \angle Y = \angle Z$, the triangle is equilateral.

Therefore, $\triangle XYZ$ is equilateral if and only if $XY = YZ$ and $\angle X = 60°$.

8.7.3 $\boxed{\text{False}}$. A rhombus is always a parallelogram, but a parallelogram can have unequal sides.

8.7.4 First, assume rhombus $\mathcal{R}$ is a square. Then $\mathcal{R}$ is also a rectangle, since all squares are rectangles.

For the other part, we assume rhombus $\mathcal{R}$ is a rectangle and wish to prove it is a square. Then $\mathcal{R}$ is a rectangle with all sides equal, so $\mathcal{R}$ is a square.

Therefore, a rhombus is a square if and only if it is also a rectangle.

8.7.5 First, we assume $PQRS$ is a parallelogram. Then its opposite sides are equal as proved in the text, so $PQ = RS$ and $SP = QR$.

Next we must show that if $PQ = RS$ and $SP = QR$ in quadrilateral $PQRS$, then $PQRS$ is a parallelogram. Since $PQ = RS$, $SP = QR$, and $PR = PR$, we have $\triangle PSR \cong \triangle RQP$ by SSS Congruence. Hence, $\angle PRS = \angle RPQ$, so $\overline{PQ} \parallel \overline{SR}$. We also have $\angle RPS = \angle PRQ$, so $\overline{PS} \parallel \overline{QR}$. Since both pairs of opposite sides are parallel, $PQRS$ is a parallelogram.

Therefore, quadrilateral $PQRS$ is a parallelogram if and only if $PQ = RS$ and $SP = QR$.

8.7.6 First, we assume trapezoid $ABCD$, with bases $\overline{AB}$ and $\overline{CD}$, is isosceles. Then $\angle DAB = \angle ABC$ and $AD = BC$ (as shown in the text), so triangles ABD and BAC are congruent by SAS. Therefore, corresponding angles $\angle ABD$ and $\angle BAC$ are equal.

Next, we assume that in trapezoid $ABCD$ with bases $\overline{AB}$ and $\overline{CD}$, we have $\angle ABD = \angle BAC$. Let K be the intersection of diagonals $\overline{AC}$ and $\overline{BD}$. Since $\angle ABD = \angle BAC$, triangle AKB is isosceles, and $AK = BK$.

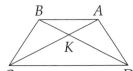

Since $\angle ACD = \angle CAB = \angle ABD = \angle BDC$, triangle CKD is also isosceles, and $CK = DK$. Therefore, $AC = AK + KC = BK + KD = BD$. Then triangles ABD and BAC are congruent by SAS. So $\angle DAB = \angle ABC$, and the trapezoid is isosceles.

Therefore, trapezoid $ABCD$ with $\overline{AB} \parallel \overline{CD}$ is isosceles if and only if $\angle ABD = \angle BAC$.

8.7.7 First, we assume $WXYZ$ is a rectangle. Then $WX = ZY$, $\angle WXY = 90° = \angle XYZ$, and $\overline{WX} \parallel \overline{ZY}$.

Next, we assume $WXYZ$ is a quadrilateral such that $WX = ZY$, $\angle WXY = \angle XYZ$, and $\overline{WX} \parallel \overline{ZY}$, and we prove $WXYZ$ is a rectangle. The first and third conditions tell us that quadrilateral $WXYZ$ is a parallelogram. The second condition along with $\angle WXY + \angle XYZ = 180°$ (since $\overline{WX} \parallel \overline{ZY}$) tells us that all four angles are right angles. Therefore, $WXYZ$ is a rectangle.

8.7.8 First, we assume triangle ABC is isosceles with $CA = CB$. Then $\angle CAM = \angle CBM$ and $AM = BM$. Hence, triangles CAM and CBM are congruent by SAS. Therefore, corresponding angles are equal, and in particular, $\angle ACM = \angle BCM$, so $\overline{CM}$ bisects $\angle ACB$.

For the other direction, we assume $\overline{CM}$ bisects $\angle ACB$ and show that $CA = CB$. By the Angle Bisector Theorem, we have $CA/AM = CB/BM$. Since $AM = BM$, we have $CA = CB$, so triangle BAC is isosceles with $CA = AB$.

8.7.9 $\boxed{\text{False}}$. The diagonals of a square are perpendicular and congruent, but the trapezoid shown also has diagonals that are perpendicular and congruent.

Exercises for Section 8.8

8.8.1 We have many 45-45-90 triangles here. Note that $\angle GCH = \angle GCE - \angle ECH = \angle BCD - \angle ECH = \angle DCE = 90° - \angle DEC = 45°$. Also, we have $\angle CHG = 90° - \angle GCH = 45°$, so $GC = HG = 3$ and $HC = \sqrt{2}HG = 3\sqrt{2}$. Hence, $BC = 2HC = 6\sqrt{2}$.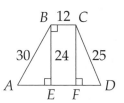

Now consider triangle FBH. We have $\angle BHF = \angle GHC = 45°$, and $\angle FBH = 90°$, so $\angle HFB = 45°$. Therefore, $FH = \sqrt{2}HB = \sqrt{2}HC = 6$, so $FG = FH + HG = 6 + 3 = 9$. Finally, in triangle CDE, we have $\angle DEC = 45°$ and $\angle EDC = 90°$, so $\angle ECD = 45°$. Therefore, $CD = CE/\sqrt{2} = FG/\sqrt{2} = 9/\sqrt{2}$.

Therefore, the area of $CEFG$ is $CG \cdot FG = 3 \cdot 9 = 27$, and the area of $ABCD$ is $CD \cdot BC = (9/\sqrt{2})(6\sqrt{2}) = 54$, and their positive difference is $54 - 27 = \boxed{27}$.

8.8.2 Let E and F be the feet of perpendiculars from B and C to $\overline{AD}$, respectively. Then $BE = CF = 24$. Since $BCFE$ is a rectangle, $EF = BC = 12$. By the Pythagorean Theorem, $AE = \sqrt{AB^2 - BE^2} = \sqrt{30^2 - 24^2} = \sqrt{324} = 18$, and $FD = \sqrt{CD^2 - CF^2} = \sqrt{25^2 - 24^2} = \sqrt{49} = 7$. Therefore, $AD = AE + EF + FD = 18 + 12 + 7 = 37$, and the area of trapezoid $ABCD$ is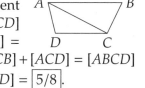

$$\frac{1}{2}(BC + AD) \cdot BE = \frac{1}{2}(12 + 37) \cdot 24 = \boxed{588}.$$

8.8.3 The height from A to $\overleftrightarrow{CD}$ equals the height from C to $\overleftrightarrow{AB}$ because both represent the distance between the parallel bases of trapezoid $ABCD$. Therefore, $[ACB]/[ACD]$ equals the ratios of the lengths of bases $\overline{AB}$ and $\overline{CD}$. So, we have $[ACB]/[ACD] = AB/CD = 5/3$. The sum of these areas is the area of the whole trapezoid. Since $[ACB] + [ACD] = [ABCD]$ and $[ACD] = (3/5)[ACB]$, we have $[ACB] + (3/5)[ACB] = [ABCD]$, so $[ACB]/[ABCD] = \boxed{5/8}$.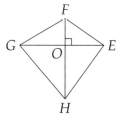

8.8.4 Since the diagonals of $EFGH$ are perpendicular, they divide the quadrilateral into four right triangles as shown. From the Pythagorean Theorem, we have

$$\begin{aligned} EF^2 &= EO^2 + FO^2 \\ GH^2 &= GO^2 + HO^2 \\ FG^2 &= FO^2 + GO^2 \\ EH^2 &= EO^2 + HO^2 \end{aligned}$$

Adding the first two gives $EF^2 + GH^2 = EO^2 + FO^2 + GO^2 + HO^2$, and adding the second two gives $FG^2 + EH^2 = EO^2 + FO^2 + GO^2 + HO^2$. Therefore, $EF^2 + GH^2 = FG^2 + EH^2$.

8.8.5 Construct point D such that $ACBD$ is a parallelogram. Then the diagonals $\overline{AB}$ and $\overline{CD}$ bisect each other, so M is also the midpoint of $\overline{CD}$. We showed in the text (Problem 8.31) that the sum of the squares of the diagonals of a parallelogram equals the sum of the squares of the sides, so $AB^2 + CD^2 = 2(AC^2 + BC^2)$. Therefore, $CD^2 = 2(AC^2 + BC^2) - AB^2 = 2(8^2 + 7^2) - 6^2 = 190$. Thus, we have $CM = CD/2 = \boxed{\sqrt{190}/2}$.

Review Problems

8.32

Fact	Parallelogram	Rhombus	Rectangle	Square
Opposite sides have equal length	Yes	Yes	Yes	Yes
All sides have equal length	Sometimes	Yes	Sometimes	Yes
Opposite sides are parallel	Yes	Yes	Yes	Yes
Opposite angles are equal	Yes	Yes	Yes	Yes
All angles are equal	Sometimes	Sometimes	Yes	Yes
Diagonals bisect each other	Yes	Yes	Yes	Yes
Diagonals have equal length	Sometimes	Sometimes	Yes	Yes
Area can be determined from sides	No	No	Yes	Yes

8.33 Let $\angle C = x$, so $\angle D = 5x - 10°$. Since $\angle A + \angle B + \angle C + \angle D = 360°$, we have $128° + 128° + x + (5x - 10°) = 360°$, so $x = 19°$. Therefore, $\angle C = x = \boxed{19°}$ and $\angle D = 5x - 10° = \boxed{85°}$.

8.34

(a) In this case, $\angle A = \angle B$, or $x = 2x - 45°$, so $x = 45°$. Then $\angle C = 180° - \angle B = \boxed{135°}$ and $\angle D = 180° - \angle A = \boxed{135°}$.

(b) In this case, $\angle A + \angle B = 180°$, or $x + (2x - 45°) = 180°$, so $x = 75°$. Then $\angle C = \angle B = 2x - 45° = \boxed{105°}$ and $\angle D = \angle A = x = \boxed{75°}$.

8.35 The trapezoid is the only quadrilateral in the list for which a diagonal does not necessarily divide the area in two equal halves. Let $ABCD$ be any of the other four types of quadrilaterals. Then $AB = CD$ and $BC = DA$, so triangles ABC and CDA are congruent by SSS. Therefore, $[ABC] = [CDA]$, and diagonal $\overline{AC}$ cuts the quadrilateral into two equal halves.

Let $ABCD$ be a trapezoid with $\overline{AB} \parallel \overline{CD}$ and $AB < CD$. Letting h be the height of the trapezoid, we have $[ABC] = (AB)(h)/2$ and $[CDA] = (CD)(h)/2$. Since $CD > AB$, we have $[CDA] > [ABC]$, so diagonal $\overline{AC}$ does not cut the trapezoid into two regions of equal area. Similarly, $\overline{BD}$ cuts $ABCD$ into $\triangle ABD$ and $\triangle CBD$, but we have $[CBD] > [ABD]$. Hence, a trapezoid that is not a parallelogram cannot be divided into two equal area triangles by a diagonal.

8.36 Let the side length of the square be x. Then the perimeter is $4x$ and the area is x^2, so $4x = x^2$, or $x = 4$. The length of a diagonal is then $\boxed{4\sqrt{2}}$.

8.37 Since $FA = AE = ED = FB$, $\triangle DEA$ and $\triangle AFB$ are 45-45-90 right triangles. Therefore, $\angle DAB = 180° - \angle EAD - \angle FAB = 90°$ and $AD = AB = AE\sqrt{2} = EF\sqrt{2}/2$. Similarly, we can show that all sides of $ABCD$ equal $EF\sqrt{2}/2$ and all angles of $ABCD$ equal 90°. Since $\triangle ABC$ is a 45-45-90 right triangle (since $ABCD$ is a square), we have $[ABC] = (AB)(BC)/2 = EF^2/4 = [EFGH]/4$. Therefore, $[ABC]/[EFGH] = \boxed{1/4}$.

8.38 In terms of x, the area of the trapezoid is $(2x + 4x)(2x)/2 = 6x^2$. Since we are given that the area is 48, we have $6x^2 = 48$, so $x = \boxed{2\sqrt{2}}$.

8.39   The shorter diagonal divides the rhombus into two equilateral triangles of side length 2. (Make sure you see why.) Hence, the area of the rhombus is $2(2^2 \sqrt{3}/4) = \boxed{2\sqrt{3}}$.

8.40   Let the two sides of the rectangle be a and b. Then $2a + 2b = 64$, so $b = 32 - a$, and $ab = 192$, so $a(32 - a) = 192$, or $a^2 - 32a + 192 = 0$. This factors as $(a - 8)(a - 24) = 0$, so the solutions are $a = 8$ and $a = 24$. Therefore, the dimensions of the rectangle are 8 by 24, so the longer side is $\boxed{24 \text{ cm}}$.

8.41   Since $\overline{WX}$ and $\overline{YZ}$ are opposite sides in a parallelogram, they are equal, so $2x - 3 = 3x - 8$, or $x = 5$. Then $WX = 2x - 3 = 7$ and $XY = x + 7 = 12$. Therefore, the perimeter of the parallelogram is $2(7 + 12) = \boxed{38}$.

8.42   Since $\overline{WZ} \parallel \overline{XY}$, $\angle W = 180° - \angle X = 90°$. Similarly, $\angle Y = 180° - \angle X = 90°$, and $\angle Z = 180° - \angle Y = 90°$. Therefore, $WXYZ$ is a rectangle. But it is also rhombus, so all sides are equal, and a rectangle with all sides equal is a square.

8.43

  (a)   Triangles ABD and CBD are congruent by SSS. Corresponding angles of these triangles are equal, so $\angle ABD = \angle CBD$.

  (b)   Since triangles ABD and CBD congruent, we have $\angle EDA = \angle EDC$. Therefore, triangles EDA and EDC are congruent by SAS. Corresponding angles of these triangles are equal, so $\angle DEA = \angle DEC$. These two angles add up to a straight angle, so both must be right angles. In other words, $\overline{AC} \perp \overline{BD}$.

8.44   A square is a rhombus. Any rhombus $ABCD$ satisfies $AB = BC$ and $CD = DA$, the definition of a kite. Therefore, squares and rhombi are kites. All the other quadrilaterals in the list do not necessarily have equal adjacent sides.

8.45   Let ABC be the triangle and $\overline{AD}$ an altitude. Then triangle ABD is a 30-60-90 triangle with $AD = (\sqrt{3}/2)(AB) = 4\sqrt{3}$. The area of the square is then $(4\sqrt{3})^2 = \boxed{48}$.

8.46

  (a)   Looking at right triangles XEH and GHF, we find $\angle XEH = 90° - \angle XHE = \angle EHG - \angle XHE = \angle GHF = \boxed{31°}$.

  (b)   Since $EY = HY$, $\angle HEY = \angle YHE$. So, $\angle YEF = 90° - \angle YEH = 90° - \angle YHE = \angle GHF = \boxed{31°}$.

  (c)   We know $\angle EFH = \angle GHF = 31°$. Hence, $\angle EFH = \angle YEF = 31°$. Therefore, $EY = YF$. Since $HY = EY$, $HY = YF$. Thus, Y is the midpoint of diagonal $\overline{HF}$ of rectangle $EFGH$. Then the other diagonal $\overline{EG}$ of the rectangle must also pass through Y. In other words, $\overrightarrow{EY}$ passes through G.

8.47   Since $EFGH$ is a rhombus, we have $FG = EF$. Therefore, $FG = EF = EG$, so EFG is equilateral with side length 6. Since $EH = HG = EF = EG$, $\triangle EHG$ is equilateral with side length 6, too. Hence, $[EFGH] = 2[EFG] = 2(6^2 \sqrt{3}/4) = \boxed{18\sqrt{3}}$.

8.48   First, we assume $ABCD$ square. Then all sides are equal and diagonals $\overline{AC}$ and $\overline{BD}$ have the same length.

Second, we must assume all sides equal and that $AC = BD$, and prove $ABCD$ is a square. Since $AB = BC = CD = DA$ and $AC = BD$, triangles BCD, CDA, DAB, and ABC are congruent by SSS.

Corresponding angles of these triangles are equal, so $\angle A = \angle B = \angle C = \angle D$. Since the four angles add up to $360°$, each angle is $90°$. Hence, $ABCD$ is a rectangle, with all sides equal. Therefore, it is a square.

We have thus proven that $ABCD$ is a square if and only if its sides are equal in length and $AC = BD$.

8.49   Let $\angle Z = a$ and $\angle Y = a + d$. Then $\angle X = a + 2d$ and $\angle W = a + 3d$. We know the four angles add up to $360°$, so $4a + 6d = 360°$.

(a) If $\angle W = 4\angle Z$, then $a + 3d = 4a$. The system of equations $4a + 6d = 360°$ and $a + 3d = 4a$ gives $a = 36°$ and $d = 36°$. So $\angle W = a + 3d = \boxed{144°}$.

(b) If $\angle W = 2\angle Z$, then $a + 3d = 2a$. The system of equations $4a + 6d = 360°$ and $a + 3d = 2a$ gives $a = 60°$ and $d = 20°$. So $\angle W = a + 3d = \boxed{120°}$.

8.50   Since E and H are the midpoints of sides $\overline{AB}$ and $\overline{AD}$ of $\triangle ABD$, we have $\overline{EH} \parallel \overline{BD}$. (This was proved in the text as the Midline Theorem in Section 7.4.) Similarly, since F and G are midpoints of sides $\overline{CB}$ and $\overline{CD}$, we have $\overline{FG} \parallel \overline{BD}$. Therefore, $\overline{EH} \parallel \overline{FG}$. By a similar argument we can show $\overline{EF} \parallel \overline{HG}$. Since its opposite sides are parallel, $EFGH$ is a parallelogram.

8.51   Let the side length of the square be $3x$. Since the perimeter of the triangle equals the perimeter of the square, the side length of the triangle is $(4 \cdot 3x)/3 = 4x$. (We pick $3x$ as the side length of the square to avoid having to use fractions too much.) The area of the square is $(3x)^2 = 9x^2$. The area of the triangle is $(4x)^2 \sqrt{3}/4 = 4\sqrt{3}x^2$. Therefore, the ratio is $9/(4\sqrt{3}) = \boxed{3\sqrt{3}/4}$.

8.52   Let a be the length of the longer base of the small trapezoids and b be the length of the shorter base. Since $\overline{EF}$ is a leg of $EFGH$ and is the shorter base of $EFBC$, the legs of the small trapezoids have length b as well. From the given information, we have $a = AB = 4$ and $a + 2b = CD = 8$, so $b = 2$. The diagram has 4 segments of length a and 7 of length b, so the total length is $4a + 7b = \boxed{30}$.

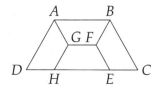

8.53   *Solution 1:* We note that the area of shaded triangle EFG equals that of $\triangle AFG$, so the shaded square has area equal to $\triangle AEG$. The large square is made up of 8 triangles congruent to $\triangle AEG$, so the shaded area is $\boxed{1/8}$ of the entire square.

Solution 2: Let the side length of the square be s, so that the area of the square is s^2 and each diagonal has length $s\sqrt{2}$. Since $EG = EA/\sqrt{2} = s/2$, the area of the shaded square is $(s/2)^2/2 = s^2/8$. Therefore, the shaded area is $\boxed{1/8}$ of the large square.

8.54   The crosswalk is a parallelogram with base 15 and height 40, so its area is $15 \cdot 40 = 600$. Let the distance between the stripes be h. Then the crosswalk is also a parallelogram with base 50 and height h, so $50h = 600$. Solving for h, we find $h = \boxed{12}$.

Challenge Problems

8.55

(a) Since $AX = XY$, we have $\angle XAY = \angle AYX$. Also, $\angle XAY + \angle XYA = 180° - \angle AXY = 180° - 70° = 110°$, so $\angle XAY = \angle AYX = 110°/2 = 55°$.

Then $\angle XAZ = 180° - \angle XAY = 180° - 55° = 125°$, so $\angle ZXA = 180° - \angle XZA - \angle XAZ = 180° - 21° - 125° = \boxed{34°}$.

(b) First, $\angle XWZ + \angle WZA = \angle XWZ + \angle WZX + \angle XZA = 125° + 34° + 21° = 180°$, so $\overline{WX} \parallel \overline{ZA}$. Since $\angle ZXA = 34° = \angle XZW$, we also have $\overline{WZ} \parallel \overline{AX}$, so $WXAZ$ is a parallelogram. Hence, $WZ = XA = \boxed{6}$.

(c) We have $\angle WXY = \angle WXZ + \angle ZXA + \angle AXY = 21° + 34° + 70° = 125° = \angle XWZ$. Also, $WZ = XY = 6$. Therefore, by SAS, triangles ZWX and YXW are congruent. Since corresponding parts are congruent, $WY = XZ$. (We could also note that $WXYZ$ is an isosceles trapezoid with $\angle Y = \angle Z$, so diagonals $\overline{WY}$ and $\overline{XZ}$ are congruent.)

8.56 Since the diagonals of a rectangle bisect each other, $AO = OC$. Since vertical angles are equal, $\angle AOF = \angle COG$. Since sides $\overline{AB}$ and $\overline{CD}$ are parallel, we have $\angle FAO = \angle GCO$. Hence, triangles AOF and GOC are congruent by ASA.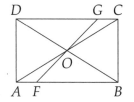

Then $[FOB] + [GOC] = [FOB] + [AOF] = [AOB]$. Right triangle ADB has area $(AD)(AB)/2 = 40$. Since $\overline{AO}$ is a median in triangle ABD, $[AOB] = [ABD]/2 = \boxed{20}$.

8.57 The angle trisectors divide each right angle of the rectangle into three $30°$ angles. Therefore $\angle PAD = \angle PDA = 30°$, so $AP = PD$. Similarly, $\angle RBC = \angle BCR = 30°$, so $BR = RC$. Since $\angle PAD = \angle RBC$, $AD = BC$ (since $ABCD$ is a rectangle), and $\angle PDA = \angle BCR$, we have $\triangle APD \cong \triangle BRC$ by ASA. Therefore, $AP = BR = RC = PD$. Similarly, we can show that $SD = SC = AQ = QB$ by showing that $\triangle SDC$ and $\triangle QAB$ are congruent isosceles triangles.

Since $\angle QAP = \angle PDS = \angle SCR = \angle RBQ = 30°$, $AP = PD = RC = RB$, and $SD = SC = QA = QB$, we have $\triangle QPA \cong \triangle SPD \cong \triangle SRC \cong \triangle QRB$ by SAS Congruence. Therefore, $QP = SP = SR = QR$, so $PQRS$ is a rhombus.

8.58 Since $\angle B = \angle C = 90°$ and $\angle QMC = 180° - \angle QMP - \angle BMP = 90° - \angle BMP = \angle MPB$, triangles QMC and MPB are similar by AA Similarity. Therefore, $PM/MQ = PB/MC = (4BC/3)/(BC/2) = \boxed{8/3}$.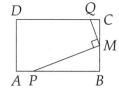

8.59

(a) Since $\overline{DA} \parallel \overline{BC}$, we have $\angle ADF = \angle ECF$ and $\angle DAF = \angle CEF$. Hence, triangles AFD and EFC are similar by AA. Since their areas are in a ratio of $4/64 = 1/16$, their corresponding sides are in a ratio of $\sqrt{1/16} = \boxed{1/4}$.

(b) Since $\overline{AB} \parallel \overline{CD}$, the distances from A and B to $\overline{CD}$ are equal. Thus, triangles BFC and ADF have equal altitudes, but their bases are in a ratio of $CF/DF = 1/4$. Hence, $[BFC] = [ADF]/4 = \boxed{16}$.

(c) Let G be the point on $\overline{AB}$ such that $\overline{AD} \parallel \overline{FG} \parallel \overline{BC}$. Then triangle ADF is half of parallelogram $AGFD$, so $[AGFD] = 2[ADF] = 128$. Similarly, triangle BFC is half of parallelogram $GBCF$, so $[GBCF] = 2[BCF] = 32$. Therefore, $[ABCD] = [AGFD] + [GBCF] = 128 + 32 = \boxed{160}$.

8.60

(a) As we showed in an earlier Exercise, the area of a quadrilateral with perpendicular diagonals is half the product of the diagonals. Hence, the area is $(10)(12)/2 = \boxed{60}$.

(b) *Solution 1:* Let F be the intersection of $\overline{AC}$ and $\overline{BD}$. Let $x = AF$, so $CF = 6\sqrt{2} - x$. Applying the Pythagorean Theorem to $\triangle BAF$ gives $BF^2 = (4\sqrt{3})^2 - x^2$, and applying it to $\triangle BCF$ gives $BF^2 = (2\sqrt{3})^2 - (6\sqrt{2} - x)^2$. Setting the two equal and solving, we get $x = 9\sqrt{2}/2$. Then $BF = \sqrt{(4\sqrt{3})^2 - x^2} = \sqrt{(4\sqrt{3})^2 - (9\sqrt{2}/2)^2} = \sqrt{30}/2$. Finally,

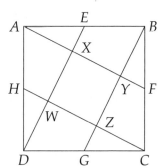

$$[ABCD] = \frac{1}{2}(AC)(BD) = (AC)(BF) = (6\sqrt{2})\left(\frac{\sqrt{30}}{2}\right) = \boxed{6\sqrt{15}}.$$

Solution 2: In triangle ABC, we have $a = 2\sqrt{3}$, $b = 6\sqrt{2}$, and $c = 4\sqrt{3}$, so the semiperimeter is $s = (a + b + c)/2 = 3\sqrt{3} + 3\sqrt{2}$. Then by Heron's formula, we have

$$[ABC]^2 = s(s-a)(s-b)(s-c) = (3\sqrt{3} + 3\sqrt{2})(\sqrt{3} + 3\sqrt{2})(3\sqrt{3} - 3\sqrt{2})(-\sqrt{3} + 3\sqrt{2})$$

$$= [(3\sqrt{3} + 3\sqrt{2})(3\sqrt{3} - 3\sqrt{2})][(\sqrt{3} + 3\sqrt{2})(-\sqrt{3} + 3\sqrt{2})] = (27 - 18)(-3 + 18) = 9 \cdot 15 = 135,$$

so $[ABC] = \sqrt{135} = 3\sqrt{15}$. Since $\triangle ABC \cong \triangle ADC$ by SSS Congruence, the area of $ABCD$ is $2[ABC] = \boxed{6\sqrt{15}}$.

8.61 Let $ABCD$ be the square, and let the midpoints of the sides of $ABCD$ and the vertices of the shaded square be labeled as shown. Since $AB = BC = CD = DA$ and $BF = CG = DH = AE$, right triangles $\triangle ABF$, $\triangle BCG$, $\triangle CDH$, and $\triangle DAE$ are congruent by SAS Congruence. Therefore, we have the following two angle equalities as corresponding parts of these triangles:

$$\angle FBY = \angle GCZ = \angle WDH = \angle XAE$$

$$\angle BFY = \angle CGZ = \angle WHD = \angle XEA.$$

Combined with $FB = GC = DH = AE$, this tells us that AXE, BYF, CZG, and DWH are congruent by ASA Congruence. Furthermore, since $\angle YFB = \angle CGZ = \angle CGB$ and $\angle FBY = \angle CBG$, these little triangles are similar to the four larger right triangles we found congruent earlier. Therefore, these little triangles are all right triangles. Moreover, $YF/BY = CG/BC = 1/2$ from $\triangle FYB \sim \triangle GCB$. Since $\overline{YF}$ and $\overline{CZ}$ are both perpendicular to $\overline{BZ}$, we have $\triangle BYF \sim \triangle BZC$ by AA Similarity (since the triangles share an angle at B). Therefore, $BY/BZ = BF/BC = 1/2$. Finally, we note that $YZ = BY$, and $GZ = YF = BY/2$, so $YZ = (BG)(2/5)$. (Similarly, we can show that all four sides of rectangle $WXYZ$ have this length, so $WXYZ$ is a square.)

Square $WXYZ$ therefore has area $YZ^2 = 4BG^2/25$. Since $BG^2 = \sqrt{CG^2 + BC^2} = \sqrt{(BC/2)^2 + BC^2} = (BC\sqrt{5})/2$, our area is $4(5BC^2/4)/25 = BC^2/5$. Because the area of $ABCD$ is 1, we know $BC = 1$, so $[WXYZ] = \boxed{1/5}$.

(Alternatively, once we have $BY = XY = AY/2$, we could note that $[AYB] = [BZC] = [CWD] = [DXA] = [XYZW]$ and these 5 pieces together make up square $ABCD$, so each must equal 1/5.)

8.62 Since $MNOP$ is a parallelogram, we have $MP = NO$. Therefore, $AM = MP/2 = NO/2 = CN$. Since $\overline{MP} \parallel \overline{ON}$, we have $\angle MCN = \angle CMA$. Since $AM = CN$, $\angle MCN = \angle AMC$, and $MC = MC$, we have

$\triangle AMC \cong \triangle NCM$. Therefore, $\angle ACM = \angle CMN$, so $\overline{AC} \parallel \overline{MN}$. Thus, $ACNM$ and $ACOP$ are parallelograms, and each has half the area of $MNOP$.

Since $\triangle AMC$ and $\triangle MNC$ are congruent, each has half the area of $ACNM$. The diagonals of parallelogram $ACNM$ bisect each other, so $BC = BM$. Therefore, $[ABC] = [ABM] = [ACM]/2$ since all three triangles share the same altitude from A and $BM = BC = CM/2$. Therefore, $[ABC] = [AMC]/2 = [ACNM]/4 = [MNOP]/8$. Similarly $[ADC] = [MNOP]/8$, so $[ABCD] = [ABC] + [ACD] = [MNOP]/4$.

8.63 Let the incenters of triangles WXA, XYA, YZA, and ZWA be P, Q, R, and S, respectively. Since P is the incenter of $\triangle XAW$, $\overline{AP}$ bisects $\angle XAW$. Similarly, $\overline{AQ}$, $\overline{AR}$, and $\overline{AS}$ bisect $\angle XAY$, $\angle YAZ$, and $\angle ZAW$, respectively. We have $\angle PAQ = \angle QAX + \angle PAX = (\angle XAY)/2 + (\angle XAW)/2 = (\angle XAY + \angle XAW)/2 = 180°/2 = 90°$ (since $\angle XAY + \angle XAW$ is straight angle $\angle YAW$). Similarly, we have $\angle PAS = \angle SAR = \angle QAR = 90°$. Since $\angle PAQ + \angle PAS = 180°$, we know that $\angle SAQ = 180°$. Therefore, S, A, and Q are on the same line. Similarly, P, A, and R are collinear. We have already shown that $\overline{PA} \perp \overline{QA}$, so we know that $\overline{PR} \perp \overline{QS}$.

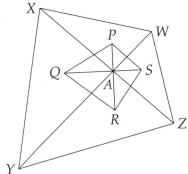

8.64 In parallelogram $EGJI$, $\overline{EG} \parallel \overline{JI}$, so the distance between $\overline{EG}$ and $\overline{JI}$ is equal to the distance from H to $\overline{EG}$. Therefore, $[EGJI] = 2[EGH]$. But triangle EGH is half of rectangle $EFGH$, so $[EGJI] = 2[EGH] = [EFGH] = \boxed{192}$. Note that the piece of information that $EF = 12$ is irrelevant.

8.65 Since $BC = 9$, we have $CH = BC - BH = 3$. Since $ABCD$ is a rectangle, $AD = 9$ also. Therefore, $AE = AD - ED = 5$. Since $\overline{CH} \parallel \overline{AE}$, we have $\angle GCH = \angle GEA$ and $\angle GHC = \angle GAE$, so $\triangle GCH \sim \triangle GEA$. $\overline{GF}$ is an altitude of $\triangle GEA$, and the corresponding altitude of $\triangle GCH$ has length $GF - CD$. Therefore, our similar triangles give us $(GF - CD)/GF = CH/EA = 3/5$. Since $ABCD$ is a rectangle, we have $CD = AB = 8$, so substitution and cross-multiplying gives $5(GF - 8) = 3GF$. Therefore, $GF = \boxed{20}$.

8.66 Let $x = BC$, and let E and F be the feet of perpendiculars from B and C to $\overline{AD}$, as shown. Triangles AEB and CFD are 45-45-90 triangles, so $AE = BE = CF = DF = AB/\sqrt{2} = 6/\sqrt{2} = 3\sqrt{2}$. Therefore, $AD = AE + EF + FD = 3\sqrt{2} + BC + 3\sqrt{2} = x + 6\sqrt{2}$. Since the area of the trapezoid is 30, we have

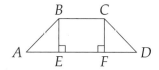

$$\frac{1}{2}(BC + AD) \cdot BE = \frac{1}{2}(x + x + 6\sqrt{2}) \cdot 3\sqrt{2} = 30.$$

Solving the above equation gives $BC = x = \boxed{2\sqrt{2}}$.

8.67 Let s be the side length of $ABCD$. Since $ABCD$ is a square, the height from X to $\overline{CD}$ is s greater than the height from X to $\overline{AB}$. Letting the height from X to $\overline{AB}$ be h, we have $hs/2 = 1$ and $(h + s)(s)/2 = 993$ from the given information. Therefore, $hs = 2$ and $hs + s^2 = 1986$, from which we find $s^2 = 1986 - hs = 1984$. The area of the square then is 1984. Now, we let the distance from X to $\overleftrightarrow{AD}$ be t. Since $ABCD$ is a square, the distance between $\overline{AD}$ and $\overline{BC}$ is s, and the height from X to $\overleftrightarrow{BC}$ has length $s - t$. Therefore, $[XAD] + [XBC] = t(AD)/2 + (s - t)(BC)/2 = ts/2 + (s - t)(s)/2 = s^2/2 = \boxed{992}$.

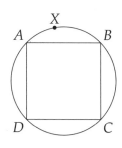

Note that our solution is a somewhat long-winded way of noting that both $[XCD] - [XAB]$ and $[XAD] + [XBC]$ equal half the area of the square, which is true because the difference in altitudes from X in the first case, and the sum of the altitudes in the second, equal the length of a side of the square.

8.68 Since $\overline{EG} \parallel \overline{AC}$, we have $\angle EFI = \angle ABI$ and $\angle FEI = \angle BAI$. Together with $AB = EF$, this gives us $\triangle IEF \cong \triangle IAB$ by ASA Congruence. Therefore, $IE = IA$, so $\overline{FB}$ bisects diagonal $\overline{AE}$. Similarly, we can show $\overline{HD}$ bisects $\overline{AE}$. Therefore, $\overline{FB}$ and $\overline{HD}$ meet at the midpoint of diagonal $\overline{AE}$, which is the center of rectangle $ACEG$. Since I is the center of the rectangle, it is $AG/2 = 4$ units from $\overline{EG}$ and $EG/2 = 7$ units from $\overline{EC}$. Therefore, the height from I to $\overline{EF}$ of $\triangle IEF$ has length 4 and the height from I to $\overleftrightarrow{ED}$ of $\triangle EID$ has length 7. Therefore, $[FEDI] = [FIE] + [DIE] = (EF)(4)/2 + (ED)(7)/2 = 18 + 7 = \boxed{25}$.

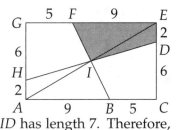

8.69 We draw altitudes from P to the sides of $WXYZ$ as shown. Since $ABCD$ is a rectangle, we have $\overline{AD} \perp \overline{AB}$. Because $\overline{PX}$ and $\overline{AD}$ are both perpendicular to $\overline{AB}$, we have $\overline{PX} \parallel \overline{AD}$. Since $\overline{AD} \perp \overline{CD}$ and $\overline{PX} \parallel \overline{AD}$, we know that $\overleftrightarrow{PX} \perp \overline{CD}$. Therefore, $\angle XPZ$ is a straight angle. Similarly, $\angle WPY$ is a straight angle as well. Hence, $AXZD$ and $BXZC$ are rectangles, so $AX = DZ$ and $BX = ZC$. We then use the Pythagorean Theorem to find

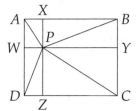

$$\begin{aligned} AP^2 &= AX^2 + PX^2 \\ CP^2 &= CZ^2 + ZP^2 \\ BP^2 &= BX^2 + PX^2 \\ DP^2 &= DZ^2 + PZ^2 \end{aligned}$$

Adding the first two equations and the last two equations, we have

$$AP^2 + CP^2 = AX^2 + PX^2 + CZ^2 + ZP^2 = DZ^2 + PX^2 + BX^2 + PZ^2 = (BX^2 + PX^2) + (DZ^2 + PZ^2) = BP^2 + DP^2.$$

Therefore, we have $PD^2 = PA^2 + PC^2 - PB^2 = 2^2 + 10^2 - 3^2 = 95$, so $PD = \boxed{\sqrt{95}}$.

8.70 Since $\overline{BC} \parallel \overline{AD}$, we have $\angle EBC = \angle EAD$, and $\angle ECB = \angle EDA$, so triangles BEC and AED are similar by AA. Let $x = BE$ and $y = CE$. From our similar triangles, $BE/BC = AE/AD$, which upon substitution becomes $x/7 = (x+6)/17$. Solving this equation gives $x = 21/5$. Similarly, $CE/CB = DE/DA$, or $y/7 = (y+8)/17$. Solving this gives $y = 28/5$.

Therefore, $BE : EC : BC = 21/5 : 28/5 : 7 = 3 : 4 : 5$. In other words, triangle BEC is similar to a 3-4-5 triangle, so it is a right triangle. Since $\angle BEC$ is opposite the longest side of this triangle, we have $\angle BEC = 90°$.

CHAPTER 9

_____Polygons

Exercises for Section 9.2

9.2.1 Using the formulas we developed in the text, we find:

Name	# of Sides	Sum of Interior $\angle$s	Int. $\angle$ Measure	Ext. $\angle$ Measure
pentagon	5	540°	108°	72°
hexagon	6	720°	120°	60°
heptagon	7	900°	$128\frac{4}{7}^\circ$	$51\frac{3}{7}^\circ$
octagon	8	1080°	135°	45°
nonagon	9	1260°	140°	40°
decagon	10	1440°	144°	36°
dodecagon	12	1800°	150°	30°
pentadecagon	15	2340°	156°	24°
icosagon	20	3240°	162°	18°
triacontagon	30	5040°	168°	12°

9.2.2 $\boxed{\text{Yes}}$, the formula for the sum of the interior angles works in any polygon. This is because any n-gon, whether convex or not, can be divided into $(n-2)$ triangles. Thus the sum of all interior angles is always $180(n-2)$ degrees.

9.2.3 *Solution 1:* Since the sum of all the angles in a polygon with n sides is $(n-2)(180°)$, we have $(n-2)(180°) = (n)(160°)$. Therefore, $9n - 18 = 8n$, so $n = \boxed{18}$.

Solution 2: Since each interior angle is $160°$, each exterior angle is $180° - 160° = 20°$. Since all the exterior angles together sum to $360°$, there must be $(360°)/(20°) = \boxed{18}$ of them.

9.2.4 From our given information, the angles have measures $3x$, $3x$, $3x$, $4x$, and $5x$ for some value of x. Since the sum of all the angles in a pentagon is $(180°)(5-2) = 540°$, we have $3x + 3x + 3x + 4x + 5x = 540°$, so $x = 30°$. Therefore, the largest angle has measure $5(30°) = \boxed{150°}$.

9.2.5 The sum of all exterior angles of any polygon is $360°$. Letting n be the number of sides, we have $180°(n-2) = 3(360°)$, so $n = \boxed{8}$.

Exercises for Section 9.3

9.3.1

(a) Draw diagonals $\overline{AD}, \overline{BE}, \overline{CF}$. Denote by O their intersection. As described in the text, these diagonals divide the regular hexagon into six congruent equilateral triangles. Thus, $AD = AO + OD = 9 + 9 = \boxed{18}$.

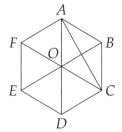

(b) The area of the hexagon is the total area of all six equilateral triangles, or $6(9^2 \sqrt{3})/4 = \boxed{243 \sqrt{3}/2}$.

(c) By symmetry, $\overline{AD}$ bisects $\angle CDE$, so $\angle CDA = \angle CDE/2 = 120/2 = 60°$. Since $CD/AD = 9/18 = 1/2$ and $\angle CDA = 60°$, triangle ACD is a 30-60-90 triangle. Therefore $AC = \sqrt{3}CD = \boxed{9\sqrt{3}}$.

9.3.2 As described in the text, a regular hexagon of side s can be divided into six equilateral triangles of side s. Thus the area is $6(s^2 \sqrt{3}/4) = \boxed{3s^2 \sqrt{3}/2}$.

9.3.3 Since $\overline{BC} \parallel \overline{DE}$ (because they are both perpendicular to $\overline{CD}$) and $BC = DE$, we know $BCDE$ is a parallelogram. Furthermore, $BCDE$ must be a square, since it is a parallelogram with a right angle and with two adjacent sides equal in length. Therefore, $BE = 4$. However, from right triangle ABE, we find $BE = \sqrt{AB^2 + AE^2} = 3$. BE can't be both 3 and 4, so this diagram is impossible.

9.3.4

(a) We show here a different method than that given in the text. Make sure you study and understand both methods. Perhaps you can also find a new method all your own. Draw $\overline{AD}$ and $\overline{EH}$, dividing the octagon into two trapezoids and a rectangle. Let the feet of perpendiculars from B and C to $\overline{AD}$ be P and Q, respectively.

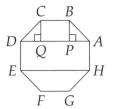

$\overline{BP}$ and $\overline{CQ}$ divide trapezoid $ABCD$ into two 45-45-90 triangles and rectangle $BPQC$. Thus $BP = AB/\sqrt{2} = 8/\sqrt{2} = 4\sqrt{2}$. The area of $ABCD$ is

$$[ABCD] = [ABP] + [BPQC] + [CDQ] = \frac{1}{2}(4\sqrt{2})^2 + (8)(4\sqrt{2}) + \frac{1}{2}(4\sqrt{2})^2 = 32 + 32\sqrt{2}.$$

This is also the area of trapezoid $EFGH$. Rectangle $ADEH$ has area $(AH)(AD) = 8(8 + 8\sqrt{2}) = 64 + 64\sqrt{2}$. The area of $ABCDEFGH$ is the sum of the areas of these three pieces, or $[ABCD] + [EFGH] + [ADEH] = \boxed{128 + 128\sqrt{2}}$. (See the text for other approaches to finding the area of a regular octagon.)

(b) In the text, we learned that the area of a regular polygon equals one-half the product of the length of the apothem and the perimeter of the polygon. In the first part, we found that the area of $ABCDEFGH$ is $128 + 128\sqrt{2}$, and the perimeter of $ABCDEFGH$ is $8(8) = 64$. Therefore, we have $(XM)(64)/2 = 128 + 128\sqrt{2}$, so $XM = \boxed{4 + 4\sqrt{2}}$.

(c) Since X is the center of the octagon, we have $XC = XB$, so $\triangle XCB$ is isosceles. Let N be the midpoint of $\overline{BC}$. Since $XC = XB$, $\overline{XN}$ is an altitude of $\triangle XCB$. It also is an apothem of the octagon, so it has length $4 + 4\sqrt{2}$ as shown in part (b). From right triangle $\triangle NXC$, we have

$$XC = \sqrt{XN^2 + NC^2} = \sqrt{(4 + 4\sqrt{2})^2 + 4^2} = \boxed{4\sqrt{4 + 2\sqrt{2}}}.$$

9.3.5 As shown earlier, a hexagon of side length s can be divided into six congruent equilateral triangles of side length s. Thus, the area of the shaded region is precisely the area of the hexagon. The answer is $\boxed{24}$.

9.3.6 Since $ABCDEFGH$ is a regular octagon, each of its interior angles has measure $135°$. Therefore, both $\angle XBC$ and $\angle XCB$ have measure $180° - 135° = 45°$, so $\angle X = 180° - 45° - 45° = 90°$. Similarly, all four little triangles in our diagram are 45-45-90 triangles. Therefore, all four angles of $WXYZ$ are $90°$. Each of our triangles has a side of the octagon as its hypotenuse. Letting each side of the octagon have length s, this means that each leg of each triangle has length $s/\sqrt{2}$. Therefore, each side of $WXYZ$ has length $s + 2s/\sqrt{2}$. Since all four angles of $WXYZ$ are the same, as are all four sides, $WXYZ$ is a square.

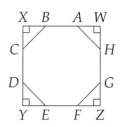

9.3.7

(a) Since each angle of a regular hexagon is $120°$, we have $\angle OAB = \angle OBA = 60°$. Therefore, $\angle AOB = 180° - \angle OAB - \angle OBA = 60°$ also, so $\triangle OAB$ is equilateral.

(b) We have $OB = AB$ from equilateral $\triangle OAB$ and $AB = BC$ because the hexagon is regular. Therefore, we have $OB = BC$, so $\angle BOC = \angle BCO$ in $\triangle OCB$. We also have $\angle OBC = 60°$ because $\overline{OB}$ bisects $\angle ABC$, so $\angle BOC = \angle BCO = (180° - \angle OBC)/2 = 60°$. Therefore, $\angle OBC = \angle BOC = \angle BCO$, so $\triangle BCO$ is equilateral.

(c) We can continue as in the previous part to show that each of $\triangle OCD$, $\triangle ODE$, $\triangle OEF$, and $\triangle OFA$ is equilateral. Since $\angle AOB + \angle BOC + \angle COD = 180°$, we have $\angle AOD = 180°$. Therefore, A, O, and D all lie on the same line.

(d) In the previous part, we showed that O is on $\overline{AD}$. Similarly, we can show that O is on $\overline{BE}$ and $\overline{CF}$. Therefore, all three long diagonals of the hexagon pass through the same point.

Exercises for Section 9.4

9.4.1 Each interior angle of a regular pentagon is $108°$. Since $ED = DC$, $\triangle EDC$ is isosceles with $\angle DEC = \angle DCE = (180° - \angle EDC)/2 = (180° - 108°)/2 = 36°$. Therefore, $\angle QEA = \angle AED - \angle DEC = 72°$. Since $\angle A = 108°$, we have $\angle A + \angle AEQ = 180°$, so $\overline{AB} \parallel \overline{EQ}$. Similarly, we can show $\overline{EA} \parallel \overline{QB}$. Therefore $AEQB$ is a parallelogram. Since $AB = EA$, $AEQB$ is a rhombus.

9.4.2

(a) As an Exercise in the previous chapter, we showed that the area of a regular hexagon with side length s is $3s^2\sqrt{3}/2$. Therefore, $[UOWABC] = 3(12^2)\sqrt{3}/2 = \boxed{216\sqrt{3}}$.

(b) Draw $\overline{VO}$. $UVWO$ consists of two equilateral triangles of sides 12. Thus, $[UVWO] = 2(12^2\sqrt{3}/4) = \boxed{72\sqrt{3}}$.

9.4.3

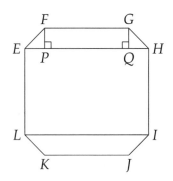

(a) The perimeter is $4(2 + 6) = \boxed{32}$.

(b) Draw $\overline{EH}$ and $\overline{IL}$, dividing the octagon into two trapezoids and a rectangle. Let the feet of perpendiculars from F and G to $\overline{EH}$ be P and Q, respectively. $\overline{FP}$ and $\overline{GQ}$ divide trapezoid $EFGH$ into two 45-45-90 triangles and rectangle $FPQG$. Thus $FP = EF/\sqrt{2} = 2/\sqrt{2} = \sqrt{2}$. The area of $EFGH$ is

$$[EFP] + [FPQG] + [GHQ] = \frac{1}{2}(\sqrt{2})^2 + (6)(\sqrt{2}) + \frac{1}{2}(\sqrt{2})^2 = \boxed{2 + 6\sqrt{2}}.$$

Similarly, $IJKL$ is a trapezoid with area $2 + 6\sqrt{2}$. $HILE$ is a rectangle with area $(HI)(EH) = (HI)(HQ+QP+PE) = 6(6+2\sqrt{2}) = 36+12\sqrt{2}$. Therefore, $[EFGHIJKL] = [EFGH]+[HILE]+[IJKL] = \boxed{40 + 24\sqrt{2}}$. (We could also have found the area by extending $\overline{FG}$, $\overline{HI}$, $\overline{JK}$, and $\overline{EL}$ to form a square.)

(c) $EHIL$ is a rectangle with side lengths 6 and $6 + 2\sqrt{2}$. Thus diagonal $EI = \sqrt{6^2 + (6 + 2\sqrt{2})^2} = 2\sqrt{3^2 + (3 + \sqrt{2})^2} = \boxed{2\sqrt{20 + 6\sqrt{2}}}$.

9.4.4 *Solution 1:* In the figure at left below, altitude $\overline{DX}$ of isosceles triangle $\triangle DEC$ divides $\triangle DEC$ into two 30-60-90 triangles. Letting the side length of the hexagon be s, we have $DX = s/2$ and $XE = XC = s\sqrt{3}/2$. Therefore, $EC = s\sqrt{3}$ and the area of $\triangle CDE$ is $(EC)(XD)/2 = s^2\sqrt{3}/4$. Similarly, the area of $\triangle AFE$ is $s^2\sqrt{3}/4$. The area of the whole hexagon (as we showed in an Exercise in the previous section) is $3s^2\sqrt{3}/2$, so the area of $ABCE$ is $[ABCDEF] - [CDE] - [AFE] = s^2\sqrt{3}$. Therefore $[ABCE]/[ABCDEF]$ is $\boxed{2/3}$.

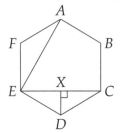 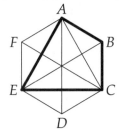

Solution 2: Dissect the hexagon as shown at right above. The hexagon is cut into 12 congruent triangles (why?). Since $ABCE$ consists of 8 of these triangles, our desired ratio is $8/12 = \boxed{2/3}$.

9.4.5 Since all the sides of the octagon have the same length and all the angles have the same measure, we have $\triangle ABC \cong \triangle CDE \cong \triangle EFG \cong \triangle GHA$ by SAS Congruence. Therefore, $AC = CE = EG = GA$. Since $\angle ABC = 135°$ and $AB = BC$, we have $\angle BAC = (180° - 135°)/2 = 22.5°$. Similarly, $\angle HAG = 22.5°$, and we have $\angle CAG = \angle BAH - \angle BAC - \angle HAG = 90°$. Similarly, all the angles of $ACEG$ are 90°. Since all its angles are the same and all its sides are the same, $ACEG$ must be a square.

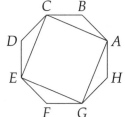

9.4.6 Triangles AHG and BCD are congruent by SAS. Thus, $AG = BD$. We also have $\angle HAG = \angle CBD$, so $\angle BAG = 135° - \angle HAG = 135° - \angle CBD = \angle ABD$. Thus triangles ABD and BAG are congruent by SAS. The altitudes from D and G to $\overline{AB}$ are therefore equal, implying $\overline{DG} \parallel \overline{AB}$. So, $ABDG$ is a trapezoid.

9.4.7 We first show $\overline{HG} \perp \overline{GD}$. Continue $\overrightarrow{BA}$ past A and $\overrightarrow{GH}$ past H to meet at Q. Since $\angle HAB = 135°$, we have $\angle HAQ = 180° - \angle HAB = 45°$. Similarly, $\angle AHQ = 45°$. Therefore, $\angle Q = 180° - \angle QAH - \angle QHA = 90°$, so $\overleftrightarrow{AB} \perp \overleftrightarrow{HG}$. In the last problem, we proved $\overline{GD} \parallel \overline{AB}$, so $\overline{GD} \perp \overleftrightarrow{HG}$ also. Therefore, we have $\angle FGX = \angle FGH - \angle XGH = 45°$. Similarly, we have $\angle XFG = 45°$, so from triangle $\triangle FXG$, we have $\angle FXG = 180° - \angle XFG - \angle XGF = 90°$.

9.4.8 Let the midpoints of the sides be as indicated. Two problems earlier we saw that $[ABC]/[ABCDEF] = 1/6$. Since triangles ABC and GBH are similar with ratio $1/2$, we have $[GBH]/[ABC] = 1/4$, so $[GBH]/[ABCDEF] = 1/24$. Similarly, each of the six small triangles outside $GHIJKL$ has area $1/24$ that of the $ABCDEF$. Thus the answer is $1 - 6(1/24) = \boxed{3/4}$.

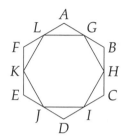

Exercises for Section 9.5

9.5.1 From the construction, we know $DA = AB = BC$ and $\angle CBA = \angle BAD = 90°$. Since $\angle BAD + \angle CBA = 180°$, we have $\overline{AD} \parallel \overline{BC}$. Because we also have $AD = BC$, $ABCD$ is a parallelogram. Since $ABCD$ has a right angle, it is a rectangle. The rectangle has equal adjacent sides $AD = AB$, so it is a square.

9.5.2 Let O be the center of the circle. Since $OZ = OS = OW$ and $\angle ZOS = 45° = \angle SOW$, isosceles triangles ZOS and SOW are congruent by SAS. Thus, $\angle ZSW = \angle ZSO + \angle WSO = \angle ZSO + \angle SZO = 180° - \angle ZOS = 135°$, and $ZS = SW$. Similarly, all adjacent sides of the octagon are equal and each of its interior angles has measure $135°$, so the octagon is regular.

9.5.3 If we connect all the vertices of a dodecagon to its center, we will form 12 congruent isosceles triangles. Each triangle has a vertex angle with measure $360°/12 = 30°$. We can use this with the circumcircle of the dodecagon to construct a regular dodecagon.

Specifically, we start with $\odot O$ and construct a $30°$ angle with vertex O. (We can construct a $30°$ angle by first constructing an equilateral triangle, then bisecting an angle of it.) Let this angle hit the circle at A and B. Since $\overarc{AB} = 30°$, we can find 12 evenly spaced points around the circle by marking off arcs using a radius of length AB. The arc with center B and length AB will give us C, then the arc centered at C will give us D, and so on around the circle. We connect each of the points thus formed in succession to construct our dodecagon. By construction, each of these chords has the same length. Each chord cuts off $30°$ of the circle, so 12 of them will bring us back to our starting point. Each angle of the dodecagon consists of two of the base angles of triangles like AOB. Therefore, each angle of the dodecagon has measure $180° - 30° = 150°$. Since all its angles are the same and all its lengths are the same, our dodecagon is regular.

Review Problems

9.18 Each exterior angle is $180° - 140° = 40°$. The sum of all exterior angles is $360°$ for any polygon. Thus, the number of sides is $360°/40° = \boxed{9}$.

9.19 Each exterior angle of a polygon with 36 sides has measure $360°/36 = 10°$. Therefore, each interior angle has measure $180° - 10° = \boxed{170°}$.

9.20 As we have seen in the text, drawing the long diagonals of a regular hexagon with side s divides the hexagon into six equilateral triangles with side length s. Therefore, the area of a hexagon with side length 4 is $6(4^2 \sqrt{3}/4) = \boxed{24\sqrt{3}}$.

9.21 The sum of all exterior angles is $360°$ for any polygon. Thus, the number of sides is $360°/6° = \boxed{60}$.

9.22 (This solution shows a different method than in the text. See text for another method.) Let the octagon be $ABCDEFGH$. Draw $\overline{AD}$ and $\overline{EH}$, dividing the octagon into two trapezoids and a rectangle. Let the feet of perpendiculars from B and C to $\overline{AD}$ be P and Q, respectively.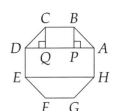

$\overline{BP}$ and $\overline{CQ}$ divide trapezoid $ABCD$ into rectangle $BPQC$ and two 45-45-90 triangles. Thus $BP = AB/\sqrt{2} = 6/\sqrt{2} = 3\sqrt{2}$. The area of $ABCD$ therefore is

$$[ABP] + [BPQC] + [CDQ] = \frac{1}{2}(3\sqrt{2})^2 + (6)(3\sqrt{2}) + \frac{1}{2}(3\sqrt{2})^2 = 18 + 18\sqrt{2}.$$

The area of trapezoid $EFGH$ likewise is $18 + 18\sqrt{2}$. Rectangle $ADEH$ has area $(AD)(DE) = 6(6 + 6\sqrt{2}) = 36 + 36\sqrt{2}$. The area of the octagon is $[ABCD] + [ADEH] + [EFGH] = \boxed{72 + 72\sqrt{2}}$.

(We could also have noted that we found the area of a regular octagon with side length 4 in the text. The area of that octagon is $32 + 32\sqrt{2}$. Because all regular octagons are similar, the area of a regular octagon with side length 6 is $(6/4)^2(32 + 32\sqrt{2}) = 72 + 72\sqrt{2}$.)

9.23 Let the measure of each exterior angle be x. Then the interior angle is $8x$. Since $x + 8x = 180°$, $x = 20°$. Therefore, each exterior angle is $20°$, so the number of sides is $360°/20° = \boxed{18}$.

9.24 By symmetry, long diagonal $\overline{EH}$ bisects $\angle FEJ$, so $\angle FEH = (1/2)(\angle FEJ) = 60°$. Since $EF = FG$ and $\angle EFG = 120°$, we have $\angle FGE = \angle FEG = (180° - 120°)/2 = 30°$. Therefore, $\angle GEH = \angle FEH - \angle FEG = 30°$ and $\angle HGE = \angle HGF - \angle FGE = 90°$. Hence, $\triangle EGH$ is a 30-60-90 triangle with hypotenuse $\overline{EH}$ and short leg $\overline{GH}$. Therefore, $EH/EG = 2/\sqrt{3} = \boxed{2\sqrt{3}/3}$.

9.25 Let the common side length be s. We are given that $s^2 = \sqrt{3}$. Since the area of a regular hexagon with side length s is 6 times the area of an equilateral triangle with side length s, the area of our regular hexagon is $6(s^2 \sqrt{3}/4) = \boxed{9/2}$.

9.26 The sum of the measures of the angles in an octagon is $(8 - 2)(180°) = 1080°$. Therefore, we have $4(x) + 2(2x) + 2(90°) = 1080°$, so $8x = 900°$, and we have $x = 112.5°$. Therefore, two of the angles have measure $2(112.5°) = 225°$, so the octagon has two angles larger than $180°$. Hence, the octagon is $\boxed{\text{concave}}$.

9.27 We showed in the text that a polygon with n sides has $n(n-3)/2$ diagonals. Solving $n(n-3)/2 = 27$ gives $n^2 - 3n = 54$, or $(n-9)(n+6) = 0$. Since n must be positive, our polygon has $n = \boxed{9}$ sides.

9.28 Let $ABCDEFGH$ be our octagon. We draw $\overline{AC}$, $\overline{AD}$, and $\overline{AE}$, as shown. We also drop altitudes from B and C to $\overline{AD}$ as shown. Since $ABCD$ is a trapezoid, we have $\angle QCB = 180° - \angle CQA = 90°$, so $\angle DCQ = \angle DCB - \angle QCB = 45°$. Therefore $\triangle CDQ$ is a 45-45-90 triangle (and similarly so is $\triangle ABP$). Therefore, $DQ = PA = AB/\sqrt{2} = 4\sqrt{2}$. Since $BCQP$ is a rectangle, $QP = BC = 8$. Hence, $AD = AP + PQ + QD = \boxed{8 + 8\sqrt{2}}$.

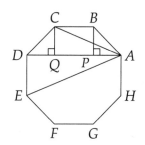

From right triangle $\triangle ACQ$ we have

$$AC = \sqrt{AQ^2 + CQ^2} = \sqrt{(8 + 4\sqrt{2})^2 + (4\sqrt{2})^2} = \boxed{8\sqrt{2 + \sqrt{2}}}.$$

Since $\angle CDQ = 45°$, we have $\angle ADE = \angle CDE - \angle CDQ = 90°$. From right triangle $\triangle ADE$, we have $AE = \sqrt{AD^2 + DE^2} = \sqrt{(8 + 8\sqrt{2})^2 + 8^2} = \boxed{8\sqrt{4 + 2\sqrt{2}}}$. (We also could have used square $ACEG$ to note that $AE = AC\sqrt{2} = 8\sqrt{4 + 2\sqrt{2}}$.)

9.29 Each interior angle of the regular hexagon has measure $120°$, and each interior angle of the regular pentagon has measure $108°$. Therefore, $\angle CAB = \angle BAD - \angle CAD = 120° - 108° = \boxed{12°}$.

9.30 Let the number of sides in the polygon be n. The number of diagonals in the polygon then is $n(n-3)/2$, so we have $n(n-3)/2 = n$. Therefore, $n - 3 = 2$, so $n = 5$. Since the polygon is a pentagon, the sum of its interior angles is $(5-2)(180°) = \boxed{540°}$.

9.31

(a) Since $BC = AJ$, $AB = IJ$, and $\angle B = \angle J$, we have $\triangle ABC \cong \triangle AJI$. Therefore, $AI = AC$.

(b) Since isosceles triangles $\triangle ABC$ and $\triangle AJI$ are congruent (from the last part), we have $\angle ACB = \angle AIJ$. Therefore, $\angle HIA = \angle HIJ - \angle AIJ = \angle DCB - \angle ACB = \angle ACD$. Since we also have $AC = AI$ and $CD = HI$, we have $\triangle ACD \cong \triangle AIH$ by SAS. Therefore, $AD = AH$.

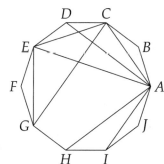

(c) Just as $\triangle ABC \cong \triangle AJI$, we have $\triangle ABC \cong \triangle AJI \cong \triangle CDE \cong \triangle EFG$. These congruent isosceles triangles give us $\angle DCE = \angle FEG = \angle DEC = \angle ACB$. Therefore $\angle CEG = \angle DEF - \angle FEG - \angle CED = \angle BCD - \angle DCE - \angle ACB = \angle ACE$. Just as we showed $AI = AC$ in the first part, we can show that $AC = CE = EG$. Combining this with $\angle ACE = \angle CEG$, we have $\triangle CEG \cong \triangle ACE$ by SAS Congruence. Therefore, $CG = AE$.

9.32 Since $ABCD$ is a square, we have $\angle ABC = 90°$. Since $ABEFG$ is a regular pentagon, we have $\angle ABE = 108°$. Therefore, $\angle CBE = 360° - \angle ABC - \angle ABE = 162°$. Since $BC = AB = BE$, $\triangle BCE$ is isosceles with $\angle BCE = \angle BEC = (180° - \angle CBE)/2 = \boxed{9°}$.

9.33

(a) Each interior angle is $180° - 360°/8 = 135°$. By symmetry, $\overline{XA}$ bisects $\angle HAB$, so $\angle XAB = \angle HAB/2 = \boxed{67.5°}$.

(b) X is the center of the octagon, meaning it is equidistant from all 8 vertices of the octagon. (Make sure you see why!) Therefore, by SSS Congruence, we see that connecting X to each of the vertices forms 8 congruent isosceles triangles. The vertex angles of these triangles are the angles at X. Since these angles are all equal, each is $360°/8 = 45°$. Since M is the midpoint of $\overline{AB}$, we have $AM = BM$. Combined with $AX = BX$ and $XM = XM$, this gives us $\triangle AMX \cong \triangle BMX$ by SSS Congruence. Therefore, $\angle AXM = \angle BXM = \angle AXB/2 = \boxed{22.5°}$.

(c) $\angle MXD = \angle MXB + \angle BXC + \angle CXD = 22.5° + 45° + 45° = \boxed{112.5°}$.

Challenge Problems

9.34

(a) Triangles $A_1A_2A_3$ and $A_2A_3A_4$ are congruent by SAS. Thus, the altitudes from A_1 and A_4 to $\overline{A_2A_3}$ (extended if necessary) have the same length. This means $\overline{A_1A_4} \parallel \overline{A_2A_3}$ since A_1 and A_4 are equidistant from $\overleftrightarrow{A_2A_3}$. Thus $A_1A_2A_3A_4$ is a trapezoid.

(b) In trapezoid $A_1A_2A_3A_4$, $\angle A_2A_1A_4 = 180° - \angle A_1A_2A_3$. Each exterior angle of the polygon also has measure $180° - \angle A_1A_2A_3$. Therefore, $\angle A_2A_1A_4$ equals an exterior angle, so it has measure $360°/n$.

9.35 We tackle both parts at once. Let our polygon be $A_1A_2A_3\cdots A_n$. Let O be the point where the angle bisectors of $\angle A_nA_1A_2$ and $\angle A_1A_2A_3$ meet. Since $\angle A_nA_1A_2 = \angle A_1A_2A_3$, we have $\angle OA_1A_2 = \angle OA_2A_1$. Therefore $OA_1 = OA_2$. Since $OA_1 = OA_2$, $\angle OA_1A_2 = \angle OA_2A_3$ (each is half one of the interior angles) and $A_1A_2 = A_2A_3$ (both are sides of the regular polygon), we have $\triangle OA_1A_2 \cong \triangle OA_2A_3$ by SAS Congruence. Therefore $OA_2 = OA_3$ and 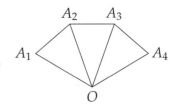 $\angle OA_3A_2 = \angle OA_2A_1 = (\angle A_2A_3A_4)/2$. Similarly, we can prove $\triangle OA_2A_3 \cong \triangle OA_3A_4$ and use this to show $OA_3 = OA_4$, and so on. Therefore, O is equidistant from all the vertices of the polygon. Moreover, as we do this, we show that each $\overline{OA_i}$ bisects the interior angle of the polygon with A_i as a vertex. (For example, we showed above that $\angle OA_3A_2 = (\angle A_2A_3A_4)/2$.)

(Note that we can quickly show there's only one point equidistant from all vertices of a regular polygon. Such a point must be on the perpendicular bisector of each side. Since the perpendicular bisectors of two adjacent sides only meet at one point, that point is the only one that can possibly be equidistant from all the vertices of the polygon.)

9.36 Since $AB = AE$ and $\angle A = 90°$, triangle ABE is a 45-45-90 triangle. Since $BE = 2$, we have $AB = AE = \sqrt{2}$, and $[ABE] = 1$. Let F be the foot of the altitude from C to $\overline{BE}$. In right triangle BCF, $BF = (BE - CD)/2 = 1/2$. Since $BC = 1$, $CF = \sqrt{3}/2$ from right triangle $\triangle BCF$. Thus $[BCDE] = (1+2)(\sqrt{3}/2)/2 = 3\sqrt{3}/4$. Finally, $[ABCDE] = [ABE] + [BCDE] = \boxed{1 + 3\sqrt{3}/4}$.

9.37 Let the number of sides of the two polygons be a and b. We are given $180°(a-2)+180°(b-2) = 1980°$, so $a + b = 15$. We are also given $a(a-3)/2 + b(b-3)/2 = 34$, so $a^2 + b^2 - 3(a+b) = 68$, which means $a^2 + b^2 = 68 + 3(a+b) = 113$. Thus, $2ab = (a+b)^2 - (a^2+b^2) = 112$. Finally, $(a-b)^2 = (a^2+b^2) - (2ab) = 1$, so the positive difference between a and b is $\boxed{1}$.

(We could also have let $a = 15 - b$ and substituted into $a^2 + b^2 = 113$ to find b, or have simply used guess-and-check to find that our polygons must have 7 and 8 sides, respectively.)

9.38   Let the measures of the seven interior angles of the heptagon be $x_1, x_2, \ldots,$ x_7 as shown. The sum of these angles is $180°(7 - 2) = 900°$. Letting the vertex at the angle with measure x_1 be X_1 and likewise for the other vertices of the interior heptagon, we can use quadrilaterals $ACEX_6$, $BDFX_7$, $CEGX_1$, $DFAX_2$, $EGBX_3$, $FACX_4$, and $GBDX_5$ to write:

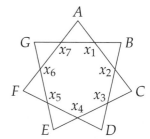

$$\angle A + \angle C + \angle E + x_6 = 360°$$
$$\angle B + \angle D + \angle F + x_7 = 360°$$
$$\angle C + \angle E + \angle G + x_1 = 360°$$
$$\angle D + \angle F + \angle A + x_2 = 360°$$
$$\angle E + \angle G + \angle B + x_3 = 360°$$
$$\angle F + \angle A + \angle C + x_4 = 360°$$
$$\angle G + \angle B + \angle D + x_5 = 360°.$$

Each angle of our star shows up in three equations. Because $x_1 + x_2 + x_3 + \cdots + x_7 = 900°$, adding all the equations gives

$$3(\angle A + \angle B + \angle C + \angle D + \angle E + \angle F + \angle G) + 900° = 7(360°).$$

Therefore, the sum of the angles in our star is $[7(360°) - 900°]/3 = \boxed{540°}$. Try figuring out the sum of the angles in stars with more points. Also, try finding the sum of the angles if you connected the points in the order *A-E-B-F-C-G-D-A*!

9.39   Let the side length of the hexagon be s, then the side length of the triangle is $2s$. As we found earlier, the area of the hexagon is $3s^2 \sqrt{3}/2$. The area of the equilateral triangle is $(2s)^2 \sqrt{3}/4 = s^2 \sqrt{3}$. Thus, the ratio is $\boxed{3/2}$

9.40   Let the side length of the hexagon be s. Then, the square and the equilateral triangles each have side length s, too. The area of each square is s^2. The area of each hexagon is $3s^2 \sqrt{3}/2$. Therefore, the ratio of the area of a square to the area of the hexagon is $(s^2)/[3s^2 \sqrt{3}/2] = 2\sqrt{3}/9$. So, the area of each square is $(2\sqrt{3}/9)(96) = \boxed{64\sqrt{3}/3}$.

As for the area of the whole figure, we note that the 6 equilateral triangles together equal the area of the hexagon, and the six squares together have area $6(64\sqrt{3}/3) = 128\sqrt{3}$. Therefore the area of the whole region is $\boxed{192 + 128\sqrt{3}}$. (Challenge: How might you use this dissection to find the area of a regular dodecagon?)

9.41   *Solution 1:* Since all the sides and all the angles of the hexagon are the same, we have $\triangle AFE \cong$ $\triangle EDC \cong \triangle CBA$ by SAS Congruence. Therefore, we have $AE = EC = CA$, so triangle $\triangle ACE$ is equilateral. We draw altitude $\overline{DX}$ of isosceles $\triangle CDE$ as shown at left below. Since $\angle EDC = 120°$, we have $\angle CDX = 60°$. Therefore, $\triangle CDX$ is a 30-60-90 triangle, so we have $DX = 6$ and $CX = 6\sqrt{3}$. Therefore, the side length of triangle ACE is $12\sqrt{3}$, so its area is $(12\sqrt{3})^2 \sqrt{3}/4 = \boxed{108\sqrt{3}}$.

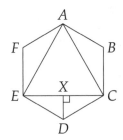

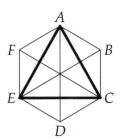

Solution 2: We could also have dissected the hexagon as shown at right above, drawing △*ACE* and the three long diagonals. This divides the hexagon into 12 congruent triangles, of which 6 are inside △*ABC*. Therefore, the area of △*ABC* is half the area of *ABCDEF*, or $(1/2)(3(12^2)\sqrt{3}/2) = \boxed{108\sqrt{3}}$.

9.42 Let the number of sides be n. Then the n angles are $172°, 172° - 4°, 172° - 2(4°), \ldots, 172 - (n-1)(4°)$. The sum of this arithmetic sequence is

$$\frac{1}{2}(172° + 172° - (4°)(n - 1))n.$$

On the other hand, we know the sum of all angles should be $180°(n - 2)$. Solving

$$\frac{1}{2}(172° + 172° - (n - 1)(4°))n = 180°(n - 2),$$

we have $86n - n(n - 1) = 180(n - 2)$, so $n^2 + 3n - 180 = 0$. Therefore $(n - 12)(n + 15) = 0$. Since n must be positive we have $n = \boxed{12}$.

Note, we could also have used the exterior angles for an even slicker solution! See if you can find it.

9.43 By the Pythagorean Theorem applied to △*ABC* and △*DEF*, we have $DF = AC = 15$. Since $AC^2 + CD^2 = AD^2$ and $DF^2 + AF^2 = AD^2$, we have $\angle AFD = \angle ACD = 90°$. Since $\overline{CD} \parallel \overline{AF}$, we have $\angle CDF = 180° - \angle AFD = 90°$. Similarly, $\angle CAF = 90°$, and $AFDC$ is a rectangle. Therefore, we have $[ABCDEF] = [ABC] + [ACDF] + [DEF] = (9)(12)/2 + (15)(20) + (9)(12)/2 = \boxed{408}$.

9.44 The interior angle is integer if and only if the exterior angle is integer. In other words, n is a factor of $360 = 2^3 3^2 5$. Any factor of 360 must have the form $2^a 3^b 5^c$. We have 4 choices for a, 3 choices for b, and 2 for c, so there are 24 factors of 360. However, we can't have a polygon with 1 or 2 sides, so we must omit those cases. There are $\boxed{22}$ values of n such that a convex regular polygon with n sides has interior angles whose measures in degrees are integers.

9.45 We'll tackle all three parts by taking care of the general case. Suppose our polygon has n sides. Then, the interior angles must sum to $180°(n - 2)$. Suppose k of these angles are right angles. If $k = n$, then we must have $(90°)k = (180°)(k - 2)$, from which we get $n = k = 4$. Therefore, in a quadrilateral, we can have up to 4 right angles.

If $k < n$, then the sum of all the angles must be less than $(90°)k + (180°)(n - k)$, since each of the remaining $n - k$ angles is less than $180°$. Since we know the sum of all the angles is $(180°)(n - 2)$, we have:

$$(180°)(n - 2) < (90°)k + (180°)(n - k).$$

Therefore, we have $k(90°) < 360°$, and $k < 4$. Therefore, no convex polygon with more than 4 sides can have more than $\boxed{3}$ right angles as interior angles. It is possible for a convex polygon with more than four sides to have 3 right angles as interior angles. (Make sure you see why!)

9.46 Let the three polygons have a, b, and c sides, respectively, where $a \le b \le c$. To surround a point, adding together the measures of one interior angle from each polygon must give a sum of $360°$. Let these angles have measures m_a, m_b, and m_c, respectively, so we have $m_a + m_b + m_c = 360°$. We can write these measures in terms of the exterior angles of the polygons. For example, $m_a = 180° - 360°/a$. Therefore, our equation becomes

$$\left(180° - \frac{360°}{a}\right) + \left(180° - \frac{360°}{b}\right) + \left(180° - \frac{360°}{b}\right) = 360°.$$

After some rearrangement, we have

$$\frac{2}{a} + \frac{2}{b} + \frac{2}{c} = 1.$$

Therefore, the smallest of a, b, and c can be at most 6 (otherwise, the sum will be less than 1). We consider four cases:

Case 1: $a = 6$. Here, we have $2/b + 2/c = 2/3$. Then, b must be at most 6, too, or else we'll have $2/b + 2/c < 2/3$. Since $b \ge a$, we must have $b = 6$, and this brings us the solution with three hexagons shown as an example in the problem.

Case 2: $a = 5$. Here, we have $2/b + 2/c = 3/5$, or $1/b + 1/c = 3/10$. Since $1/7 + 1/7 = 2/7 < 3/10$, we see that b cannot be greater than 6. We find that $b = 6$ makes c a fraction (impossible) and $b = 5$ gives $c = 10$, so that two pentagons and a decagon completely surround a point.

Case 3: $a = 4$. Here, we have $2/b + 2/c = 1/2$, or $1/b + 1/c = 1/4$. Since $1/8 + 1/8 = 1/4$, we either have $b = c = 8$ (giving us the solution of a square and two octagons) or we have $4 \le b < 8$. We find that $b = 7$ and $b = 4$ do not give solutions, but $b = 5$ gives us $c = 20$ (so a square, a pentagon and a 20-gon surround a point), and $b = 6$ gives us $c = 12$ (so a square, a hexagon, and a dodecagon surround a point as shown as an example in the problem).

Case 4: $a = 3$. Here, we have $2/b + 2/c = 1/3$, so $1/b + 1/c = 1/6$. Since $1/12 + 1/12 = 1/6$, we either have $b = c = 12$ (giving us the solution of a hexagon and two dodecagons), or $3 \le b < 12$. If $b < 7$, we quickly see that c either doesn't exist or must be negative. Trying the other possibilities, we see that $b = 11$ fails, but everything else works, giving us four more possible sets $\{a, b, c\}$: $\{3, 10, 15\}$; $\{3, 9, 18\}$; $\{3, 8, 24\}$; $\{3, 7, 42\}$.

For an extra challenge, figure out which of these patterns could be used to tile a whole floor (meaning each vertex is 'surrounded' with the same three types of polygons). Then try it with 4 polygons around each point!

9.47 All regular pentagons are similar to each other. Since the ratio of the areas of two similar figures equals the square of the ratio of corresponding sides of the figures, we have $[P_A]/[P_C] = (BC/AB)^2$ and $[P_B]/[P_C] = (AC/AB)^2$. Since $\triangle ABC$ is right with $\angle ACB = 90°$, we have $AC^2 + BC^2 = AB^2$. Therefore,

$$[P_A] + [P_B] = \left(\frac{BC}{AB}\right)^2 [P_C] + \left(\frac{AC}{AB}\right)^2 [P_C] = \left(\frac{BC^2 + AC^2}{AB^2}\right) [P_C] = \left(\frac{AB^2}{AB^2}\right) [P_C] = [P_C].$$

Perhaps you aren't satisfied with our simply claiming that $[P_A]/[P_C] = (BC/AB)^2$ – good for you! We'll run through the proof for pentagons. (The proof for other regular polygons is very similar.) Take any two regular pentagons and connect each vertex of each pentagon to the center of its pentagon. We thus form five isosceles triangles in each pentagon. Each of the base angles in these isosceles triangles has measure equal to half an interior angle of the pentagon, or $54°$. (We proved this in an earlier Challenge Problem.) Therefore, all of the isosceles triangles of one pentagon are similar to all of the isosceles triangles of the other. Thus, the ratio of the area of an isosceles triangle in one pentagon to the area of an isosceles triangle in the other is the square of the ratio of the sides of the pentagons (since the corresponding bases of these triangles are sides of the pentagons). Because the area of each pentagon is the sum of five of these isosceles triangles, the ratio of the areas of the pentagons equals the ratio of the areas of an isosceles triangle of one pentagon to that of the other. Therefore, the ratio of the areas of the pentagons equals the square of the ratio of the sides of the pentagons.

9.48 *Solution 1:* Because it is usually much easier to find the area of a right triangle than it is to find the area of funky octagons, we try to find the area of the octagon by subtracting the area of four of the little right triangles on the outside from one of our unit squares. First, we note that the 8 triangles are all congruent (by symmetry), so we only have to find the area of one of the right triangles.

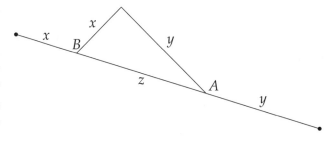

We therefore focus on one of the little triangles, letting its side lengths be x, y, and z as shown. We use the fact that all these little triangles are congruent to see that $x + y + z = 1$ since together they make up a side of one of our squares. We are given $z = 43/99$, so we have $x + y = 56/99$. The Pythagorean Theorem gives us $x^2 + y^2 = z^2 = (43/99)^2$. Before pounding away to find x and y, we note that we don't need x and y; we only need $xy/2$, the area of the triangle. As we've seen before, we can get $2xy$ by squaring $(x + y)$ and subtracting $x^2 + y^2$. We know what these both equal, so we can find our triangle area:

$$(x^2 + 2xy + y^2) - (x^2 + y^2) = \frac{56^2}{99^2} - \frac{43^2}{99^2}.$$

Therefore,

$$2xy = \left(\frac{56}{99} + \frac{43}{99}\right)\left(\frac{56}{99} - \frac{43}{99}\right) = \frac{13}{99}.$$

We note that we want to subtract 4 of our triangles from the square, so our desired area is $1 - 4(xy/2) = 1 - 2xy = \boxed{86/99}$.

We could also have jumped past all this algebra by dividing the octagon into 8 congruent triangles (where O is the center of the octagon):

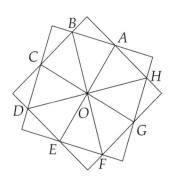

The altitude from O in each of these triangles is $1/2$, since the center of the squares is $1/2$ away from the sides of the squares. We are given that the 'base' of each of these triangles is $AB = 43/99$. Hence, each of the these triangles has area $(1/2)(43/99)(1/2)$. Our octagon is eight of these triangles, for a total area of $\boxed{86/99}$.

Challenge: Can you rigorously prove that those little triangles are all congruent without invoking symmetry?

CHAPTER 10

Geometric Inequalities

Exercises for Section 10.1

10.1.1

(a) Since $\angle A = 180° - \angle B - \angle C = 30°$, we have $\angle B > \angle C > \angle A$. Therefore, we have $\boxed{AC > AB > BC}$.

(b) Since $\angle Q = 180° - \angle R - \angle P = 87°$, we have $\angle Q > \angle P > \angle R$. Therefore, we have $\boxed{PR > QR > PQ}$.

(c) From $\triangle WXY$, we have $\angle XWY = 90° - 20° = 70°$. From $\triangle WYZ$, we have $\angle YWZ = 90° - 40° = 50°$. Therefore, $\angle X > \angle XWY > \angle WYX$, so $WY > XY > WX$. $\angle WYZ > \angle YWZ > \angle Z$, so $WZ > YZ > WY$. Combining our side inequalities gives $\boxed{WZ > YZ > WY > XY > WX}$.

10.1.2 Since $\angle P = 54°$ and $PQ = PR$, we have $\angle Q = \angle R = (180° - \angle P)/2 = 63°$. Because $\angle R > \angle P$, we have $PQ > QR$, so $\boxed{PQ}$ is longer than $\overline{QR}$.

10.1.3 Since $\angle AXC = 100°$, we know that AC is the longest side of $\triangle AXC$, so $AX < AC$. Similarly, since $\angle ACB = 100°$, we have $AC < AB$. Therefore, $AX < AC < AB$, which means that X cannot be beyond B on $\overrightarrow{AB}$.

Now we turn our attention to $\triangle CXB$ to show that X cannot be beyond A on $\overrightarrow{BA}$. If X is beyond A on $\overrightarrow{BA}$ (as shown at right), then we have $\angle BXC = \angle AXC = 100°$ and $\angle XCB > \angle ACB$. Since $\angle ACB = 100°$, we have $\angle XCB > 100°$. However, if we have $\angle BXC = 100°$ and $\angle XCB > 100°$, then the angles of $\triangle XCB$ add to more than $180°$, which is impossible. Therefore, X cannot be beyond A on $\overrightarrow{BA}$ if $\angle ACB = \angle AXC = 100°$. (Note that in our diagram, we do not have $\angle AXC = 100°$.)

Since X cannot be beyond B on $\overrightarrow{AB}$, nor beyond A on $\overrightarrow{BA}$, X must be on $\overline{AB}$.

Exercises for Section 10.2

10.2.1 Let $a \le b \le c$ be the lengths of a side of a triangle. The triangle is acute if $a^2 + b^2 > c^2$, right if $a^2 + b^2 = c^2$, and obtuse if $a^2 + b^2 < c^2$. Note that we only have to compare the sum of the squares of the smallest two sides to the square of the largest side. Make sure you see why!

(a) $6^2 + 8^2 = 10^2$, so the triangle is right.

(b) $6^2 + 8^2 = 100 < 11^2$, so the triangle is obtuse.

(c) $6^2 + 8^2 = 100 > 9^2$, so the triangle is acute.

(d) First, $3\sqrt{3} = \sqrt{27}$, and $\sqrt{13} < \sqrt{15} < \sqrt{27}$. Then $(\sqrt{13})^2 + (\sqrt{15})^2 = 28 > (\sqrt{27})^2 = 27$, so the triangle is acute.

(e) $1.8^2 + 2.1^2 = 3.24 + 4.41 = 7.65 < 9.61 = 3.1^2$, so the triangle is obtuse.

(f) First, $9/14 < 5/7 < 1$. Then $\left(\frac{9}{14}\right)^2 + \left(\frac{5}{7}\right)^2 = \frac{81}{196} + \frac{25}{49} = \frac{181}{196} < 1^2$, so the triangle is obtuse.

10.2.2 Let $a = BC = 27$, $b = AC$, and $c = AB = 17$. Triangle ABC is acute if and only if

$$a^2 + b^2 > c^2,$$
$$a^2 + c^2 > b^2,$$
$$b^2 + c^2 > a^2.$$

The first inequality becomes $27^2 + b^2 > 17^2$, or $b^2 > 17^2 - 27^2 = -440$, which is automatically satisfied. The second inequality becomes $27^2 + 17^2 > b^2$, or $b^2 < 17^2 + 27^2 = 1018$. This is satisfied by $b = 1, 2, 3,$ $\ldots, 31$. The third inequality becomes $b^2 + 17^2 > 27^2$, or $b^2 > 27^2 - 17^2 = 440$. This is satisfied by $b \geq 21$.

Therefore, triangle ABC is acute for $b = 21, 22, 23, \ldots, 31$, which consists of $\boxed{11}$ values.

10.2.3

(a) Since $\triangle ABX$ is a right triangle with hypotenuse $\overline{AB}$, we have $BX^2 + AX^2 = AB^2$.

(b) If C is between B and X, then we have $\angle ACB = 180° - \angle ACX$. Since $\angle ACX$ is one of the acute angles of right triangle $\triangle ACX$, we have $\angle ACX < 90°$, which means that $\angle ACB > 90°$, so $\triangle ABC$ is obtuse.

(c) If X is on $\overline{BC}$ we have $BX + XC = BC$.

(d) If X is on $\overline{BC}$, we have $BX = BC - CX$ from part (c). Substituting this into part (a) gives $(BC - CX)^2 + AX^2 = AB^2$, from which we have $BC^2 - 2(BC)(CX) + CX^2 + AX^2 = AB^2$. Since $\triangle AXC$ is right no matter where on $\overleftrightarrow{BC}$ point X is, we have $AX^2 + CX^2 = AC^2$. Therefore, our equation for AB^2 becomes $AB^2 = BC^2 + AC^2 - 2(BC)(CX)$. Since BC and CX are nonnegative, this means that $AB^2 \leq BC^2 + AC^2$. So, it is not possible for AB^2 to be greater than $AC^2 + BC^2$.

(e) In part (d) we showed that if X is on $\overline{AB}$, then it is impossible for AB^2 to be greater than $AC^2 + BC^2$. Similarly, if X is beyond B on $\overrightarrow{CB}$, we have $CX - BX = BC$, so $BX = CX - BC$. We can then follow the same steps as in (d) to show that in this case, we have $AB^2 < BC^2 + AC^2$. Therefore, if we are given that $AB^2 > AC^2 + BC^2$, we know that X, the foot of the altitude from A, is on $\overleftrightarrow{BC}$ such that C is between X and B. From part (b), we know that this means $\angle ACB$ is obtuse.

Exercises for Section 10.3

10.3.1 Let $a \leq b \leq c$ be three positive real numbers. Note that the inequalities $b + c > a$ and $a + c > b$ are then automatically satisfied, so to check that a, b, and c form the sides of a triangle, we only have to check that $a + b > c$.

(a) We have $4 \leq 5 \leq 6$ and $4 + 5 = 9 > 6$, so 4, 5, and 6 can be sides of a triangle.

(b) We have $7 \leq 9 \leq 20$, but $7 + 9 = 16 < 20$, so 7, 9, and 20 cannot be the sides of a triangle.

(c) We have $1/6 \leq 1/3 \leq 1/2$, but $1/6 + 1/3 = 1/2$, so 1/6, 1/3, and 1/2 cannot be the sides of a triangle.

(d) We have $3.4 \leq 9.8 \leq 11.3$ and $3.4 + 9.8 = 13.2 > 11.3$, so 3.4, 9.8, and 11.3 can be the sides of a triangle.

(e) We have $\sqrt{5} \leq \sqrt{14} \leq \sqrt{19}$. We must compare $\sqrt{5} + \sqrt{14}$ and $\sqrt{19}$. Let $x = \sqrt{5} + \sqrt{14}$. Then $x^2 = 5 + (2\sqrt{5})(\sqrt{14}) + 14 > 5 + 14 = 19$, so $x > \sqrt{19}$. Therefore, $\sqrt{5}$, $\sqrt{14}$, and $\sqrt{19}$ can be the sides of a triangle.

10.3.2   Due to the Triangle Inequality, the third side must be less than $3 + 7 = 10$ cm. Therefore, the largest the third side can be is 9 cm. If we let the third side of the triangle be x, then we must have $x + 3 > 7$, so $x > 4$ cm. Therefore, the smallest the third side can be is 5 cm. Then the greatest possible perimeter is $3 + 7 + 9 = 19$, and the least possible perimeter is $3 + 7 + 5 = 15$, and their difference is $19 - 15 = \boxed{4}$.

10.3.3   Let the vertices of the quadrilateral be A, B, C, and D, in that order. Then by the Triangle Inequality applied to triangles ABC, ACD, ABD, and BCD, we have

$$AC < AB + BC,$$
$$AC < DA + CD,$$
$$BD < AB + DA,$$
$$BD < BC + CD.$$

Adding, we get $2AC + 2BD < 2AB + 2BC + 2CD + 2DA$, so $AC + BD < AB + BC + CD + DA$.

10.3.4   From $\triangle ACO$ we have $AC + CO > AO$. From $\triangle BDO$, we have $BO + DO > BD$. Adding these gives $AC + CO + BO + DO > AO + BD$. Since $ABCD$ is a square, we have $AC = BD$, so subtracting these from our inequality, we have the desired $BO + CO + DO > AO$.

10.3.5   First we must show that given $a + b > c$, $a + c > b$, and $b + c > a$, we have $\sqrt{a} + \sqrt{b} > \sqrt{c}$, $\sqrt{a} + \sqrt{c} > \sqrt{b}$, and $\sqrt{b} + \sqrt{c} > \sqrt{a}$. Since $(\sqrt{a} + \sqrt{b})^2 = a + b + 2\sqrt{ab}$, and $a + b > c$, we know that $(\sqrt{a} + \sqrt{b})^2 > c + 2\sqrt{ab}$, so $(\sqrt{a} + \sqrt{b})^2 > c$. Taking the square root of both sides gives the desired $\sqrt{a} + \sqrt{b} > \sqrt{c}$. We can prove the other two desired inequalities similarly. Therefore, if a, b, and c can be the sides of a triangle, so can $\sqrt{a}$, $\sqrt{b}$, $\sqrt{c}$.

For the second part, we provide a counterexample to show that a^2, b^2, and c^2 do not necessarily have to be the sides of a triangle. Note that 3, 5, and 7 can be the sides of a triangle because $3 + 5 > 7$, $3 + 7 > 5$, and $5 + 7 > 3$. However, 3^2, 5^2, and 7^2 cannot be, since $3^2 + 5^2 < 7^2$.

Review Problems

10.18

(a) Since $2 + 3 = 5 > 4$, the numbers are the sides of a triangle. Since $2^2 + 3^2 = 13 < 4^2$, the triangle is obtuse.

(b) Since $1.7 + 2.1 = 3.8 < 3.9$, the numbers are not the sides of a triangle.

(c) Let $x = 2 + \sqrt{3}$. Then $x^2 = 4 + 4\sqrt{3} + 3 = 7 + 4\sqrt{3} > 5$, so $x = 2 + \sqrt{3} > \sqrt{5}$. Therefore, the numbers are the sides of a triangle. Furthermore, $2^2 + (\sqrt{3})^2 = 4 + 3 = 7 > (\sqrt{5})^2$, so the triangle is acute.

(d) Since $199 + 297 = 496 > 401$, the numbers are the sides of a triangle. Now, $199^2 + 297^2 < 200^2 + 300^2 = 40,000 + 90,000 = 130,000 < 160,000 = 400^2 < 401^2$. Therefore, the triangle is obtuse.

(e) Since $1/3 + 1/2 = 5/6 < 1$, the numbers are not the sides of a triangle.

(f) Note that $24 = 12 \cdot 2$, $48 = 12 \cdot 4$, and $60 = 12 \cdot 5$, so a triangle with sides 24, 48, and 60 is similar to a triangle with sides 2, 4, and 5 (by SSS Similarity). Therefore, the analysis for the numbers 24, 48, and 60 will be the same as the analysis for the numbers 2, 4, and 5. Since $2 + 4 = 6 > 5$, the numbers are the sides of a triangle. Since $2^2 + 4^2 = 20 < 5^2$, the triangle is obtuse.

10.19 Since $119 + 120 < 261$, it is impossible for a triangle to have side lengths 119, 120, and 261.

10.20 Let $a = BC = 11$, $b = AC$, and $c = AB = 5$. First, we determine which values of b make $a, b,$ and c the sides of a triangle. By the Triangle Inequality, we have $5 + b > 11$, so $b > 6$, and $5 + 11 > b$, so $b < 16$. Therefore, we must have $6 < b < 16$.

Then triangle ABC is obtuse if and only if either $5^2 + b^2 < 11^2$ or $5^2 + 11^2 < b^2$. The first inequality gives us $b^2 < 96$. This, combined with $b > 6$, gives $b = 7, 8,$ and 9 as possibilities. The second inequality gives us $b^2 > 146$. Combining this with $b < 16$ gives $b = 13, 14,$ and 15 as possibilities. Therefore, triangle ABC is obtuse for $b = \boxed{7, 8, 9, 13, 14, \text{ and } 15}$.

10.21 If $n = 4$, then $\triangle A_1A_2A_3$ is a right triangle with hypotenuse $\overline{A_1A_3}$. Therefore $A_1A_3 > A_1A_2$.

For $n > 4$, we consider the exterior angles of the polygon, each of which measures $360°/n$. Since $n > 4$, we have $360°/n < 360°/4 = 90°$, so the exterior angles are acute. Since each interior angle is supplementary to each exterior angle, we know that each interior angle is obtuse. Specifically, $\angle A_1A_2A_3$ is obtuse, so it is the largest angle in $\triangle A_1A_2A_3$. Therefore, $\overline{A_1A_3}$, which is opposite the obtuse angle in $\triangle A_1A_2A_3$, is the longest side of $\triangle A_1A_2A_3$. Thus, $A_1A_3 > A_1A_2$.

10.22 In any triangle $\triangle ABC$, the Triangle Inequality gives us $AB + AC > BC$. Adding BC to both sides gives $AB + AC + BC > 2BC$, so $BC < (AB + AC + BC)/2$. Therefore, BC is less than half the perimeter of $\triangle ABC$. There's nothing special about $\triangle ABC$ or side $\overline{BC}$ of $\triangle ABC$ in this proof, so we've shown that each side of a triangle must be less than half the triangle's perimeter.

10.23 There are two possibilities: The sides are of the form $3x, 3x,$ and $8x$ for some number x, or they are of the form $3x, 8x,$ and $8x$ for some number x. In the first case, the sides do not satisfy the Triangle Inequality because $3x + 3x = 6x < 8x$. In the second case, the sum of the sides is $3x + 8x + 8x = 19x = 38$, so $x = 2$. Hence, the side lengths are 6, 16, and 16, so the shortest side length is $\boxed{6 \text{ cm}}$.

10.24 Since $6x + 7$ is greater than both $2x + 3$ and $3x + 8$ for all positive integers x, we only have to check that $(2x + 3) + (3x + 8) > (6x + 7)$ in order to be sure the Triangle Inequality is satisfied. From this inequality, we find $x < 4$, which is true for $x = 1, 2,$ and 3. Therefore, the positive integers x for which the given expressions can be the side lengths of a nondegenerate triangle are $\boxed{1, 2, \text{ and } 3}$.

10.25 Suppose we have a triangle with side lengths $a, b,$ and c such that these are all positive integers and $a \le b \le c$. We wish to find all such groups of $a, b,$ and c such that $a + b + c = 20$, $a + b > c$ and $a^2 + b^2 < c^2$. First, we note that $c < 10$, since otherwise we will have $a + b \le c$. Next, since c is the largest side of the triangle, we must have $c \geq 7$, since no three integers that are each less than 7 can sum to 20. Therefore, we only have three cases to check.

Case 1: $c = 9$. If $c = 9$, we have $a + b = 11$. We also need $a \leq b$ and $a^2 + b^2 < 81$. We thus find the solutions $a = 5, b = 6$; $a = 4, b = 7$; and $a = 3, b = 8$.

Case 2: $c = 8$. If $c = 8$, we have $a + b = 12$. We also need $a \leq b$ and $a^2 + b^2 < 64$. Testing $b = 6$ and $b = 7$, we find that there are no solutions to this case.

Case 3: $c = 7$. If $c = 7$, we have $a + b = 13$. We also need $a \leq b$ and $a^2 + b^2 < 49$. Since $a + b = 13$ and $a \leq b$, b must be at least 7, so we cannot have $a^2 + b^2 < 49$. So, there are no solutions to this case.

Therefore, there are $\boxed{3}$ distinct obtuse triangles with integer side lengths and perimeter of 20.

10.26 The sum of the legs of the triangle must be greater than the base, so $2(x + 1) > 3x - 2$, or $x < 4$. Since the legs have the same length, clearly the base plus a leg is greater than the other leg. Also, the length of the base must be positive, so $3x - 2 > 0$, or $x > 2/3$. Therefore, all possible values of x are given by $\boxed{2/3 < x < 4}$.

10.27 Since $YZ > XY$, we have $\angle X > \angle Z$. Since $\angle Z$ and $\angle Y$ are the base angles of isosceles $\triangle XYZ$, we have $\angle Y = \angle Z$. From $\triangle XYZ$, we have $\angle X + \angle Y + \angle Z = 180°$. We also have $\angle X > \angle Z$ and $\angle Y = \angle Z$, so $\angle X + \angle Y + \angle Z < 3\angle X$. Hence, $180° < 3\angle X$, so $\angle X > 60°$.

Challenge Problems

10.28 Let P be the point where we place the stake. We wish to make $WP + XP + YP + ZP$ as small as possible. Therefore, we use the Triangle Inequality to create inequalities involving sums of these lengths. Applying the Triangle Inequality to $\triangle WPY$ gives $WP + YP \geq WY$, and applying it to $\triangle XPZ$ gives $XP + ZP \geq XZ$. Adding these two inequalities gives $WP + XP + YP + ZP \geq WY + XZ$. We get equality if and only if P lies on segments $\overline{WY}$ and $\overline{XZ}$, or in other words, if P is the intersection of $\overline{WY}$ and $\overline{XZ}$. Hence, this is where we should place the stake.

10.29 Let Orion's number be o, Michelle's number be m and Joshua's be j. If we can form a triangle with these numbers as side lengths, then we must have $o + j > m$ and $m + j > o$. Therefore, we must have $j > m - o$ and $j > o - m$. So, if o and m are not the same, then it is possible for Joshua to choose a j that violates one of these inequalities, since he can always choose a number that is smaller than whichever of $o - m$ and $m - o$ is positive. Since we are told it is impossible for Joshua to choose a number such that the three numbers violate the Triangle Inequality, we must conclude that it is impossible for Orion's and Michelle's numbers to be different. Finally, if o and m are the same, then any positive j satisfies $o + j > m$ and $m + j > o$. Furthermore, since we are told that Joshua's number is smaller than Orion's, we have $m + o > j$.

10.30 Since $AB > BC$, we have $\angle ACB > \angle BAC$. Since $\angle ACB = \angle ECD$, we have $\angle ECD > \angle BAC$. Since $CD > DE$, we have $\angle DEC > \angle ECD$; therefore, $\angle DEC > \angle BAC$. Since $\angle GEF = \angle DEC$, we have $\angle GEF > \angle BAC$. Since $EF > FG$, we have $\angle EGF > \angle GEF$, so $\angle EGF > \angle BAC$. Continuing in this manner, $\angle IGH = \angle EGF$, so $\angle IGH > \angle BAC$. Since $GH > HI$, we have $\angle GIH > \angle IGH$, so $\angle GIH > \angle BAC$. Because $\angle AIJ = \angle GIH$, we find $\angle AIJ > \angle BAC$. Since $IJ > JA$, we have $\angle IAJ > \angle AIJ$, so $\angle IAJ > \angle BAC$. However, $\angle IAJ = \angle BAC$. Since it is impossible for $\angle IAJ$ to be both equal to and greater than $\angle BAC$ at the same time, it is impossible for the side lengths to be as described in the problem.

10.31

(a) Let $\odot O$ be the circle that is externally tangent to the three given circles. Also, let the radii of $\odot O$, $\odot P$, $\odot Q$, and $\odot R$ be r_O, r_P, r_Q, and r_R. As we learned in the text, if two circles are externally tangent, then the point of tangency and the centers of the two circles are all collinear. Therefore, the distance between the centers of these circles is the sum of the radii of the two circles.

   Applying the Triangle Inequality to $\triangle OPR$, we have $OP + OR > PR$, so $(r_O + r_P) + (r_O + r_R) > PR$. Since our given circles do not intersect, we have $PR > r_P + r_R + 2r_Q$, so $(r_O + r_P) + (r_O + r_R) > PR > r_P + r_R + 2r_Q$. Therefore, $2r_O > 2r_Q$, or $r_O > r_Q$, as desired.

(b) Define $\odot O$ and r_O, r_P, r_Q, and r_P as in the previous part. Either $\angle OQR$ or $\angle OQP$ must be at least $90°$. Suppose without loss of generality that $\angle OQR \geq 90°$. Then, $\overline{OR}$ must be the longest side of $\triangle OQR$, so $OR > QR$. Therefore, $r_O + r_R > QR$. Since $\odot Q$ and $\odot R$ don't intersect, we have $QR > r_Q + r_R$. Therefore, $r_O + r_R > QR > r_Q + r_R$, so $r_O > r_Q$.

10.32 Since $WX > WY$, we have $\angle WYX > \angle WXY$. Since $WXYZ$ is convex, we have $\angle ZXY < \angle WXY$ and $\angle XYZ > \angle WYX$. Therefore, we have $\angle XYZ > \angle WYX > \angle WXY > \angle ZXY$. Since $\angle XYZ > \angle ZXY$, we have $XZ > ZY$, as desired.

10.33 We must show that $a + b > c$. Since $(a + b)^2 = a^2 + b^2 + 2ab$, and we are given $a^2 + b^2 > c^2$, we have $(a + b)^2 = a^2 + b^2 + 2ab > a^2 + b^2 > c^2$. Taking the square root of both sides of $(a + b)^2 > c^2$, we have $a + b > c$. (Note that several of these steps require a and b to be positive, as we are given.) Similarly, we can show that $a + c > b$ and $b + c > a$, so a, b and c can be the sides of a triangle.

10.34 Let P be the intersection of diagonals $\overline{AC}$ and $\overline{BD}$. Then by the Triangle Inequality applied to triangles ABP, BCP, CDP, and DAP, we have

$$AP + BP > AB,$$
$$BP + CP > BC,$$
$$CP + DP > CD,$$
$$DP + AP > DA.$$

Adding, we get $2(AP + BP + CP + DP) = 2(AC + BD) > AB + BC + CD + DA$, so

$$AC + BD > \frac{1}{2}(AB + BC + CD + DA).$$

Therefore, the sum of the diagonals of a quadrilateral is greater than half its perimeter.

10.35 We'll prove $3YZ > XY$ and $XY > 2YZ$ separately. For the first, the $20°$ and the 3 in what we want to prove, together with the observation $3(20°) = 60°$, inspire us to try to combine three copies of $\triangle XYZ$ as shown in our first diagram at left below. Specifically, we draw $\triangle XAY$ and $\triangle XBZ$ such that each is congruent to $\triangle XYZ$. Therefore, $XA = XY = XZ = XB$. Moreover, $\angle AXY = \angle YXZ = \angle ZXB = 20°$, so $\angle AXB = 60°$. Since we also have $XA = XB$, we know that $\triangle XAB$ is equilateral. So, we have $AB = XA = XY$. From $\triangle AYB$, we have $AY + YB > AB$. From $\triangle YZB$ we have $YZ + ZB > YB$. Combining these gives $AY + YZ + ZB > AY + YB > AB$. Since $AB = XY$ and $AY = ZB = YZ$, this inequality is equivalent to the desired $3YZ > XY$.

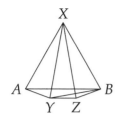

Figure 10.1: Diagram for $3YZ > XY$

Figure 10.2: Diagram for $XY > 2YZ$

Tackling $XY > 2YZ$ calls for a different strategy. We locate point Q on $\overline{XY}$ such that $YQ = YZ$. Since $\angle Y = (180° - 20°)/2 = 80°$, we have $\angle YQZ = \angle YZQ = (180° - 80°)/2 = 50°$. Therefore, we have $\angle XQZ = 180° - \angle YQZ = 130°$ and $\angle XZQ = \angle XZY - \angle QZY = 30°$. Since $\angle QZX > \angle QXZ$ in $\triangle QXZ$, we have $QX > QZ$. Also, we have $\angle QYZ > \angle ZQY$ in $\triangle QZY$, so $QZ > YZ$. Therefore, we have $QX > QZ > YZ$. Finally, we can write $XY = XQ + QY = XQ + YZ > YZ + YZ$, so $XY > 2YZ$, as desired.

10.36 Let $AB \le AC$. (The case $AC < AB$ can be proved similarly.) Arrange the two triangles so that $\overline{AB}$ and $\overline{A'B'}$ coincide, as shown in the diagram. Since $\angle BAC > \angle B'A'C'$, $\overrightarrow{AC}$ in our diagram must be inside $\angle BAC$. Since $AB < AC$, we have $\angle ACB < \angle ABC$. Therefore, $\angle ACB$ must be acute. Since $\angle ACB$ is acute and $AC' = AC$, $\overrightarrow{AC'}$ must intersect $\overline{BC}$ as shown. (We can prove this by letting D be the foot of the altitude from A to $\overleftrightarrow{BC}$ and E be the point where $\overrightarrow{AC'}$ hits $\overline{BC}$.

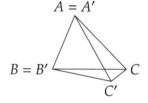

Suppose E is on $\overline{DC}$. Since $\angle ACB$ is acute, we have $DE < DC$, so $AD^2 + DE^2 < AD^2 + DC^2$. Therefore, $AE < AC$, so C' is beyond E on $\overrightarrow{AE}$. If E is on $\overline{DB}$ instead of $\overline{DC}$, we can note $AE < AB \le AC$.)

Since $AC' = AC$, we have $\angle AC'C = \angle ACC'$. In $\triangle BCC'$, we therefore have $\angle BC'C > \angle AC'C = \angle ACC' > \angle BCC'$. Since $\angle BC'C > \angle BCC'$, we have $BC > BC'$. Therefore, if we have two triangles ABC and $A'B'C'$ such that $AB = A'B'$, $AC = A'C'$ and $\angle ABC > \angle A'B'C$, then we have $BC > B'C'$.

10.37 Let P and Q be the two points inside the triangle. Extend $\overline{PQ}$ to the sides of the triangle at points P' and Q', as shown. WLOG, assume that P' is on side $\overline{AB}$ and that Q' is on side $\overline{AC}$.

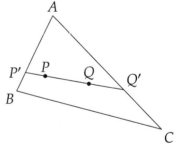

Clearly, $PQ \le P'Q'$. Applying the Triangle Inequality to $\triangle P'BQ'$ gives $P'Q' \le P'B + BQ'$. Applying the Triangle Inequality to $\triangle BQ'C$ gives $BC + Q'C \ge BQ'$, which combined with our relationship from $\triangle P'BQ'$ gives $P'Q' \le P'B + BQ' \le P'B + BC + CQ'$. Adding this to $P'Q' < P'A + AQ'$ (from the Triangle Inequality applied to $\triangle P'AQ'$) gives $2P'Q' < P'A + AQ' + Q'C + BC + BP' = AB + AC + BC$ (since $AQ' + Q'C = AC$ and $BP' + P'A = AB$), which implies that

$$PQ \le P'Q' < \frac{1}{2}(AB + AC + BC).$$

CHAPTER 11

Circles

Exercises for Section 11.1

11.1.1 Circumference = $2\pi(\text{Radius}) = 2\pi(4) = \boxed{8\pi}$.

11.1.2 Let r be the radius. Then $2\pi r = 12\pi$, so $r = \boxed{6}$.

11.1.3

(a) $\overline{OC}$ is a radius, so $OC = \boxed{18}$.

(b) The circumference is $2\pi(18) = 36\pi$. Since $\overset{\frown}{AC}$ has length 1/3 the circumference, $\angle AOC = 360°/3 = \boxed{120°}$.

(c) Let $\overline{OB}$ be the altitude from O to $\overline{AC}$. Then triangles ABO and CBO are congruent 30-60-90 triangles with hypotenuse $AO = CO = 18$. Thus, $AB = BC = 9\sqrt{3}$. Therefore, $AC = AB + BC = \boxed{18\sqrt{3}}$.

11.1.4 Since $\overset{\frown}{WX} = \overset{\frown}{YZ}$, we have $\angle WQX = \angle YQZ$. Since $\overline{QX}$, $\overline{QW}$, $\overline{QY}$, and $\overline{QZ}$ are all radii, they have equal length. Therefore, we have $\triangle WQX \cong \triangle YQZ$ by SAS Congruence, so $WX = YZ$.

Exercises for Section 11.2

11.2.1 Since the diameter is 18, the radius is 9. Therefore, the area is $(9^2)\pi = \boxed{81\pi}$.

11.2.2 Let r be the radius. Then $2\pi r = 12\pi$, so $r = 6$. Therefore, our area is $\pi r^2 = \pi(6)^2 = \boxed{36\pi}$.

11.2.3 Let r be the radius. Then $2\pi r = \pi r^2$. Since $r \neq 0$, $r = \boxed{2}$.

11.2.4 The ratio of the areas of the pizza is $(20/12)^2 = 25/9$. Since the small pizza provides four meals, the large one provides $(25/9) \cdot 4 = \boxed{100/9}$ meals.

11.2.5

(a) The radius of the circle is 12, so its area is $(12^2)\pi = \boxed{144\pi}$.

(b) Sector AOB is $120°/360° = 1/3$ of the circle, so its area is $(144\pi)(1/3) = \boxed{48\pi}$.

11.2.6 The area of sector XQZ is $(30\pi)/(100\pi) = 3/10$ of the entire circle. Therefore, $\angle XQZ$ cuts off $3/10$ of the circle, so it has measure $(3/10)(360°) = \boxed{108°}$. Since $QX = QZ$ (both are radii of $\odot Q$), $\triangle XQZ$ is isosceles with $\angle X = \angle Z$. Since $\angle XQZ = 108°$, we have $\angle Z + \angle X + \angle XQZ = \angle Z + \angle X + 108° = 180°$ from $\triangle XQZ$. Therefore, $\angle XZQ = \angle Z = (180° - 108°)/2 = \boxed{36°}$.

11.2.7 As the man walks in a circle, the point on his hat closest to the center is $5 - 3 = 2$ feet from the center. This point traces out a circle. Inside this circle, the grass lives. The point on the hat that is farthest from the center is $5 + 3 = 8$ feet from the center. This point also traces out a circle as the man walks. All the grass that is inside this larger circle, but outside the aforementioned small circle, dies. This larger circle has area $(8^2)\pi = 64\pi$ and the smaller has area $(2^2)\pi = 4\pi$, so the area of dead grass is $64\pi - 4\pi = \boxed{60\pi}$.

Exercises for Section 11.3

11.3.1 The area of circular segment AB equals (Area of sector AOB) $-$ $[AOB]$. Since sector AOB is $120°/360° = 1/3$ of the circle, its area is $(1/3)(6^2)\pi = 12\pi$. To find the area of $\triangle AOB$, we draw altitude $\overline{OX}$ of $\triangle AOB$. Since $OA = OB$, this altitude is also a median and an angle bisector. Therefore, $\angle AOX = \angle BOX = 60°$, and $\triangle BOX$ and $\triangle AOX$ are 30-60-90 triangles. Thus, $OX = OB/2 = 3$ and $BX = AX = AO\sqrt{3}/2 = 3\sqrt{3}$. Hence, $[AOB] = (AB)(OX)/2 = (6\sqrt{2})(3)/2 = 9\sqrt{3}$. Subtracting this from the area of sector AOB, the area of our shaded region is $\boxed{12\pi - 9\sqrt{3}}$.

11.3.2 To find the desired area, we must find the area of the circumcircle and subtract the area of the equilateral triangle. Since the triangle has side length 9, it has area $9^2\sqrt{3}/4 = 81\sqrt{3}/4$. To find the area of the circumcircle, we need the circumradius. We draw altitude $\overline{XA}$, which is also a median and an angle bisector. Since $\triangle XYA$ is a 30-60-90 triangle, we have $XA = XY\sqrt{3}/2 = 9\sqrt{3}/2$. The circumcenter, O, of $\triangle XYZ$ is the centroid of the triangle, so it is on median $\overline{XA}$ such that $XO/XA = 2/3$. Therefore, $XO = (2/3)(XA) = 3\sqrt{3}$. Since O the center of $\triangle XYZ$, we know that the circumradius of $\triangle XYZ$ is $XO = 3\sqrt{3}$. Hence, the area of the circumcircle of $\triangle XYZ$ is $(3\sqrt{3})^2\pi = 27\pi$. Finally, our desired area is $\boxed{27\pi - 81\sqrt{3}/4}$.

(See if you can figure out how we could have used the answer from the previous problem to solve this problem.)

11.3.3 The area of our shaded region equals

(Area of $\triangle ABC$) $-$ (Areas of sectors centered at A, B, and C).

Since $\triangle ABC$ is equilateral with side length 6, it has area $(6^2)\sqrt{3}/4 = 9\sqrt{3}$. Since $\angle A = \angle B = \angle C = 60°$, each of our sectors is $60°/360° = 1/6$ of a circle. Each sector has radius $6/2 = 3$ since each has its center at a vertex of $\triangle ABC$ and passes through the midpoint of a side connected to that vertex. Therefore, the area of each sector is $(1/6)(3^2)\pi = 3\pi/2$. Finally, our shaded region has area $9\sqrt{3} - 3(3\pi/2) = \boxed{9\sqrt{3} - 9\pi/2}$.

11.3.4 Our shaded region is formed by subtracting $\triangle ABD$ from sector DAB and combining this with the region that is formed by subtracting $\triangle BCD$ from sector DCB. These two regions are congruent, so what we seek is $2[$(Area of sector DAB) $-$ (Area of $\triangle BAD$)$]$. Since $\angle A = 90°$ and $AB = AD = 6$, we find that the area of sector DAB is $(90°/360°)(6^2)\pi = 9\pi$ and $[BAD] = (BA)(AD)/2 = 18$. Therefore, our shaded area is $2(9\pi - 18) = \boxed{18\pi - 36}$.

11.3.5 Let S be the point besides O where the semicircles meet, and let R and Q be the centers of the semicircles as shown. Since $OQ = QS = RS = RO$ (each is a radius of one of our congruent semicircles), and $\angle ROQ = 90°$, $ORSQ$ is a square. If we start with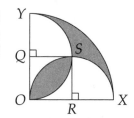

(Area of sector XOY) − (Area of semicircle Q) − (Area of semicircle R),

the result is the outer shaded region (the region with $\overset{\frown}{XY}$ as a boundary) minus the inner shaded region. Since we want the sum of these shaded regions, we must find the inner shaded region separately and add it back twice to the value we find above. Since the area of sector XOY is $(90°/360°)(4^2)\pi = 4\pi$, and the area of each semicircle is $(180°/360°)(2^2)\pi = 2\pi$, we find that the area of the outer shaded region minus the area of the inner shaded region is 0! Therefore, the two shaded regions have the same area.

To find the area of the inner region, we follow the same process as in the previous problem.

(Area of inner region) $= 2[(\text{Area of sector } SRO) - [SRO]] = 2[(90°/360°)(2^2)\pi - (2^2)/2] = 2\pi - 4$.

Since the area of the outer region equals the area of the inner region, the desired area is $2(2\pi - 4) = \boxed{4\pi - 8}$.

We could also solve this problem by drawing $\overline{SO}$, dividing the inner shaded region into two circular segments. Circular segment SO of $\odot R$ is congruent to circular segment SX of $\odot R$. Similarly, circular segment SO of $\odot Q$ is congruent to circular segment SY of $\odot Y$. Therefore, the total shaded region equals the area of circular segment XY of $\odot O$, which is $OX^2\pi/4 - OX^2/2 = \boxed{4\pi - 8}$.

11.3.6 We start by expressing the area in terms of pieces we can handle. We do so by starting with sector ABC of $\odot B$ and subtracting pieces.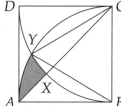

Desired area = (Area of sector ABC) − (Area of sector YBC)

−(Area of segment BY of $\odot C$) − (funky area AXB).

The area of circular segment BY of $\odot C$ is (Area of sector BCY)−$[BCY]$. Furthermore, since $CB = BY = CY$ (each is a radius of a circle with radius BC), $\triangle BCY$ is equilateral and $\angle BCY = \angle YBC = 60°$. Therefore, sectors YCB and YBC have the same area, $(60°/360°)(BC^2)\pi = 8\pi/3$. Since $\triangle BYC$ is equilateral, we have $[BYC] = 4^2\sqrt{3}/4 = 4\sqrt{3}$. Also, $\angle ABC = 90°$ tells us the area of sector ABC is $(90°/360°)(BC)^2\pi = 4\pi$, so we have:

$$\begin{aligned} \text{Desired area} &= (\text{Area of sector } ABC) - 2(\text{Area of sector } YBC) + [BYC] - (\text{funky area } AXB) \\ &= 4\pi - 2(8\pi/3) + 4\sqrt{3} - (\text{funky area } AXB) \\ &= -4\pi/3 + 4\sqrt{3} - (\text{funky area } AXB) \end{aligned}$$

Funky area AXB is the result of subtracting sector BCX from $\triangle ABC$, so its area is $(AB)(BC)/2 - (45°/360°)(BC^2)\pi = 8 - 2\pi$. Therefore, our desired area is $-4\pi/3 + 4\sqrt{3} - (8 - 2\pi) = \boxed{2\pi/3 + 4\sqrt{3} - 8}$.

Review Problems

11.14

(a) The area is $(3\sqrt{3})^2\pi = \boxed{27\pi}$.

(b) The circumference is $2\pi(3\sqrt{3}) = \boxed{6\pi\sqrt{3}}$.

(c) Since $\angle AOZ = 90°$, $\overset{\frown}{AZ}$ is $90°/360° = 1/4$ of the whole circle. Therefore, the length of $\overset{\frown}{AZ}$ is $(1/4)(6\pi\sqrt{3}) = \boxed{3\pi\sqrt{3}/2}$.

(d) As in the previous part, sector AOZ is $1/4$ the entire circle. Therefore, its area is $(1/4)(27\pi) = \boxed{27\pi/4}$.

(e) Since $[AOZ] = (AO)(OZ)/2 = 27/2$, the area of circular segment is (Area of sector AOZ) $- [AOZ] = \boxed{27\pi/4 - 27/2}$.

11.15 Six miles equals $6(5280) = 31680$ feet. In each full revolution of a tire, the earth-mover moves a distance equal to the circumference of the tire, which is 11.5π feet. Therefore, to cover 31680 feet, the tires must turn $31680/(11.5\pi) \approx \boxed{877}$ times.

11.16 Since the circle has radius 14, its circumference is $2\pi(14) = 28\pi$. Our arc is $78°/360° = 13/60$ of the circle, so its length is $(28\pi)(13/60) = \boxed{91\pi/15}$.

11.17 Let the radius be r. Then, we have $A = \pi r^2$ and $C = 2\pi r$. Therefore, $r = C/(2\pi)$, so $A = \pi(C/(2\pi))^2 = \boxed{C^2/(4\pi)}$.

11.18

(a) Let r_O be the radius of $\odot O$, r_B be the radius of $\odot B$, and r_C be the radius of $\odot C$. We are given that $r_O = 2r_B$ since a diameter of $\odot B$ is a radius of $\odot O$, and $r_B = 2r_C$ because a diameter of $\odot C$ is a radius of $\odot B$. Therefore, $r_O = 2(2r_C) = 4r_C$, so the area of $\odot O$ is $\pi r_O^2 = \pi(4r_C)^2 = 16\pi r_C^2$. The area of $\odot C$ is πr_C^2, so the ratio of the area of $\odot C$ to the area of $\odot O$ is $\boxed{1/16}$.

(b) Since $\overline{OA}$ is a diameter of $\odot B$, B is the midpoint of $\overline{OA}$. So, $AB = BO = AO/2$. Since $\overline{OB}$ is a diameter of $\odot C$, C is the midpoint of $\overline{OB}$. Therefore, $BC = CO = BO/2 = AO/4$. Finally, we have $CA = OA - CO = 3AO/4$, so $CA/OA = \boxed{3/4}$.

11.19 A $36°$ arc is $36°/360° = 1/10$ of the entire circle. Since the arc is 24π units long, the entire circumference is $(24\pi)(10) = \boxed{240\pi}$. A circle with circumference 240π has diameter 240 and radius $240/2 = 120$ and therefore has area $(120)^2\pi = \boxed{14400\pi}$.

11.20 Since the radius of the circle is 6, its circumference is $2(6)\pi = 12\pi$. The hexagon can be dissected into six equilateral triangles. As we see in the diagram at right, each of these triangles has side length equal to the radius of the circle. Therefore, the perimeter of the hexagon is $6(6) = 36$. The ratio of the circumference of the circle to the perimeter of the hexagon is $12\pi/36 = \boxed{\pi/3}$.

11.21 *Solution 1:* Let the legs of the triangle have lengths a and b, where $a < b$. We have $(1/2)(a/2)^2\pi = 36$, so $a = 12\sqrt{2}/\sqrt{\pi}$. Similarly, $(1/2)(b/2)^2\pi = 64$, so $b = 16\sqrt{2}/\sqrt{\pi}$. Therefore, the hypotenuse has length

$c = \sqrt{a^2 + b^2} = 20\sqrt{2}/\sqrt{\pi}$ (which we could have figured out quickly by noting that $a/b = 3/4$, so the triangle is a 3-4-5 right triangle). So, the area of the semicircle with the hypotenuse as diameter is $(1/2)(c^2)\pi = \boxed{100 \text{ square centimeters}}$.

Solution 2: Again, let the sides of the triangle be a, b, and c, as usual. The Pythagorean Theorem gives us $a^2 + b^2 = c^2$. Multiplying this equation by $\pi/8$ and noting that $a^2/4 = (a/2)^2$ gives: $(1/2)(a/2)^2\pi + (1/2)(b/2)^2\pi = (1/2)(c/2)^2\pi$. Therefore, the sum of the areas of the semicircles on our two legs equals the area of the semicircle on the hypotenuse. So, our final semicircle has area $36 + 64 = \boxed{100 \text{ square centimeters}}$.

11.22 To find the unshaded region, we subtract the shaded region from the entire region. The entire region consists of a right triangle with legs 3 cm and 2(1) = 2 cm, and a semicircle with radius 1 cm. Therefore, the area of the entire region is $(3)(2)/2 + (1^2)\pi/2 = 3 + \pi/2$. The shaded region is a circle with radius 0.5 cm, so its area is $(0.5)^2\pi = \pi/4$. Therefore, the unshaded region has area $3 + \pi/2 - \pi/4 = \boxed{3 + \pi/4 \text{ cm}^2}$.

11.23 Let $\angle AOD = x$. Therefore, the area of sector AOD is $(x/360°)(AO)^2\pi$ and the area of sector BOC is $(x/360°)(BO)^2\pi$. So, the ratio of the area of sector BOC to the area of sector AOD is BO^2/AO^2. We are given that $AO/BO = 1.5$, so $BO/AO = 2/3$ and the ratio of the area of sector BOC to the area of sector AOD is $(2/3)^2 = 4/9$. Partial ring $ABCD$ is what's left of sector AOD when sector BOC is removed, so the ratio of the area of partial ring $ABCD$ to the area of sector AOD is $1 - 4/9 = 5/9$. Therefore, our desired ratio of areas is $(5/9)/(4/9) = \boxed{5/4}$.

11.24 To find the desired area, we must subtract the area of the rectangle from the area of the circle. Since the midpoint of the hypotenuse of a right triangle is the center of the circumcircle of the right triangle, each diagonal of the rectangle is a diameter of the circle. Therefore, the circle has diameter $\sqrt{5^2 + 12^2} = 13$ and area $\pi(13/2)^2 = 169\pi/4$. The area of the rectangle is $(5)(12) = 60$, so the area of the region inside the circle but outside the rectangle is $\boxed{169\pi/4 - 60}$.

11.25 To find the area of the region Charlyn can see, we must find its boundary. In other words, we must find all points that are 1 km from the square. The 5 km square is dashed in our diagram, and the boundary is solid. Inside the 5 km square, the boundary forms a square that has side length 3 km, since we must take away 1 km from each side of the original square to reach the limit of where Charlyn can see. Outside the square is a little trickier. Along the sides, the boundary is still 1 km away, but when we reach a corner of our 5 km square, Charlyn can see a full quarter circle with radius 1 km when looking away from the square, as shown.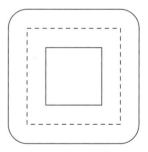

The area of the visible interior region is therefore the difference between a square with side length 5 and a square with side length 3, or $5^2 - 3^2 = 16$. The exterior region consists of four 5×1 rectangles and four quarter circles that together form a circle with radius 1 km and area $1^2\pi = \pi$ square kilometers. Therefore the total visible area is $16 + 4(5) + \pi = \boxed{36 + \pi}$ square kilometers.

Challenge Problems

11.26 Let the centers of the circles be A, B, and C. Since the circles all have the same radii and each of $\overline{AB}$, $\overline{AC}$, and $\overline{BC}$ is a radius of two of the circles, we have $AB = BC = AC = 12$. Therefore, $\triangle ABC$ is equilateral. The area of the shaded region then consists of an equilateral triangle plus three congruent circular segments, as shown. Each circular segment has an area equal to the difference between the area of a $60°$ sector of the circle and the area of $\triangle ABC$. Therefore, the shaded area equals

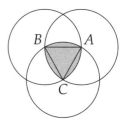

$$[ABC] + 3(\text{Area of circular segment } AB) = [ABC] + 3(\text{Area of sector } ACB - [ABC])$$
$$= 3(\text{Area of sector } ACB) - 2[ABC].$$

The triangle has area $(12^2)\sqrt{3}/4 = 36\sqrt{3}$ and the sector has area $(60°/360°)(12^2)\pi = 24\pi$, so the shaded area is $\boxed{72\pi - 72\sqrt{3}}$.

11.27 When my wheels make one full revolution, my speedometer thinks the car moves 24π inches, since the circumference of the old 24 inch diameter wheels is 24π. However, with my new wheels, the car really moves 28π inches. Therefore, the car moves $(28\pi)/(24\pi) = 7/6$ as fast as my speedometer thinks it moves. So, when my speedometer registers 40 miles per hour, the car is really moving at a rate that covers $(40)(7/6) = \boxed{140/3 \text{ miles per hour}}$.

11.28 Let the point on $\mathcal{Z}$ that is touching $\mathcal{A}$ initially be point P. When the coin has moved a quarter turn, the point on $\mathcal{Z}$ that is touching $\mathcal{A}$ is the one that is one quarter of the circumference of $\mathcal{Z}$ from P. As we can see in the diagram, in this quarter turn, P goes from being the bottom point of $\mathcal{Z}$ to being the top point. Therefore, it has revolved a full half-turn about the center of $\mathcal{Z}$ as $\mathcal{Z}$ is moved $1/4$ of the way around $\mathcal{A}$. After four of these quarter-turns, P will have moved 4 half-turns about the center of $\mathcal{Z}$. Therefore, coin $\mathcal{Z}$ revolves $\boxed{\text{twice}}$ about its center.

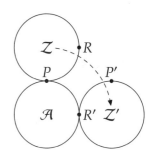

11.29 Let the radius of the semicircle be r. The perimeter of the semicircle is the diameter plus the semicircular arc, or $2r + r\pi$. The area of the semicircle is $\pi r^2/2$. Therefore, we have $2r + r\pi = \pi r^2/2$. Since the radius can't be 0, we can divide by r, giving $2 + \pi = \pi r/2$, so $r = \boxed{(4 + 2\pi)/\pi}$.

11.30 The total amount of water flowing through a cross-section at a given time equals the area of the cross-section times the rate of the water flowing through it. (Make sure you see why!) The total amount of water flowing through any section of the river must be the same throughout the river. Therefore, the area of the initial cross-section times the rate of water there equals the area of the downstream cross-section times the rate of water there.

We find our initial area by using our tactics from the quadrilaterals chapter. We draw altitudes from the short side of the trapezoid to the long side, dividing the trapezoid into a rectangle and two right triangles. These right triangles have one leg of length $(16 - 10)/2 = 3$ ft, and a hypotenuse of length 5 ft. Therefore, the other leg, which is the height of the trapezoid, is 4 ft. The area of the trapezoid then is $4(10 + 16)/2 = 52$ ft. Letting the holes downstream have radius r, the total cross-sectional area downstream is $4\pi r^2$ ft^2. Now we equate the product of the cross-sectional area and the rate of the water at both points to find $(52)(\pi) = (4\pi r^2)(16)$. Therefore, $r^2 = 13/16$ ft^2, so $r = \boxed{\sqrt{13}/4 \text{ ft}}$.

11.31 Since $\triangle ABD$ is a right triangle, its circumcenter is the midpoint of hypotenuse $\overline{AD}$. We can use the Pythagorean Theorem or note that $\triangle ABD$ is a 30-60-90 triangle since $AB/BD = \sqrt{3}$ to find $AD = 16$. The shaded area equals the area of the semicircle containing the triangle minus $[ABD]$. The area of the semicircle is $(8^2)\pi/2 = 32\pi$ and the area of $\triangle ABD$ is $(AB)(BD)/2 = 32\sqrt{3}$. Therefore, the shaded area is $\boxed{32\pi - 32\sqrt{3}}$.

11.32 If we subtract the area of the square from the area of the circle, the overlapping area 'cancels out' and we are left with the difference between the nonoverlapping area of the circle and the nonoverlapping area of the square. This is exactly the quantity we seek! Therefore, the answer is $(3^2)\pi - 2^2 = \boxed{9\pi - 4}$.

11.33 Let the center of the hexagon be point O. Connecting each vertex of the hexagon to O forms six equilateral triangles. The shaded piece of the diagram inside $\triangle AOF$ is the result of removing circular segment OF of $\odot A$ and circular segment OA of $\odot F$ from $\triangle OAF$. To find the area of circular segment OF, we subtract $[OAF]$ from the area of sector OAF. Since $\triangle OAF$ is equilateral, $[OAF] = (AF)^2\sqrt{3}/4 = 9\sqrt{3}$ and the area of sector OAF is $(60°/360°)(OA^2)\pi = 6\pi$. Therefore, the area of circular segment OF of $\odot A$ is $6\pi - 9\sqrt{3}$. Similarly, the area of circular segment OA is also $6\pi - 9\sqrt{3}$, so the area of one of our shaded pieces is $[AOF] - 2(6\pi - 9\sqrt{3}) = 27\sqrt{3} - 12\pi$. Our total shaded region consists of six of these shaded pieces, so the desired area is $\boxed{162\sqrt{3} - 72\pi}$.

11.34 Let each side of the triangle have length s. The perimeter of the triangle is $3s$. The circumradius of an equilateral triangle equals 2/3 an altitude of the triangle. Drawing an altitude of the triangle forms two 30-60-90 triangles, of which the altitude is the leg opposite the 60° angle and a side of triangle is the hypotenuse. Therefore, the altitude has length $s\sqrt{3}/2$, so the circumradius has length $(2/3)(s\sqrt{3}/2) = s\sqrt{3}/3$. So, the area of the circumcircle is $s^2\pi/3$. Since this must equal our perimeter, we have $s^2\pi/3 = 3s$, or $s = 9/\pi$, since s cannot be 0. Therefore, the radius of the circle is $s\sqrt{3}/3 = \boxed{3\sqrt{3}/\pi}$.

11.35 We start by drawing our chords, $\overline{AB}$ and $\overline{CD}$, and connecting their endpoints to the center of the circle, O. We draw $\overline{XY}$ perpendicular to both chords (since they are parallel) and passing through O, as shown. Since $\overline{OX}$ is part of a radius and it is perpendicular to chord $\overline{AB}$, it bisects $\overline{AB}$. Similarly, Y is the midpoint of $\overline{CD}$. Moreover, we have $\triangle OAB \cong \triangle OCD$ by SSS, so corresponding altitudes $\overline{OX}$ and $\overline{OY}$ of these triangles have the same length. Therefore, we have $AX = BX = OX = OY = CY = DY = x/2$, and $\triangle AOX$, $\triangle BOX$, $\triangle COY$, and $\triangle DOY$ are all 45-45-90 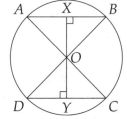 triangles. Since $\angle AOX = \angle BOX = \angle DOY = \angle COY = 45°$, we have $\angle COD = \angle AOB = 90°$. Therefore, $\angle AOD + \angle BOC = 360° - \angle AOB - \angle COD = 180°$. Hence, sectors AOD and BOC together make a semicircle. The radius of this semicircle is $AO = x\sqrt{2}/2$, so its area is $AO^2\pi/2 = x^2\pi/4$. Each of $\triangle ABO$ and $\triangle CDO$ has area $(x)(x/2)/2 = x^2/4$, so the total area of the region between $\overline{AB}$ and $\overline{CD}$ is $x^2\pi/4 + 2(x^2/4)$. We are given that this equals $2 + \pi$, so we have $x^2\pi/4 + 2(x^2/4) = 2 + \pi$. Therefore, $x^2(\pi + 2) = 4(2 + \pi)$, so $x = \boxed{2}$.

Note that we can also solve this problem by drawing $\overline{BC}$ and $\overline{AD}$ and noting that $ABCD$ is a square. The area between $\overline{AB}$ and $\overline{CD}$ then is one-half the difference between the area of the circle and the area of the square (since two of the circular segments outside the square are between the chords), plus the area of the square.

11.36 Let X be the intersection of $\overset{\frown}{AC}$ of $\odot B$ and $\overset{\frown}{BD}$ of $\odot C$. Consider the region bound by $\overline{AB}$, $\overset{\frown}{AX}$, and $\overset{\frown}{BX}$. We'll call this 'funky region ABX'. As suggested by the diagram, four of these funky regions plus the desired area equals the whole square, so we can solve the problem if we find the area of funky region ABX. Funky region ABX equals sector ABC minus the region bound by $\overline{BC}$, $\overset{\frown}{BX}$, and $\overset{\frown}{CX}$. Since $\overline{BC}$, $\overline{BX}$ and $\overline{CX}$ are radii of congruent circles, $\triangle BCX$ is equilateral. Therefore, the region bound by $\overline{BC}$, $\overset{\frown}{BX}$, and $\overset{\frown}{CX}$ has area $[BCX] + 2$(Area of circular segment CX of $\odot B$). The area of the 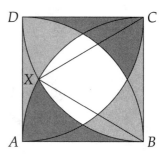 circular segment is the difference between the area of sector CBX and the area of $\triangle CBX$. Therefore, the area of the region bound by $\overline{BC}$, $\overset{\frown}{BX}$, and $\overset{\frown}{CX}$ is

$$\frac{BC^2\sqrt{3}}{4} + 2\left[\left(\frac{60°}{360°}\right)\pi BC^2 - \frac{BC^2\sqrt{3}}{4}\right] = 48\pi - 36\sqrt{3}.$$

We subtract this from sector ABC to get the area of funky region ABX:

$$\left(\frac{90°}{360°}\right)BC^2\pi - (48\pi - 36\sqrt{3}) = 36\sqrt{3} - 12\pi.$$

The area of the shaded region equals the area of the square minus four of these funky regions, or

$$[ABCD] - 4(36\sqrt{3} - 12\pi) = \boxed{144 + 48\pi - 144\sqrt{3}}.$$

11.37 Let $OX = x$. To find the area of the region bound by $\overline{AX}$, $\overline{XD}$, and $\overset{\frown}{AD}$, 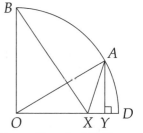 we subtract $[AOX]$ from the area of sector AOD. Since $\angle AOD = 30°$, the area of sector AOD is $(30°/360°)(OD^2) = \pi/12$. To find the area of $\triangle AOX$ in terms of x, we draw the altitude from A to $\overleftrightarrow{OX}$ as shown. Since $\angle AOY = 30°$, we have $AY = AO/2 = 1/2$. Therefore, the area of $\triangle AOX$ is $(OX)(AY)/2 = x/4$ and the area of the region bound by $\overline{AX}$, $\overline{XD}$, and $\overset{\frown}{AD}$ is $\pi/12 - x/4$. This plus the area of $\triangle BOX$ must equal half the area of the quarter circle. The area of $\triangle BOX$ is $(BO)(OX)/2 = x/2$ and the area of the quarter circle is $\pi/4$ so we have $\pi/12 - x/4 + x/2 = (\pi/4)/2$, so $x/4 = \pi/8 - \pi/12 = \pi/24$. Therefore, $x = \boxed{\pi/6}$.

12

CHAPTER

Circles and Angles

Exercises for Section 12.1

12.1.1

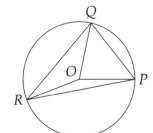

(a) The given information tells us that the points are situated on the circle as shown at right. $\angle QPR$ is inscribed in $\overset{\frown}{QR}$, so its measure is $\overset{\frown}{QR}/2 = \boxed{61.5°}$.

(b) $\angle PQR$ is inscribed in $\overset{\frown}{PR}$, which has measure $360° - 201° = 159°$. Therefore, $\angle PQR = \overset{\frown}{PR}/2 = \boxed{79.5°}$.

(c) $\angle PRQ$ is inscribed in $\overset{\frown}{PQ}$, so $\angle PRQ = \overset{\frown}{PQ}/2 = \boxed{39°}$.

(d) $\angle POQ = \overset{\frown}{PQ} = \boxed{78°}$.

(e) Since $PO = QO$ (radii of the circle), $\triangle POQ$ is isosceles with $\angle PQO = \angle QPO$. From $\triangle PQO$, we have $\angle POQ + \angle PQO + \angle QPO = 180°$, so $\angle PQO = (180° - \angle POQ)/2 = \boxed{51°}$.

(f) $\angle POR = \overset{\frown}{PR} = \boxed{159°}$. (Make sure you see why the answer is not 201°.)

12.1.2 If point T is on minor arc $\overset{\frown}{RS}$, then $\angle RTS$ is inscribed in major arc $\overset{\frown}{RS}$, which has measure $360° - 50° = 310°$. In this case, $\angle RTS = (310°)/2 = 155°$. If T is on major arc $\overset{\frown}{RS}$, then $\angle RTS$ is inscribed in minor arc $\overset{\frown}{RS}$, so its measure is $(50°)/2 = 25°$. Therefore, our possibilities for $\angle RTS$ are $\boxed{25° \text{ and } 155°}$.

12.1.3

(a) From $\triangle EFG$, we have $\angle EGF = 180° - \angle EFG - \angle GEF = \boxed{54°}$.

(b) Since $\angle EFG$ is inscribed in $\overset{\frown}{EG}$, we have $\overset{\frown}{EG} = 2\angle EFG = \boxed{96°}$.

(c) Since $\angle EFG < 90°$, we know that F is on major arc $\overset{\frown}{EG}$ (instead of the shorter minor arc $\overset{\frown}{EG}$). In the previous part, we found that minor arc $\overset{\frown}{EG} = 96°$, so $\overset{\frown}{EFG} = 360° - 96° = \boxed{264°}$.

12.1.4

(a) $\angle D$ is inscribed in the same arc as $\angle C$. Since both equal $\overset{\frown}{AB}/2$, the two have equal measure. Therefore, $\angle D = \angle ACB = \boxed{36°}$.

(b) $\overset{\frown}{AB} = 2\angle D = \boxed{72°}$.

(c) Since $\angle ABD = \widehat{AD}/2 = (180° - \widehat{AB})/2 = 54°$, we have $\angle ABF = 180° - \angle ABD = \boxed{126°}$.

(d) From right triangle $\triangle EDF$, we have $\angle EFD = 90° - \angle D = \boxed{54°}$.

(e) $\widehat{CD} = \angle COD = \angle AOB = \widehat{AB} = \boxed{72°}$.

(f) Combining $\angle ABD = 54°$ from part (c) and $\angle EFD = 54°$ from part (d), we have $\angle ABD = \angle EFD$. Therefore, $\overline{EF} \parallel \overline{AB}$. (We also could have noted $\angle BAD = \angle FED = 90°$, so $\overline{EF} \parallel \overline{AB}$.)

(g) Since $BO = AO$, $CO = DO$, and $\angle BOC = \angle AOD$, we have $\triangle AOD \cong \triangle BOC$ by SAS Congruence. Therefore, corresponding lengths AD and BC are equal.

12.1.5

(a) Let $\widehat{AC} = x$ and $\widehat{BC} = y$. Therefore, $\angle AOC = x$, $\angle BOC = y$, and $\angle AOB = x + y$. Isosceles triangles $\triangle OAB$ and $\triangle OBC$ give us $\angle OBA = (180° - \angle AOB)/2 = (180° - x - y)/2$ and $\angle OBC = (180° - \angle BOC)/2 = (180° - y)/2$. So,

$$\angle ABC = \angle OBC - \angle OBA = \frac{180° - x - y}{2} - \frac{180° - y}{2} = \frac{x}{2} = \frac{\widehat{AC}}{2}.$$

(b) Using our result from the first part, we have $\angle BAC = \widehat{BC}/2 = y/2$. Therefore, $\triangle ABC$ gives us $\angle ACB = 180° - \angle ABC - \angle BAC = 180° - x/2 - y/2$. Major arc $\widehat{AB}$ has measure $360° - \widehat{AC} - \widehat{CB} = 360° - x - y$, so $\angle ACB$ has measure one-half the measure of major arc $\widehat{AB}$.

12.1.6 Let $\widehat{BC} = x$, so $\angle COB = x$ and $\widehat{AC} = 180° - \widehat{BC} = 180° - x$. Therefore $\angle AOC = 180° - x$. From isosceles triangle $\triangle AOC$, we have $\angle CAB = \angle CAO = (180° - \angle AOC)/2 = x/2 = \widehat{BC}/2$.

12.1.7 We consider two cases. The first is shown at left below, in which the arcs do not overlap. Since $\widehat{XW} = \widehat{YZ}$, we have $\angle WZX = \widehat{WX}/2 = \widehat{YZ}/2 = \angle YWZ$. Since $\angle YWZ = \angle WZX$, we have $\overline{WY} \parallel \overline{XZ}$. (Essentially the same proof holds if we relabel the points so that Y is on $\widehat{WZ}$.)

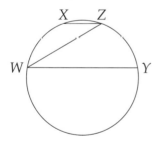
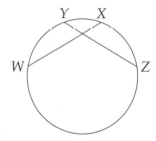

For our second case, suppose the arcs overlap as shown in the second diagram. Then, we have $\widehat{WY} = \widehat{WX} - \widehat{XY} = \widehat{YZ} - \widehat{XY} = \widehat{XZ}$. From this point, we can follow essentially the same proof as in the first case to show that $\overline{XY} \parallel \overline{WZ}$.

12.1.8 Consider the circumcircle of $ABCDEFGHIJKLMNO$. Because the polygon is regular, each side subtends and arc of equal measure. Since there are 15 such little arcs, each has measure $360°/15 = 24°$. $\angle ACD$ is inscribed in major arc $\widehat{AED}$, which consists of 12 of these little arcs (all but $\widehat{AB}$, $\widehat{BC}$, and $\widehat{CD}$). Therefore, $\angle ACD = \widehat{AED}/2 = (12 \cdot 24°)/2 = \boxed{144°}$. Similarly, $\angle ADE$ is inscribed in an arc that consists of 11 of these smaller arcs, so it has measure $(11 \cdot 24°)/2 = \boxed{132°}$. (Perhaps you'll agree that this is much easier than the method we used earlier!)

Exercises for Section 12.2

12.2.1 $\widehat{AD} = 360° - \widehat{AB} - \widehat{BC} - \widehat{CD} = 134°$. Therefore, $\angle AED = (\widehat{BC} + \widehat{AD})/2 = \boxed{118.5°}$.

12.2.2

(a) Since $\angle YWZ = (\widehat{YZ} - \widehat{XV})/2$, we have $30.5° = (\widehat{YZ} - 32°)/2$, so $\widehat{YZ} = \boxed{93°}$.

(b) $\widehat{VZ} = 360° - \widehat{YZ} - \widehat{XY} - \widehat{XV} = \boxed{157°}$.

12.2.3 We have $\widehat{QR} = \widehat{PQ}/2$, $\widehat{RS} = \widehat{PQ}/3$, and $\widehat{SP} = \widehat{PQ}/4$. Therefore, we have $360° = \widehat{PQ} + \widehat{QR} + \widehat{RS} + \widehat{SP} = \widehat{PQ}\left(1 + \frac{1}{2} + \frac{1}{3} + \frac{1}{4}\right)$, so $25\widehat{PQ}/12 = 360°$. Thus, $\widehat{PQ} = (12/25)(360°) = 172.8°$. Therefore, $\angle QZR = (\widehat{QR} + \widehat{PS})/2 = (\widehat{PQ}/2 + \widehat{PQ}/4)/2 = (3/8)(\widehat{PQ}) = \boxed{64.8°}$.

12.2.4 We have $\widehat{YZ} = \widehat{XYZ} - \widehat{XY} = 117° - 75° = 42°$ and $\widehat{XW} = 360° - \widehat{YZW} - \widehat{XY} = 112°$.

(a) We have $\angle YPZ = (\widehat{XW} - \widehat{YZ})/2 = \boxed{35°}$.

(b) We have $\angle YOZ = (\widehat{XW} + \widehat{YZ})/2 = \boxed{77°}$.

(c) $\angle XYW = \widehat{XW}/2 = \boxed{56°}$.

(d) $\angle OWZ = \widehat{YZ}/2 = \boxed{21°}$.

12.2.5 Since $\angle ROS = \widehat{RS}/2$, we have $\angle POR = 180° - \angle ROS = 180° - \widehat{RS}/2$. Also, note that $\widehat{OR} + \widehat{RS} + \widehat{OTS} = 360°$, so $\widehat{RS}/2 = 180° - \widehat{OR}/2 - \widehat{OTS}/2$, from which we find

$$\angle POR = 180° - \left(180° - \frac{\widehat{OR}}{2} - \frac{\widehat{OTS}}{2}\right) = \frac{\widehat{OR} + \widehat{OTS}}{2}.$$

Exercises for Section 12.3

12.3.1

(a) Let minor arc $\widehat{UI} = x$. Therefore, major arc $\widehat{UI}$ has measure $2x$. Together these make up the whole circle, so $3x = 360°$, from which we find $x = \boxed{120°}$.

(b) As described in the text, $\angle UPI$ is half the difference between major arc $\widehat{UI}$ and minor arc $\widehat{UI}$. Therefore, $\angle UPI = (240° - 120°) = \boxed{60°}$.

(c) $\overline{IO}$ is a radius and $\overline{PI}$ is tangent to the circle. Therefore, $\angle PIO = \boxed{90°}$.

(d) Just as $\angle PIO = 90°$, we have $\angle PUO = 90°$. From $PUOI$ we have $\angle IOU = 360° - \angle PIO - \angle IPU - \angle OUP = \boxed{120°}$. (We could also have noted $\angle IOU = \widehat{UI} = 120°$.)

(e) $\triangle PIU$ is isosceles with $\angle PIU = \angle PUI = (180° - \angle IPU)/2 = \boxed{60°}$.

12.3.2 Since $\overline{TX}$ is tangent to $\odot O$ at X and $\overline{OX}$ is a radius of the circle, we have $\overline{TX} \perp \overline{OX}$. The Pythagorean Theorem then gives us $TO = \sqrt{TX^2 + OX^2} = \sqrt{144 + 36} = \boxed{6\sqrt{5}}$.

12.3.3 We mimic our exploration of the specific case given in the text to build our proof. If the chord is a diameter (such as $\overline{AC}$ in the diagram), then it forms a right angle with the tangent. Since the two arcs $\overparen{AC}$ are semicircles, we have $\angle ZAC = \overparen{ABC}/2$, as required.

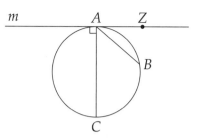

Consider chord $\overline{AB}$, which is not a diameter. We have $\angle BAC = \overparen{BC}/2$, and $\overparen{AB} = 180° - \overparen{BC}$. Therefore, $\angle ZAB = 90° - \angle BAC = 90° - \overparen{BC}/2 = \overparen{AB}/2$. Finally, the obtuse angle formed by chord $\overline{AB}$ and the tangent has measure $180° - \angle ZAB = 180° - \overparen{AB}/2$. Since $\overparen{ACB} = 360° - \overparen{AB}$, the obtuse angle between the chord and the tangent is also one-half the arc it intercepts. Therefore, the angle formed by a tangent and a chord with the point of tangency as an endpoint equals one-half the arc intercepted by the angle.

12.3.4 Let minor arc $\overparen{YZ} = x$, so major arc $\overparen{YZ} = 360° - x$ and $\angle YXZ = (360° - x - x)/2$. Since we are given $\angle YXZ = 51°$, we have $51° = 180° - x$, so $x = \boxed{129°}$.

12.3.5

(a) As we draw our circle with diameter $\overline{OP}$, we must draw two semicircles, one on either side of $\overline{OP}$. Each of these semicircles connects a point outside $\odot O$ to the center of $\odot O$. Therefore, each of these semicircles must hit $\odot O$. Since each semicircle stays wholly on its 'side' of $\overleftrightarrow{OP}$, the only points they have in common are O and P. Therefore, the two points where the semicircles hit $\odot O$ must be different.

(b) Suppose C and $\odot O$ met at three points. The only circle through the three points is the circumcircle of the triangle with these three points as vertices. Since both C and $\odot O$ go through the three points, each is the aforementioned circumcircle. This is clearly impossible because the two circles are different (O is on one circle but not the other). Therefore, C and $\odot O$ cannot meet at three different points.

(c) Each of $\angle PAO$ and $\angle PBO$ is inscribed in a semicircle of circle C. Therefore, each is a right angle.

(d) We showed in the text that if a line passes through a point on a circle such that it is perpendicular to the radius drawn to that point, then the line is tangent to the circle. In the previous part, we found that $\overline{PA}$ is perpendicular to radius $\overline{OA}$ of $\odot O$, so $\overline{PA}$ must be tangent to $\odot O$. Similarly, $\overline{PB}$ must be tangent to $\odot O$ as well.

(e) Since D is on $\odot O$ such that $\overline{PD}$ is tangent to $\odot O$, we must have $\angle PDO = 90°$. The circumcircle of $\triangle PDO$ therefore is the circle with $\overline{PO}$ as diameter. This circle is C! Therefore, D must be on C.

(f) From the previous part, we have found that if a point D is on $\odot O$ such that $\overline{PD}$ is tangent to $\odot O$, then D must be on C. Since we have seen that the only points on C that are also on $\odot O$ are points A and B, we know that D must be A or B. In other words, $\overleftrightarrow{PA}$ and $\overleftrightarrow{PB}$ are the only lines through P that are tangent to $\odot O$.

12.3.6

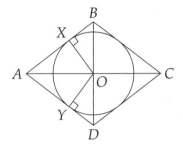

(a) Let the foot of the altitude from O to $\overline{AB}$ be X and the foot of the altitude from O to $\overline{AD}$ be Y. Since $\odot O$ is tangent to both $\overline{AB}$ and $\overline{AD}$, we have $OX = OY$ and $\angle OXA = \angle OYA = 90°$. Since we also have $OA = OA$, we have $\triangle OXA \cong \triangle OYA$ by HL Congruence. Therefore, we have $\angle OAX = \angle OAY$, so O is on the angle bisector of $\angle BAD$. Similarly, O must be on the angle bisector of all four angles of the rhombus. Since the diagonals of a rhombus bisect the angles of a rhombus, we conclude that O must be on both diagonals of the rhombus.

(b) Since $ABCD$ is a rhombus, we know that $\overline{AC} \perp \overline{BD}$. Therefore, $\triangle AOB$ is a right triangle. Since $ABCD$ is a parallelogram, $\overline{AC}$ and $\overline{BD}$ bisect each other at O. Therefore, $AO = AC/2 = 12$. From right triangle AOB, we have $OB = 9$. So, we have $[AOB] = (9)(12)/2 = 54$. We can now find the distance from O to $\overline{AB}$. Let the foot of the altitude from O to $\overline{AB}$ be X. Since $(AB)(OX)/2 = [ABO]$, we have $OX = 2[ABO]/AB = 36/5$. Therefore the area of $\odot O$ is $OX^2\pi = \boxed{1296\pi/25}$. (We could also use similar right triangles to find OX.)

Exercises for Section 12.4

12.4.1 $\boxed{\text{No}}$. To see this, pick three points A, B, and C, such that they don't all lie on the same line. The circumcircle of $\triangle ABC$ is the only circle through these three points. Now, if we pick any point D not on this circle, it's impossible for a circle to go through all four vertices of $ABCD$. We know this because there's only one circle through all three of A, B, and C, and point D is not on this circle.

12.4.2 In the text we showed that the opposite angles of a cyclic quadrilateral must add to $180°$.

(a) The opposite angles of a parallelogram are equal. In a cyclic parallelogram, these angles add to $180°$. Therefore, each must measure $90°$. Since all the angles of the parallelogram are $90°$, the parallelogram is a rectangle.

(b) A rhombus is a parallelogram. By the previous part, a cyclic rhombus is a rectangle. A rhombus that is also a rectangle is a square.

(c) Let the trapezoid be $ABCD$ with $\overline{AB} \parallel \overline{CD}$. Since $ABCD$ is cyclic, we have $\angle A + \angle C = 180°$. Since $\overline{AB} \parallel \overline{CD}$, we have $\angle B + \angle C = 180°$. Combining these gives $\angle A + \angle C = \angle B + \angle C$, so $\angle A = \angle B$, which means the trapezoid is isosceles.

12.4.3 Since $XM = YM = ZM$, the circle with center M and radius XM passes through all three vertices of $\triangle XYZ$ and is therefore the circumcircle of $\triangle XYZ$. The midpoint of $\overline{YZ}$ is the center of the circle, so $\overline{YZ}$ is a diameter of this circle. Therefore, $\angle YXZ$ is inscribed in a semicircle, so $\angle YXZ = 180°/2 = 90°$.

12.4.4 Both $\angle ACB$ and $\angle ADB$ are inscribed in the same arc $(\overset{\frown}{AB})$ of the circle through the vertices of $ABCD$, so $\angle ACB = \angle ADB$.

12.4.5 Let the points where the circle touches the quadrilateral be W, X, Y, and Z, as shown. Because tangents from the same point to the same circle have the same length, we have $AW = AZ$, $BW = BX$, $CY = CX$, and $DY = DZ$. We assign variables to these lengths as shown. We then have

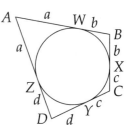

$$a + b = AB$$
$$b + c = BC$$
$$c + d = CD$$
$$d + a = AD$$

Therefore, $AB + CD = a + b + c + d = BC + AD$, as desired.

12.4.6 $\boxed{\text{No}}$. Consider a rectangle with length 100000 and width 1. Try drawing a circle that's inscribed in that!

12.4.7 We have tangents, and we're looking for a length, so we build right triangles by connecting O and P to each other and to A and B, respectively. Let T be the point where $\overline{AB}$ and $\overline{OP}$ meet. Since $\angle ATO = \angle PTB$ and $\angle OAB = \angle PBA = 90°$, we have $\triangle AOT \sim \triangle BPT$ by AA Similarity. If we let $OT = x$, then we have $PT = 36 - x$. From our similarity we have $OT/OA = PT/PB$. Therefore, $x/8 = (36 - x)/4$. Solving this equation yields $x = 24$. Therefore, $AT = \sqrt{OT^2 - AO^2} = 16\sqrt{2}$ and $PT = \sqrt{PT^2 - PB^2} = 8\sqrt{2}$, so $OP = OT + TP = \boxed{24\sqrt{2}}$.

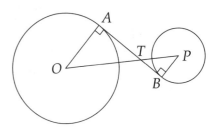

12.4.8 Let our right triangle be $\triangle ABC$ with $\angle C = 90°$, $AB = c$, $AC = b$, and $BC = a$. Let our incircle have center I and meet $\overline{AC}$ and $\overline{BC}$ at X and Y as shown. Since $\overline{IX}$ and $\overline{IY}$ are radii drawn to points of tangency, we have $\overline{IX} \perp \overline{AC}$ and $\overline{IY} \perp \overline{BC}$. We also have $IX = IY$, $\angle C = 90°$, and $CX = CY$. Therefore, $\triangle XCY$ and $\triangle XIY$ are 45-45-90 triangles with the same hypotenuse. Hence, they are congruent. Therefore, all the sides of $IXCY$ are equal, as are all the angles. Therefore, $IXCY$ is a square. Hence, the inradius of $\triangle ABC$ equals CX. In our final problem in the text, we proved that CX (the distance from a vertex to an adjacent point of tangency of the incircle) is $s - c = (a + b - c)/2$ (where s is the semiperimeter of the triangle).

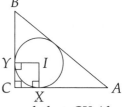

See if you can also solve this problem using the Pythagorean Theorem and the fact that the inradius of a triangle equals its area divided by its semiperimeter!

12.4.9 By symmetry, the center of the middle circle is equidistant from the vertices of the square. Therefore, this center is on both diagonals. The center of the little circle nearest A (point E in the diagram) is equidistant from $\overline{AB}$ and $\overline{AD}$, so it is on the angle bisector of $\angle BAD$. Diagonal $\overline{AC}$ bisects $\angle BAD$, so it must pass through E. Similarly, $\overline{AC}$ must pass through G, the center of the little circle nearest C. Drawing altitudes from E and G to $\overline{AB}$ and $\overline{CD}$, respectively, we form 45-45-90 triangles. Letting the radius of each little circle be r, we have $GX = EY = r$, so $AE = CG = r\sqrt{2}$. We also have $EG = 4r$ because this segment includes one radius of each of $\odot E$ and $\odot G$, and the diameter of the center circle. Therefore, we have $AC = 4r + 2r\sqrt{2}$. Since we also have $AC = AD\sqrt{2} = 4\sqrt{2}$, we have $4r + 2r\sqrt{2} = 4\sqrt{2}$. Therefore, we have $r(2 + \sqrt{2}) = 2\sqrt{2}$, so $r = (2\sqrt{2})/(2 + \sqrt{2}) = \boxed{2\sqrt{2} - 2}$.

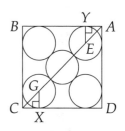

Exercises for Section 12.5

12.5.1   Let our starting lines be m and n, and their intersection point be Y. Since the circle must be tangent to both lines, its center must be equidistant from the lines. Therefore, the center of the circle must be on an angle bisector of one of the angles formed by the lines. So, we start by constructing an angle bisector. Then, we choose a point O on the angle bisector. We construct a line through O perpendicular to m. Let X be the point where this perpendicular hits m. Since $\overline{OX} \perp m$, the circle with radius OX must be tangent to m. Since

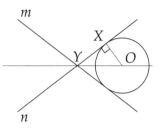

O is on an angle bisector of lines m and n, it must also be a distance of OX from n. Therefore, the circle with radius OX and center O is tangent to n as well.

12.5.2   We know $\overleftrightarrow{XY}$ is tangent to $\odot O$ by construction, because we construct this line by drawing a line through a point on $\odot O$ that is perpendicular to the radius of $\odot O$ drawn to that point. Point Z in our construction is on the circle with diameter $\overline{OP}$, so $\angle OZP$ is inscribed in a semicircle. Therefore, $\angle OZP = 90°$, so $\angle XZP = 180° - \angle OZP = 90°$. Let the radius of $\odot O$ be r_O and the radius of $\odot P$ be r_P. Then, we have $OZ = r_O - r_P$ by construction, so we have $ZX = OX - OZ = r_P$. Therefore, if we take the point Y on our tangent line through X such that $XY = ZP$

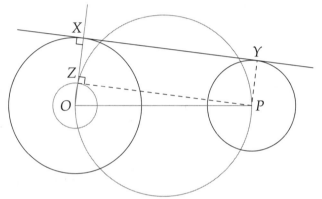

and Y is on the same side of $\overline{XZ}$ as P is, then we have $XY = PZ$ and $\overline{XY} \parallel \overline{PZ}$ so $XYPZ$ is a parallelogram. Furthermore, since $\angle YXZ = \angle XZP = 90°$, we see that $XYPZ$ is a rectangle. Therefore, $YP = XZ = r_P$, so Y is on $\odot P$, and $\angle PYX = 90°$, so $\overline{XY}$ is tangent to $\odot P$.

12.5.3   Our construction is shown below.

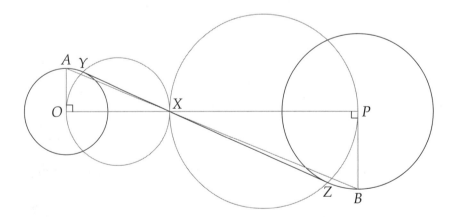

We start by drawing $\overline{OP}$, then draw radii $\overline{OA}$ and $\overline{PB}$ of our given circles such that each is perpendicular to $\overline{OP}$. We then draw $\overline{AB}$; we let point X be where this segment meets $\overleftrightarrow{OP}$. Point X is also the point where our common tangent hits $\overline{OP}$, as we'll prove on the next page. We then construct tangents from X to $\odot O$ and $\odot P$ using the construction described in the text for constructing a tangent to a circle from a point outside the circle. Let $\overline{XY}$ be one of the tangents to $\odot O$ and $\overline{XZ}$ be one of the tangents to $\odot P$. We choose these on opposite sides of $\overline{OP}$, as shown. To prove that $\overline{XZ}$ is the common tangent to both circles, we must show that $\angle YXZ = 180°$, i.e., that Y, X, and Z are collinear.

Since $\angle OXA = \angle PXB$ and $\angle AOP = \angle BPO$, we have $\triangle XAO \sim \triangle XBP$ by AA Similarity. Therefore, $XO/XP = AO/BP = r_O/r_P$, where r_O is the radius of $\odot O$ and r_P is the radius of $\odot P$. Turning our attention to $\triangle XYO$ and $\triangle XZP$, we see that $\angle XYO = \angle XZP = 90°$, and $OY/ZP = XO/XP = r_O/r_P$, so $\triangle XYO \sim \triangle XZP$ by HL Similarity. Therefore, $\angle YXO = \angle ZXP$, so $\angle ZXO = 180° - \angle ZXP = 180° - \angle YXO$. This tells us that $\angle YXO + \angle ZXO = 180°$, so Y, X, and Z lie on a straight line. Therefore, $\overleftrightarrow{YZ}$ is our desired common internal tangent.

Review Problems

12.28

(a) $\widehat{AC} = 360° - 120° - 190° = 50°$, so $x = \widehat{AC}/2 = \boxed{25°}$.

(b) $\angle J = 180° - \angle I - \angle H = 40°$, so $x = 2(\angle J) = \boxed{80°}$.

(c) Since $RS = RT$, we have $\widehat{RS} = \widehat{RT}$. Furthermore, $\widehat{ST} = 2(\angle R) = 80°$, so adding all the arcs of the circle gives us $x + x + 80° = 360°$, so $x = 280°/2 = \boxed{140°}$.

(d) Since $OA = OB = OC$, we deduce that O is the center of the circle (why?). $\widehat{AB} = 2(\angle ACB) = 20°$, so $\angle AOB = \widehat{AB} = \boxed{20°}$.

(e) Because $WXYZ$ is a cyclic quadrilateral, we have $\angle W + \angle Y = 180°$. Therefore, $x = 180° - 120° = \boxed{60°}$.

(f) $\angle QPR$ and $\angle RSQ$ are inscribed in the same arc and are therefore equal. Hence, $\angle QPR = \boxed{20°}$.

12.29

(a) Since $BC = DE$, we have $\widehat{DE} = \widehat{BC} = x$. Therefore, $\widehat{CE} = 360° - x - 31° - x = 329° - 2x$. Since $\angle A = (\widehat{CE} - \widehat{BC})/2$, we have $24° = (329° - 2x - 31°)/2 = 149° - x$. Therefore, $x = 149° - 24° = \boxed{125°}$.

(b) Since $\angle WZX = (\widehat{WX} + \widehat{VY})/2$, we have $90° = (155° + 2x)/2$, from which we find $x = \boxed{12.5°}$.

(c) $\widehat{PS} = 360° - 61° - 133° = 166°$, so $x = \angle Q = (\widehat{PS} - \widehat{PR})/2 = \boxed{52.5°}$.

(d) Since minor arc $\widehat{GI}$ equals x, major arc $\widehat{GI}$ equals $360° - x$. Therefore, we have $\angle GHI = (360° - x - x)/2 = 180° - x$. Since $\angle GHI = 45°$, we have $x = \boxed{135°}$.

(e) $\angle K = \widehat{JL}/2 = \angle JLI = 43°$. (Angles $\angle K$ and $\angle JLI$ are inscribed in the same arc.) Therefore, from $\triangle JKL$, we have $x = 180° - 43° - 31° = \boxed{106°}$.

12.30

(a) $\widehat{AB} = 2(\angle ACB) = \boxed{116°}$.

(b) $\widehat{BC} = 180° - 116° = 64°$, so $\angle BAC = \widehat{BC}/2 = \boxed{32°}$.

(c) $\angle D = \angle A = 32°$. Since $\triangle BDC$ is isosceles with $DB = DC$, we have $\angle DBC = (180° - 32°)/2 = \boxed{74°}$.

12.31 Since $\angle BEC = (\overarc{BC} + \overarc{AD})/2$ and $\overarc{BC} = 360° - 34° - 102° - 109° = 115°$, we have $\angle BEC = (115° + 109°)/2 = \boxed{112°}$.

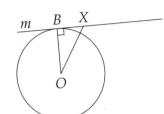

12.32 Let X be a point on m besides point B. Since $\triangle OBX$ is a right triangle, we have $OX = \sqrt{OB^2 + BX^2}$. Therefore, $OX > OB$, so X must be outside $\odot O$. This is true for all X on m besides point B, so m must meet $\odot O$ in exactly one point.

12.33 Let our triangle be $\triangle ABC$ with obtuse angle at $\angle C$. Since $\angle C > 90°$, it is inscribed in an arc that is greater than $2(90°) = 180°$. Therefore, the arc of the circumcircle connecting A to B that does not pass through C must be more than $180°$. Hence, point C and the center of our circle must be on opposite sides of $\overline{AB}$. Since the triangle lies entirely on the same side of $\overline{AB}$ as point C, this means that the circumcenter of the triangle cannot be inside the triangle.

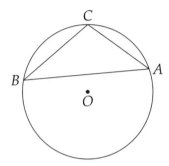

12.34 Since $\angle ROS = 53°$, minor arc $\overarc{RS} = 53°$. Therefore, major arc $\overarc{RS} = 360° - 53° = 307°$, so $\angle RTS = (\text{major arc } \overarc{RS})/2 = \boxed{153.5°}$.

12.35 Let P be on $\odot T$ such that $\overline{KP}$ is tangent to the circle. Since $\overline{TP}$ is a radius drawn to this point of tangency, we have $\angle KPT = 90°$. Since $YT = 2$, the radius of $\odot T$ is 2, so $TP = 2$. Moreover, $KT = KY + YT = 8$, so the Pythagorean Theorem applied to $\triangle KPT$ gives us $KP = \sqrt{KT^2 - YT^2} = \boxed{2\sqrt{15}}$.

12.36

(a) Since $\overline{AB}$ is a diameter of $\odot F$, we have $\overarc{YA} = \overarc{AYB} - \overarc{YB} = 180° - 134° = \boxed{46°}$.

(b) $\angle YXF = (\overarc{YB} - \overarc{YA})/2 = \boxed{44°}$.

(c) $\angle XYA = \overarc{YA}/2 = \boxed{23°}$.

12.37 Since $\overline{TA}$ and $\overline{TB}$ are tangent to the circle, both $\angle A$ and $\angle B$ are right angles. We can divide $OATB$ into two right triangles by drawing $\overline{OT}$. Since $\overline{AT}$ and $\overline{BT}$ are tangents to the same circle from the same point, we have $AT = BT$. We also have $OA = OB$ and $OT = OT$, so $\triangle OAT \cong \triangle OBT$ by SSS Congruence. Therefore, $\angle AOT = \angle BOT = (\angle AOB)/2 = 60°$, so both of our right triangles are 30-60-90 triangles. Since $AT = 6$, we have $OA = AT/\sqrt{3} = 2\sqrt{3}$ and $OT = 2OA = \boxed{4\sqrt{3}}$. We also have $\angle ATB = \angle ATO + \angle OTB = \boxed{60°}$. (We also could have used the angles of $AOBT$ to find this.) Finally, $[OATB] = [OAT] + [OBT] = (OA)(AT)/2 + (OB)(BT)/2 = \boxed{12\sqrt{3}}$.

12.38 Since $\angle A = 50°$, $\overarc{BD} = 2(50°) = 100°$. Since $\overline{AD}$ is a diameter, we have $\overarc{AB} = 180° - \overarc{BD} = 80°$. Therefore, $\overarc{AB}$ is $(80°/360°) = 2/9$ of the whole circle. Since the circumference of the circle is $AD\pi = 36\pi$, the length of $\overarc{AB}$ is $(2/9)(36\pi) = \boxed{8\pi}$.

12.39 The sides of a regular dodecagon divide its circumcircle into 12 equal arcs. Therefore, each of these small arcs has measure $(360°/12) = 30°$.

(a) $\angle ABC$ is inscribed in an arc that consists of 10 of these smaller arcs. Therefore, major arc $\overarc{AC}$ has measure $10(30°) = 300°$, so $\angle ABC = (300°)/2 = \boxed{150°}$.

(b) ∠ACD is inscribed in an arc that consists of 9 of these little arcs, so it has measure $(9 \cdot 30°)/2 = \boxed{135°}$.

(c) ∠ADJ is inscribed in an arc that consists of three of these little arcs ($\widehat{JK}$, $\widehat{KL}$, and $\widehat{LA}$), so it has measure $(3 \cdot 30°)/2 = \boxed{45°}$.

(d) The acute angles between the chords each equal $(\widehat{AB} + \widehat{FI})/2$. Since $\widehat{AB} = 30°$ and $\widehat{FI}$ consists of three of our little arcs of measure 30°, our angle has measure $(30° + 3 \cdot 30°)/2 = \boxed{60°}$.

12.40

(a) $∠ZWX = 180° - ∠VWZ = 99°$. △WXZ then gives $∠VXZ = 180° - ∠WZX - ∠XWZ = \boxed{43°}$.

(b) $∠WYZ = \widehat{WZ}/2 = ∠WXZ = \boxed{43°}$.

(c) Let $\overline{XZ}$ and $\overline{WY}$ meet at O. Right triangle △OZY gives us $∠XZY = 90° - ∠OYZ = \boxed{47°}$.

(d) $∠WYX = \widehat{WX}/2 = ∠WZX = 38°$ and $∠WYZ = 43°$, so $∠XYZ = ∠WYX + ∠WYZ = \boxed{81°}$. (Is it a coincidence that $∠XYZ = ∠VWZ$?)

(e) Since $WXYZ$ is a cyclic quadrilateral, we have $∠WXY + ∠WZY = 180°$. Since $∠WZY = ∠WZX + ∠XZY = 85°$, we have $∠WXY = 180° - ∠WZY = \boxed{95°}$.

12.41 (Figure at left below.) Since $ABCD$ has area 100, $AB = 10$. Since △ABD is a 45-45-90 triangle, we have $BD = 10\sqrt{2}$. Point O is the midpoint of $\overline{BD}$, so $OD = BD/2 = 5\sqrt{2}$. Finally, if we connect O to the points E and F, where the circle meets sides $\overline{AB}$ and $\overline{CD}$ of the square, we see that $EFDA$ is a rectangle, so $EF = AD = 10$. Therefore, the radius of the circle inscribed in the square is $10/2 = 5$, so $DX = OD - OX = \boxed{5\sqrt{2} - 5}$.

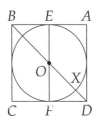
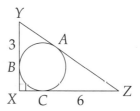

12.42 Label the points of tangency as shown at right above. Since $YA = YB$ and $ZA = ZC$, we have $YZ = YA + AZ = YB + ZC = \boxed{9}$.

12.43 We have already shown in the text that there is a circle centered at the intersection of the angle bisectors of a triangle that is tangent to all three sides of the triangle. We must now show that this is the only such circle. Let ⊙I be a circle that is tangent to all three sides of △ABC. Since ⊙I is tangent to all three sides of △ABC, it is equidistant from the sides of △ABC. Since I is equidistant from $\overline{AB}$ and $\overline{AC}$, I is on the angle bisector of ∠BAC. Similarly, I must be on the bisectors of ∠CBA and ∠ACB. Therefore, any circle that is tangent to all three sides of a triangle must be centered at the intersection of the angle bisectors of the triangle. Only one circle centered at this point is tangent to all three sides of the triangle, so each triangle has exactly one incircle.

12.44 We follow the same steps that allowed us to solve the similar problem in the text. We find the point X on $\overline{BQ}$ such that $QX = PA = 4$. Since $\angle BQP = \angle QPA = 90°$, we have $\angle BQP + \angle QPA = 180°$, so $\overline{QB} \parallel \overline{PA}$. Since we also have $AP = QX$, we know that $QPAX$ is a parallelogram. Since one of the angles of this parallelogram is $90°$, we know $QPAX$ is a rectangle. Since $QPAX$ is a rectangle, we have $AX = PQ$ and $\angle AXB = 90°$. Therefore, we have $AX^2 + XB^2 = AB^2$. Since $BQ = 8$ and $QX = 3$, we have $BX = BQ - QX = 5$. We also have $AB = AC + CB = 11$, so $AX = \sqrt{AB^2 - XB^2} = \sqrt{96} = \boxed{4\sqrt{6}}$.

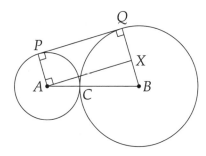

Note that $PQ = 2\sqrt{(AP)(BQ)} = 2\sqrt{24} = 4\sqrt{6}$. Is this a coincidence?

12.45 Since the inner circle is tangent to all six sides of the hexagon, a radius of the inner circle is the altitude from the center to a side ($\overline{OX}$ in the figure). The circumcircle of the hexagon goes through all six vertices, so its radius is $\overline{OB}$ in the figure. The desired ratio of areas then is the square of the ratio of the radii. Since $\triangle OBC$ isosceles and $\overline{OB}$ and $\overline{OC}$ both bisect angles of the hexagon, we have $\angle OBC = \angle OCB = 120°/2 = 60°$. Therefore $\triangle OXB$ is a 30-60-90 triangle and we have $OX/OB = \sqrt{3}/2$. Because the ratio of the radii of the circles is $\sqrt{3}/2$, the ratio of the areas of the circles ($\sqrt{3}/2)^2 = \boxed{3/4}$.

Challenge Problems

12.46 We have $\overset{\frown}{AB} = 2(40°) = 80°$ and $\overset{\frown}{BC} = 2(50°) = 100°$. Therefore, $\overset{\frown}{ABC}$ is a semicircle and $\overline{AC}$ is a diameter of the circle. The angle formed by chord $\overline{AC}$ and tangent ℓ is therefore $\overset{\frown}{ABC}/2 = 90°$, as is the angle formed by $\overline{AC}$ and m. Since $\overline{AC}$ is perpendicular to both ℓ and m, we must have $\ell \parallel m$.

12.47 Since $ABCD$ is a parallelogram, we have $\angle A = \angle C$. Since $\angle E$ and $\angle A$ are inscribed in the same arc ($\overset{\frown}{BD}$), we have $\angle E = \angle A$. Therefore, $\angle E = \angle A = \angle C$, so $\triangle EDC$ is isosceles with $ED = DC$.

12.48 Let the sides of the triangle be a, b, and c, where c is the hypotenuse. As we found in an Exercise, the inradius of the triangle is $s - c$, where s is the semiperimeter. Therefore, we have $s - c = 5$. Doubling this gives us $2s - 2c = 10$, or $p - 2c = \boxed{10}$, where p is the perimeter of the triangle.

12.49 Because $\angle QCD = \angle QCB$, we have $\overset{\frown}{QB} = \overset{\frown}{QD}$. Similarly, $\angle PAB = \angle PAD$ gives us $\overset{\frown}{PB} = \overset{\frown}{PD}$. Therefore, we have $\overset{\frown}{QBP} = \overset{\frown}{QB} + \overset{\frown}{BP} = \overset{\frown}{QD} + \overset{\frown}{DP} = \overset{\frown}{QDP}$. Since $\overset{\frown}{QBP}$ and $\overset{\frown}{QDP}$ together are the whole circle and the two arcs are equal, they must both be semicircular arcs. Because $\overset{\frown}{QBP} = 180°$, we know that $\overline{PQ}$ is a diameter.

12.50 $\angle EDF$ is inscribed in $\overset{\frown}{EF}$ of the incircle (where $\overset{\frown}{EF}$ is the arc connecting E to F that does not include D). Therefore, $\angle EDF = \overset{\frown}{EF}/2$. Since $\angle A$ is formed by two sides of $\triangle ABC$ that are tangent to the incircle, we have $\angle A = (\overset{\frown}{EDF} - \overset{\frown}{EF})/2 = (360° - \overset{\frown}{EF} - \overset{\frown}{EF})/2 = 180° - \overset{\frown}{EF}$. Since $\angle A = 32°$, we have $\overset{\frown}{EF} = 180° - \angle A = 148°$. Therefore, $\angle EDF = \overset{\frown}{EF}/2 = \boxed{74°}$.

12.51 Since $AX = AY$, we have $\overarc{AX} = \overarc{AY}$. Therefore, $\overarc{BX} = \overarc{BXA} - \overarc{AX} = 180° - \overarc{AY} = \overarc{BYA} - \overarc{AY} = \overarc{BY}$. So, we have $\angle XAB = \overarc{BX}/2 = \overarc{BY}/2 = \angle YAB$. Let $\overline{XY}$ meet $\overline{AB}$ at M. Since $AX = AY$, $\angle XAB = \angle YAB$ and $AM = AM$, we have $\triangle AXM \cong \triangle AYM$ by SAS Congruence. Therefore, $XM = YM$ and $\angle AMX = \angle AMY$. Since $\angle AMX + \angle AMY = 180°$ also, we must have $\angle AMX = \angle AMY = 90°$, so $\overleftrightarrow{AB}$ is the perpendicular bisector of $\overline{XY}$.

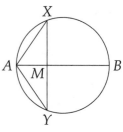

12.52 Earlier, we found that $YZ = YA + ZA = YB + ZC = 9$. Let $XC = x$. Since $\overline{XC}$ and $\overline{XB}$ are tangents from the same point to the same circle, we have $XB = XC = x$. Now we can apply the Pythagorean Theorem to $\triangle XYZ$: $(x + 3)^2 + (x + 6)^2 = 9^2$. Therefore, $2x^2 + 18x - 36 = 0$, so $x = (-9 \pm 3\sqrt{17})/2$. Since x clearly must be positive, we have $x = (-9 + 3\sqrt{17})/2$, so $XZ = 6 + x = \boxed{(3 + 3\sqrt{17})/2}$.

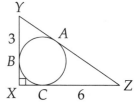

12.53 We label the centers of the circles as shown, and build right angles by drawing radii from points of tangency, and by completing rectangle $XIWV$. Since the area of the square on the left is 256, its side length is $\sqrt{256} = 16$, so $TV = 16/2 = 8$. Similarly, the side length of the square on the right is 4, so $IW = 2$. Since VW equals the sum of the radii of the circles (make sure you see why) and $XV = TV - XV = TV - IW$ (since $XIWV$ is a rectangle), we have $TX = 6$ and $XI = 10$. Therefore, $TI = \sqrt{TX^2 + IX^2} = \sqrt{136} = \boxed{2\sqrt{34}}$.

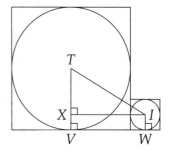

12.54 Let O be the center of our circle and $\overline{AB}$ be one of our tangents. Since $\overline{AB}$ is tangent to the circle at its midpoint, the radius from O to the midpoint of $\overline{AB}$ is perpendicular to $\overline{AB}$. From right triangle OTB, we have $OB = \sqrt{OT^2 + TB^2} = \sqrt{5}$. Similarly, $OA = \sqrt{5}$, and the endpoints of all of our tangent segments are $\sqrt{5}$ away from O. Therefore, the endpoints of our tangent segments trace out a circle with radius $\sqrt{5}$. The segments therefore fill out the region between this circle with radius $\sqrt{5}$ and the original circle, which has radius 2. So, the area of our desired region is $(\sqrt{5})^2\pi - (2)^2\pi = \boxed{\pi}$.

12.55 The two cases we must consider are shown below.

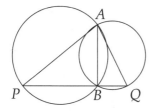
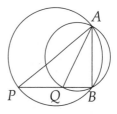

In both cases, since $\overline{AP}$ and $\overline{AQ}$ are diameters of their respective circles, both $\angle ABP$ and $\angle ABQ$ are inscribed in semicircles, so $\angle ABP = \angle ABQ = 90°$. Therefore, $\overline{AB} \perp \overline{BP}$ and $\overline{AB} \perp BQ$, so both P and Q lie on the line through B that is perpendicular to $\overline{AB}$. (Notice that this proof takes care of both cases.)

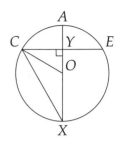

12.56 Since $\overline{CE}$ is the perpendicular bisector of radius $\overline{AO}$, we have $OY = OA/2 = OC/2$. Therefore, $\triangle OYC$ is a 30-60-90 triangle. Since $\angle COY = 60°$, we have $\widehat{AC} = 60°$. Therefore, we have $\angle CXA = \widehat{AC}/2 = 30°$. From isosceles triangle $\triangle OCX$ (since $\overline{OC}$ and $\overline{OX}$ are radii), we have $\angle OCX = \angle OXC = \boxed{30°}$.

12.57 Let $\overline{AB}$ touch the circle at Z. We are given $BZ + AZ = 6$. We also have $BZ = BY$ and $AZ = AX$ because tangents from the same point to the same circle are equal. Therefore, we have $CY + DX = (BC - BY) + (DA - AX) = BC + DA - BY - AX = 7 + 9 - BZ - AZ = 16 - (AB) = \boxed{10}$.

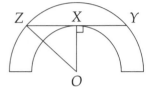

12.58 The longest line of sight will be tangent to the smaller semicircle. One such line of sight is shown in the figure, $\overline{YZ}$. Since $\overline{OX}$ is a radius of the smaller semicircle, it is perpendicular to tangent $\overline{YZ}$. It is also part of a radius of the larger semicircle. Since $\overline{OX}$ is perpendicular to chord $\overline{YZ}$ of the larger semicircle it thus bisects the chord, so $XY = XZ$. We therefore have $XZ = ZY/2 = 6$. Furthermore, from right triangle $\triangle OXZ$, we have $OZ^2 - OX^2 = XZ^2 = 36$. The area of the room is the difference in the areas of the semicircles, or $(OZ^2)\pi/2 - (OX^2)\pi/2 = (\pi/2)(OZ^2 - OX^2) = \boxed{18\pi}$.

(Note also that since OX and OZ are fixed no matter which chord we choose of the larger semicircle that is tangent to the smaller, we can use the Pythagorean Theorem as above to show the length of such a tangent chord is fixed.)

12.59 Let I be the incenter of $\triangle ABC$ and X be the point where the incircle is tangent to $\overline{AC}$. We have $IX = IZ$ because they are both inradii, and we have $AZ = AX$ because they are tangents to the same circle from the same point. Since we are given $AZ = IZ$, $IXAZ$ is a rhombus. Furthermore, since the incircle is tangent to $\overline{AB}$ at Z, we have $\overline{IZ} \perp \overline{AZ}$. Therefore, rhombus $IXAZ$ is a square with $\angle A = 90°$.

12.60 Let the legs of the triangle have length a and b, and let the hypotenuse have length c. From the Pythagorean Theorem we have $a^2 + b^2 = c^2$, or $a^2 + b^2 - c^2 = 0$. The quantity $a^2 + b^2 - c^2$ can only be even if either all three of a, b, and c are even, or if exactly one of them is even. Since this quantity must equal zero for our triangle, we know that either all three of a, b, and c are even, or if exactly one of them is even. As offered as an Exercise in the text, the inradius of our right triangle has length $(a + b - c)/2$. Since $(a + b - c)$ is an even integer if all three of a, b, and c are even, or if exactly one of them is even, we know that our inradius must be an integer for any right triangle with integer side lengths.

12.61

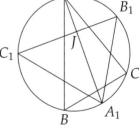

(a) Since $\widehat{A_1 B} = \widehat{A_1 C}$, we have $\angle A_1 AB = \widehat{A_1 B}/2 = \widehat{A_1 C}/2 = \angle CAA_1$. Therefore, $\overrightarrow{AA_1}$ bisects $\angle BAC$, so it must pass through the incenter of $\triangle ABC$.

(b) Let J be the point where $\overline{AA_1}$ meets $\overline{B_1 C_1}$. We have $\angle C_1 J A_1 = (\widehat{AB_1} + \widehat{A_1 BC_1})/2 = (\widehat{AB_1} + \widehat{BC_1} + \widehat{BA_1})/2 = (\widehat{AC}/2 + \widehat{BC}/2 + \widehat{AB}/2)/2 = (\widehat{AC} + \widehat{BC} + \widehat{AB})/4 = 360°/4 = 90°$. Therefore, $\overline{AA_1} \perp \overline{B_1 C_1}$.

(c) Following the logic in part (a), we can show that each of $\overleftrightarrow{AA_1}$, $\overleftrightarrow{BB_1}$, and $\overleftrightarrow{CC_1}$ passes through the incenter of $\triangle ABC$. Following the logic in part (b), we can show that each of these lines contain the altitudes of $\triangle A_1 B_1 C_1$, so these lines must meet at the orthocenter of $\triangle A_1 B_1 C_1$. Therefore, the incenter of $\triangle ABC$ is the orthocenter of $\triangle A_1 B_1 C_1$.

12.62   We start by drawing the circumcircle of $\triangle ABC$. Point D can be either inside, outside, or on this circle. We have shown in the text that if D is on the circle, then we have $\angle A + \angle C = 180°$.

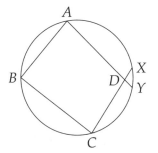   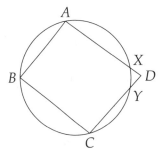

If D is inside the circumcircle of $\triangle ABC$ above, as in the diagram at left above, we have $\angle A + \angle C = \overset{\frown}{BCY}/2 + \overset{\frown}{BAX}/2 = (\overset{\frown}{BCY} + \overset{\frown}{BAX})/2 = (360° - \overset{\frown}{XY})/2 = 180° - \overset{\frown}{XY}/2$. Therefore, if D is inside the circumcircle, then $\angle A + \angle C < 180°$.

If D is outside the circumcircle, as in the diagram at right above, we have $\angle A + \angle C = \overset{\frown}{BCX}/2 + \overset{\frown}{BAY}/2 = (\overset{\frown}{BCX} + \overset{\frown}{BAY})/2 = (360° + \overset{\frown}{XY})/2 = 180° + \overset{\frown}{XY}/2$. Therefore, if D is outside the circumcircle, then $\angle A + \angle C > 180°$.

Since the only case in which we can possibly have $\angle A + \angle C = 180°$ is when D is on the circumcircle of $\triangle ABC$, then if $\angle A + \angle C = 180°$, we know that D is on the circumcircle of $\triangle ABC$, so $ABCD$ is a cyclic quadrilateral.

12.63   Since $BC = AC = DC = 10$, the circle with center C and radius 10 goes through A, B, and D as shown at left below. Since $AB = BC = AC$, $\triangle ABC$ is equilateral, so $\angle ACB = 60°$. Since $\angle ACB = 60°$, $\overset{\frown}{AB}$ of our circle is also $60°$. Since $\angle ADB$ is inscribed in this arc, we have $\angle ADB = \overset{\frown}{AB}/2 = \boxed{30°}$.

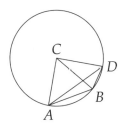   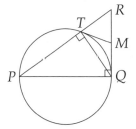

12.64   Our diagram is at right above. Since $\overline{PQ}$ is a diameter, $\angle PTQ = 90°$. Since $\overline{MT}$ and $\overline{MQ}$ are tangents from the same point to the same circle, we have $MT = MQ$. Therefore, we have $\angle MTQ = \angle MQT$. From right triangle QTR we have $\angle R = 90° - \angle MQT$. From right angle $\angle QTR$, we have $\angle MTR = 90° - \angle MTQ = 90° - \angle MQT = \angle R$. Therefore, $\triangle MTR$ is isosceles with $MT = MR$. We now have $MR = MT = MQ$, so M is the midpoint of $\overline{RQ}$.

CHAPTER 13

Power of a Point

Exercises for Section 13.1

13.1.1

(a) The power of point C gives $(CB)(CE) = (AC)(CD)$, so $BC = (AC)(CD)/(CE) = \boxed{16/3}$.

(b) The power of point T gives us $(TS)(TQ) = (TP)(TR)$, so we have $(TQ)(9 - TQ) = 20$. Therefore, we have $TQ^2 - 9TQ + 20 = 0$, so $(TQ - 5)(TQ - 4) = 0$, which gives us $TQ = 4$ or $TQ = 5$. Since $TQ < TS$ and $TS = 9 - TQ$, we cannot have $TQ = 5$. Therefore, our answer is $TQ = \boxed{4}$.

(c) By the power of point V, we have $(VY)(WV) = (VX)(VZ)$, so $VY = (VX)(VZ)/VW = 9$. Applying the Pythagorean Theorem to $\triangle YVZ$ gives $YZ = \sqrt{VY^2 + VZ^2} = \boxed{3\sqrt{13}}$.

13.1.2

(a) The power of point A gives $(AB)(AC) = (AD)(AE)$, so $AE = (AB)(AC)/(AD) = 27$. Therefore, $DE = AE - AD = \boxed{23}$.

(b) The power of point X gives $XY^2 = (XW)(XZ)$, so $XW = XY^2/XZ = 16/3$. Therefore, $WZ = XZ - XW = \boxed{20/3}$.

13.1.3 The power of point Q gives $(AQ)(BQ) = (XQ)(YQ)$. Therefore, $YQ = 3$, so $XY = XQ + YQ = \boxed{11}$.

13.1.4 Since M is the midpoint of $\overline{ST}$, we have $SM = MT = 6$. The power of point M gives $(MS)(MT) = (MU)(MV)$. Let $MU = x$, so that $MV = 15 - x$. Therefore, we have $36 = x(15 - x)$, so $x^2 - 15x + 36 = 0$, from which we have $(x - 12)(x - 3) = 0$. So, the two possible values of UM are $\boxed{3 \text{ and } 12}$.

13.1.5

(a) Since the radius of the circle is c and $OX = b$, we have $CX = OC - OX = \boxed{c - b}$ and $XD = XO + OD = \boxed{b + c}$.

(b) The power of point X gives us $(AX)(XB) = (CX)(XD)$. Since radius $\overline{OC}$ is perpendicular to chord $\overline{AB}$, it bisects the chord. Therefore, $AX = XB = a$, and our Power of a Point relationship is $a^2 = (c - b)(c + b)$. Rearranging gives us $a^2 + b^2 = c^2$.

(c) Suppose we have a right triangle $\triangle TUV$ with right angle at U. We draw a circle with center T and radius TV. We extend $\overrightarrow{VU}$ to meet the circle again at point W, and extend $\overleftrightarrow{TU}$ to meet the circle at R and S as shown. Since radius $\overline{TR}$ is perpendicular to chord $\overline{VW}$, we have $VU = UW$. We also have $RU = TR - TU$ and $US = TU + TS$. The power of point U gives $(VU)(UW) = (RU)(US)$, or $VU^2 = (TR - TU)(TS + TU)$. Since $TR = TS = TV$ (radii of the same circle), we have $VU^2 = (TV - TU)(TV + TU)$. Rearranging this gives us $TU^2 + VU^2 = TV^2$, and we have proved the Pythagorean Theorem.

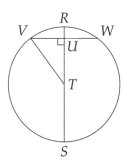

13.1.6 The power of point A gives us $(AW)(AX) = (AY)(AZ)$, or $(AW)(AW + WX) = (AY)(AY + YZ)$. Since $WX = YZ$, we have $(AW)(AW + YZ) = (AY)(AY + YZ)$, or $AW^2 + (YZ)(AW) = AY^2 + (YZ)(AY)$. Rearranging gives $AW^2 - AY^2 = (YZ)(AY) - (YZ)(AW)$. Therefore, we have $(AW - AY)(AW + AY) = (-YZ)(AW - AY)$, or $(AW - AY)(AW + AY) + (YZ)(AW - AY) = 0$. One more bit of factoring gives $(AW - AY)(AW + AY + YZ) = 0$. Dividing by $AW + AY + YZ$ and rearranging gives $AW = AY$, as desired.

Exercises for Section 13.2

13.2.1 Let the radius of the circle be r and let $\overline{BO}$ meet the circle at X. The power of point B gives $AB^2 = (BX)(BD)$, or $16 = (12 - 2r)(12)$. Solving for r, we find $r = 16/3$. Our area then is $\pi r^2 = \boxed{256\pi/9}$.

13.2.2

(a) The power of point P gives $(PT)(PQ) = (PS)(PR)$, so $PR = (PT)(PQ)/(PS) = 9$. Therefore, the Pythagorean Theorem gives us $QR = \sqrt{PQ^2 - PR^2} = 3\sqrt{7}$ and we have $[PQR] = (PR)(QR)/2 = \boxed{27\sqrt{7}/2}$.

(b) Since $\angle R$ is a right angle, $\overline{SQ}$ is a diameter of the circle. Therefore, $\angle STQ$ is inscribed in a semicircle, so $\angle STQ = 90°$. We then have $\angle STP = \angle PRQ = 90°$ and $\angle TPS = \angle QPR$, so $\triangle STP \sim \triangle QRP$ by AA. Our similarity gives us $ST/TP = QR/PR$, so $ST = (QR/PR)(TP) = \boxed{\sqrt{7}}$. (We could also have used the Pythagorean Theorem on $\triangle STP$.)

(c) The Pythagorean Theorem gives us $SQ = \sqrt{QR^2 + SR^2} = 2\sqrt{22}$. As we saw in the previous part, $\overline{SQ}$ is a diameter of the circle, so the circle's radius is $\sqrt{22}$ and its area is $(\sqrt{22})^2\pi = \boxed{22\pi}$.

(d) From part (a), we have $[PQR] = 27\sqrt{7}/2$. Since $\triangle TQR$ and $\triangle PQR$ share an altitude from R, we have $[TQR]/[PQR] = TQ/PQ$. Therefore, we have $[TQR] = (TQ/PQ)[PRQ] = \boxed{81\sqrt{7}/8}$.

13.2.3 We have $XA^2 = (XB)(XC)$ from the power of point X. Multiplying $XB < XC$ by XB gives $XB^2 < (XB)(XC)$, and multiplying $XB < XC$ by XC gives $(XB)(XC) < XC^2$. Since $(XB)(XC) = XA^2$, our two inequalities become $XB^2 < XA^2$ and $XA^2 < XC^2$. Taking square roots gives the desired $XB < XA < XC$.

13.2.4 The power of point A gives us $(AW)(AX) = (AY)(AZ)$. Since $AW = AY$, we have $AX = AZ$, so $WX = AX - AW = AZ - AY = YZ$. Since $\triangle ABC$ is equilateral, we have $AB = AC$. Therefore, $BX = AB - AX = AC - AZ = CZ$ and $BW = BX + XW = CZ + ZY = CY$. Finally, we use the powers of points B and C to find: $BQ^2 = (BX)(BW) = (CZ)(CY) = CQ^2$, so $BQ = CQ$. Point Q is therefore the midpoint of $\overline{BC}$.

Review Problems

13.11

(a) The power of point E gives $(AE)(EC) = (BE)(ED)$, so $EC = \boxed{28/3}$.

(b) The power of point T gives $(PT)(TR) = (QT)(TS)$, so $2x^2 = 24$. Therefore, $x = 2\sqrt{3}$, so $PR = 3x = \boxed{6\sqrt{3}}$.

13.12

(a) The power of point W gives $(WX)(WY) = (WV)(WZ)$. Therefore, $WZ = (WX)(WY)/WV = 6$, so $VZ = WZ - WV = \boxed{3}$.

(b) The power of point A gives $(AB)(AC) = AD^2$, so $AC = AD^2/AB = 18$. Therefore, $BC = AC - AB = \boxed{16}$. The Pythagorean Theorem applied to $\triangle ACD$ gives us $CD = \sqrt{AC^2 + AD^2} = \sqrt{18^2 + 6^2} = \sqrt{6^2(3^2 + 1^2)} = \boxed{6\sqrt{10}}$.

(c) The power of point P gives $(PS)(PT) = (PQ)(PR) = 48$. Since $PT = PS+2$, we have $PS^2+2PS-48 = 0$, so $(PS - 6)(PS + 8) = 0$. Since PS must be positive, we have $PS = \boxed{6}$.

13.13 By the power of point M we have $(GM)(MH) = (IM)(MJ)$. Therefore, $MJ = (GM)(MH)/IM = (12)(6)/2 = 36$. Since $\overline{GM} \perp \overline{MJ}$, $\triangle GMJ$ is a right triangle. Therefore $GJ = \sqrt{GM^2 + MJ^2} = \sqrt{12^2 + 36^2} = \sqrt{12^2(1 + 3^2)} = \boxed{12\sqrt{10}}$.

13.14 If P is inside $\odot O$, then P is on $\overline{AB}$ and $\overline{CD}$. Then, we would have $PB = AB - PA = 6$ and $PD = CD - PC = 1$. From Power of a Point we must have $(PA)(PB) = (PC)(PD)$, but $(PA)(PB) = 18$ and $(PC)(PD) = 4$. Therefore, P must not be inside $\odot O$. With A on $\overline{PB}$ and C on $\overline{PD}$, we have $PB = PA+AB = 12$ and $PD = PC + CD = 9$, so $(PA)(PB) = 36 = (PC)(PD)$, as required by Power of a Point.

13.15 The distance from you to the horizon is the length of a tangent segment. The distance from you to the surface of the Earth is the given distance, and the distance from you to the opposite end of the Earth (i.e., if you drew a line through yourself and the center of the Earth) is 8000 miles plus your distance from the surface of the Earth. Therefore we can use Power of a Point (applied to you!) to give:

$$(\text{Distance to horizon})^2 = (\text{Your Distance from Earth})(8000 + \text{Your Distance from Earth}).$$

(a) We have $x^2 = (1)(8001)$. Therefore, x is $\boxed{\text{approximately 90 miles}}$.

(b) We have $x^2 = (6)(8001) = 48006$. Therefore, x is $\boxed{\text{approximately 220 miles}}$.

(c) We have $x^2 = (100)(8001) = 800100$. Therefore, x is $\boxed{\text{approximately 900 miles}}$.

13.16 The power of point Z gives $(ZY)(ZX) = (ZW)(ZV)$, or $ZY/ZW = ZV/ZX$. Since $ZY < ZW$, we have $ZY/ZW < 1$. Since $ZY/ZW = ZV/ZX$, we have $ZV/ZX < 1$ also, so $ZV < ZX$.

13.17   The power of point P applied to the top circle gives $PQ^2 = (PA)(PB)$, and applied to the bottom circle gives $PR^2 = (PB)(PA)$, so we have $PQ^2 = (PA)(PB) = PR^2$. Therefore, $PQ = PR$.

13.18 $\boxed{\text{Yes.}}$ For example, we could have $AX = XB = 6$, $CX = 4$, and $DX = 9$. This would still give us $(AX)(BX) = (CX)(DX)$, which Power of a Point requires. (To visualize this, consider a circle with diameter $\overline{CD}$ of length 13 bisecting a chord $\overline{AB}$ of length 12.)

13.19 $\boxed{\text{No.}}$ Power of a point requires that $(AX)(BX) = (CX)(DX)$. Since $AX = BX$, they both must equal $AB/2$. Since $CX = 2DX$ and $CD = AB$, we have $CX = CD/3 = AB/3$ and $DX = 2CD/3 = 2AB/3$. Therefore $(AX)(BX) = AB^2/4$ and $(CX)(DX) = 2AB^2/9$. These are clearly not the same, so the situation described in the problem is impossible.

13.20   The power of point A gives $(AC)(AB) = (AD)(AF)$. From the diagram, we have $AB = 2AC$ and $AF = 3AD$, so we have $(AC)(2AC) = (AD)(3AD)$. Therefore, $AC^2/AD^2 = 3/2$. Taking the positive square root of both sides gives $AC/AD = \sqrt{3/2} = \boxed{\sqrt{6}/2}$.

13.21 $\boxed{\text{Yes.}}$ Suppose that $\overleftrightarrow{PB}$ meets the circle at both B and C. From the power of point A, we have $PA^2 = (PB)(PC)$. Since $PA = PB$, our equation becomes $PA^2 = PC^2$. Taking the square root of both sides gives $PA = PC$. Combined with $PA = PB$, this gives $PB = PC$. Since C is on $\overleftrightarrow{PB}$, and P is not inside the circle, point C must be the same as point B. Therefore, $\overleftrightarrow{PB}$ cannot possibly meet the circle at a second point.

13.22   Let $\overleftrightarrow{PO}$ meet the circle at X and Y, with X on $\overline{PY}$. The power of point P gives $(PA)(PB) = (PX)(PY)$. Since $PX = PO - r$ and $PY = PO + r$, we have the desired $(PA)(PB) = (PO - r)(PO + r) = PO^2 - r^2$. If point P is inside the circle, we still have $(PA)(PB) = (PX)(PY)$. However, when P is inside the circle, PX and PY are $r - PO$ and $r + PO$, so our equation becomes $(PA)(PB) = (r - PO)(r + PO) = r^2 - PO^2$, where the right hand side is just the negative of the right hand side when P is outside the circle.

13.23   For any right triangle, we can draw a circle centered at the vertex of one of the acute angles and generate a diagram like the one given in the problem, by extending the hypotenuse to meet the circle a second time. Let our triangle be $\triangle XAO$ as in the problem. Let the radius of the circle be a, $XA = b$, $XO = c$, and let the circle meet $\overleftrightarrow{XO}$ at B and C as shown in the problem. The power of point X gives $XA^2 = (XB)(XC)$, or $b^2 = (c - a)(c + a)$. A little algebra then gives the desired $a^2 + b^2 = c^2$.

Challenge Problems

13.24   Let the radius of the circle be r, let $\overline{PO}$ meet the circle at X, and let ray $\overrightarrow{PO}$ hit the circle past O at Y. Since $PO = 2r$ and $OB = r$, the Pythagorean Theorem applied to $\triangle POB$ gives us $PB = r\sqrt{5}$. The power of point P gives us $(PX)(PY) = (PA)(PB)$. Since $OX = OY = r$, we have $PX = r$ and $PY = 3r$. Therefore, $(r)(3r) = (PA)(r\sqrt{5})$, so $PA = 3r/\sqrt{5} = 3r\sqrt{5}/5$. Finally, $AB = PB - PA = 2r\sqrt{5}/5$, so $PA/AB = \boxed{3/2}$.

(Note, we could also have solved this problem by extending $\overrightarrow{BO}$ to hit the circle again at T, then noted that $\triangle POB$ and $\triangle TAB$ are similar right triangles.)

13.25   Power of a point requires $(PX)(QX) = (RX)(SX)$. Therefore, we have $(PX)(PQ - PX) = (RX)(RS - RX) = (RX)(PQ - RX)$. Expanding and rearranging gives $RX^2 - PX^2 - PQ(RX - PX) = 0$, or $(RX -$

$PX)(RX + PX) - PQ(RX - PX) = 0$. Therefore, $(RX + PX - PQ)(RX - PX) = 0$. So, we must have either $RX = PX$ or $RX + PX = PQ$. Since $PX + QX = PQ$, the latter possibility is equivalent to $RX = QX$. Thus, RX must equal either PX or QX.)

13.26 Power of a Point correctly applied to point P gives $(PA)(PB) = (PC)(PD)$. Since $PA = PB + AB$ and $PC = PD + CD$, we have $(PB + AB)(PB) = (PD + CD)(PD)$, so $PB^2 + (AB)(PB) = PD^2 + (CD)(PD)$. Since Jake still gets the right answer using $(AB)(PB) = (CD)(PD)$, this equation must be true for this problem. Therefore, our correct Power of a Point relationship tells us that $PB^2 = PD^2$, so $PB = PD$. Since $PB = PD$, our Power of a Point equation $(PA)(PB) = (PC)(PD)$ tells us that $PA = PC$, so $AB = PA - PB = PC - PD = CD$.

13.27 The power of point C is $(CB)(CA) = 28(52)$. We found in Problem 13.22 that since C is outside the circle, the power of point C equals $CO^2 - r^2$. Therefore, we have $CO^2 = 28(52) + 15^2 = 1681 = 41^2$, so $CO = \boxed{41}$. (We can also solve this problem by letting M be the midpoint of $\overline{AB}$ and considering right triangles $\triangle OAM$ and $\triangle OCM$.)

13.28

(a) From Problem 13.22, if point P is outside both circles, the power of P with respect to C_1 is $OP^2 - r_1^2$ and with respect to C_2 is $OP^2 - r_2^2$. Since $r_1 \neq r_2$, these powers cannot be the same.

(b) Similar to the first part, if P is inside both circles, we must have $r_1^2 - OP^2 = r_2^2 - OP^2$, which is impossible because $r_1 \neq r_2$.

(c) If P is inside C_1 and outside C_2, our powers are $r_1^2 - OP^2$ and $OP^2 - r_2^2$. We find that these are equal when $OP = \sqrt{(r_1^2 + r_2^2)/2}$. Therefore, the points that are on the circle with center O and radius $OP = \sqrt{(r_1^2 + r_2^2)/2}$ have the same power with respect to both circles.

13.29 Let the point where $\overrightarrow{XP}$ hits the circumcircle of $\triangle XYZ$ be Q. We will show that Q is W. The power of point P with respect to our circle gives us $(QP)(PX) = (YP)(PZ)$, so $QP = (YP)(PZ)/PX$. Since we also know that W is on $\overrightarrow{XP}$ past point P such that $WP = (YP)(PZ)/PX$ (from the given equation), we know that W and Q are the same point, because there is only one point past P on $\overrightarrow{XP}$ that is a distance of $(YP)(PZ)/PX$ from point P. Therefore, the circumcircle of $\triangle XYZ$ goes through W.

13.30 Let T be the point of intersection of the three chords. The power of point T with respect to C_1 gives us $(PT)(TQ) = (AT)(TB)$, and the power of T with respect to C_2 gives us $(RT)(TS) = (AT)(TB)$. Therefore, $(PT)(TQ) = (RT)(TS)$, so by the previous problem, the four points P, Q, R, and S lie on a circle.

13.31 Construct a segment $\overline{PS}$ with length a, then extend the segment past S to T such that $ST = b$. We therefore have a segment of length $a + b$. Find the midpoint M of this segment and construct the circle with center M and radius MP. Then, we construct a line perpendicular to $\overline{PT}$ through S. Let the points where this line hits our circle be X and Y. Since $\overline{XY}$ is a chord perpendicular to diameter $\overline{PT}$ of the circle, the diameter must bisect the chord. Therefore, $XS = SY$. The power of point S gives us $(XS)(YS) = (PS)(TS)$, so we have $XS^2 = ab$, or $XS = \sqrt{ab}$.

13.32 Since $\angle APB = \widehat{AB}/2 = \angle ACB$ and $\angle PAC = \widehat{PC}/2 = \angle CBP$, we have $\triangle PBQ \sim \triangle CAQ$ by AA Similarity. Therefore, $CQ/PQ = AC/PB$. Similarly, we have $\triangle PCQ \sim \triangle BAQ$, from which we have $BQ/PQ = AB/PC$. Adding these and noting that $CQ + BQ = BC$, we have $AC/PB + AB/PC = (CQ + BQ)/PQ = BC/PQ$. Since $\triangle ABC$ is equilateral, we have $AB = BC = AC$, so we can divide our equation by AC to get the desired $1/PQ = 1/PB + 1/PC$.

13.33

(a) Since $\angle ABE = \overset{\frown}{AC}/2 = \angle CDE$ and $\angle BEA = \angle CED$, we have $\triangle AEB \sim \triangle CED$ by AA. The ratio of the areas of these triangles is therefore the square of the ratio of corresponding sides, or $[ABE]/[CDE] = (AB/CD)^2$.

(b) We have $\angle ABC = \angle ADC$ and $\angle BPC = \angle DPA$, so $\triangle BPC \sim \triangle DPA$ by AA. The ratio of the areas of these triangles equals the square of the ratio of corresponding sides, or $[PBC]/[PAD] = (PC/PA)^2$.

(c) The power of point P gives $(PC)(PD) = (PA)(PB)$, so $PD = (PA)(PB)/PC = 5$, so $CD = PD - PC = 1$. Since $\triangle PAE$ and $\triangle AEB$ share an altitude from E, we have $[PAE]/[AEB] = PA/AB = 1/4$. Similarly, $[PEC]/[CDE] = PC/CD = 4$. Dividing $[PEC]/[CDE] = 4$ by the relationship we found in part (a), $[ABE]/[CDE] = (AB/CD)^2 = 64$, we have $[PEC]/[ABE] = 1/16$. Therefore,

$$\frac{[PAEC]}{[BAE]} = \frac{[PAE] + [PEC]}{[BAE]} = \frac{[PAE]}{[BAE]} + \frac{[PEC]}{[BAE]} = \frac{1}{4} + \frac{1}{16} = \boxed{\frac{5}{16}}.$$

13.34 Since $BD/BA = BE/BC$ and $\angle DBE = \angle ABC$, we have $\triangle ABC \sim \triangle DBE$ by AA Similarity. Therefore, $\triangle BDE$ is equilateral, so we have $DE = EB = DB = AD$. Extend $\overline{DE}$ past D and E to meet the circle at points G and F as shown. The power of point D gives us $(DG)(DF) = (AD)(DB)$. By symmetry, we have $DG = EF$, and we found earlier that $AD = DB = DE$. Therefore, our power of a point relationship becomes $(EF)(DF) = DE^2$, or $DE/EF = DF/DE$. Since $DF = DE + EF$, we have $DE/EF = DF/DE = (DE + EF)/DE = 1 + EF/DE$. Letting $x = DE/EF$, we have $x = 1 + 1/x$, so $x^2 - x - 1 = 0$, from which we find $x = (1 \pm \sqrt{5})/2$. Since our ratio must be positive, we have $x = \boxed{(1 + \sqrt{5})/2}$. (Notice that our answer is the golden ratio!)

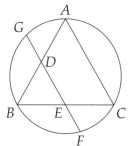

13.35 The power of point F gives $(FR)(FG) = IF^2$, so $FG = IF^2/FR = 49$. Therefore, $RG = FG - FR = 31$. The power of point O then gives us $(IO)(ON) = (OR)(OG)$. Since $OR + OG = GR = 31$, we have $OG = 31 - OR$, so we have $(OR)(31 - OR) = (IO)(ON) = 240$. Therefore, we have $OR^2 - 31OR + 240 = 0$, so $(OR - 15)(OR - 16) = 0$. Since $OR > GO$ and $OR = 15$ gives $GO = 16$, we discard this solution to our equation. Therefore, our answer is $OR = \boxed{16}$.

13.36

(a) From the power of point P we have $BP = (CP)(DP)/(AP) = 4$. We then connect O to the midpoints of $\overline{AB}$ and $\overline{CD}$. We call these points M and N as shown. Since $\overline{OM}$ is part of a radius that bisects chord $\overline{AB}$, we must have $\overline{OM} \perp \overline{AB}$. Similarly, $\overline{ON} \perp \overline{CD}$. Therefore, $MPNO$ is a rectangle. Since $AM = AB/2 = 7/2$, we have $PM = MA - AP = 1/2$. Similarly, we find $NP = CN - CP = CD/2 - CP = 2$. Since $\overline{OP}$ is a diagonal of rectangle $MPNO$, we have $OP = MN = \sqrt{MP^2 + NP^2} = \sqrt{17}/2$.

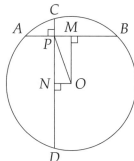

(b) We extend $\overline{OP}$ to meet the circle at X and Y, with X closer to P. The power of point P gives us $(XP)(PY) = (AP)(PB)$. Letting our radius be r, this means $(r - \sqrt{17}/2)(r + \sqrt{17}/2) = 12$, so $r^2 - 17/4 = 12$. Therefore, $r = \boxed{\sqrt{65}/2}$.

CHAPTER 14

Three-Dimensional Geometry

Exercises for Section 14.1

14.1.1 By definition, if line m is perpendicular to plane $\mathcal{P}$ at X, it is perpendicular to every line in $\mathcal{P}$ through X. Therefore, m and k must be perpendicular.

14.1.2 Since $ABCD$ is a square, $DA = AB = 5$. Because $\overline{DA}$ is perpendicular to plane ABP and $\overleftrightarrow{AP}$ is a line in plane ABP that passes through A, $\angle DAP$ is a right angle. Therefore, by the Pythagorean Theorem, $PD^2 = DA^2 + PA^2 = 5^2 + 3^2 = 34$, so $PD = \boxed{\sqrt{34}}$.

14.1.3 Let Q be any point on line m, and R be the foot of the perpendicular from P to m. Then triangle PQR is a right triangle with hypotenuse $\overline{PQ}$ and right angle at R. Therefore, PQ is always greater than or equal to PR. The answer is $\boxed{\text{yes}}$, the foot of the perpendicular from P to m is the closest point on m to P.

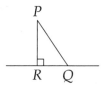

14.1.4 Let Q be any point on plane $\mathcal{P}$, and R be the foot of the perpendicular from X to $\mathcal{P}$. Since $\mathcal{P} \perp m$ at point R, we must have $\overline{RQ} \perp m$. Therefore, triangle XQR is a right triangle with hypotenuse $\overline{XQ}$ and right angle at R, so $XR < XQ$ when Q and R are different points. Since all points in $\mathcal{P}$ besides R must be more than XR from X, the foot of the perpendicular from X to $\mathcal{P}$ is the point in $\mathcal{P}$ closest to X.

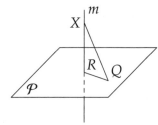

14.1.5 $\boxed{\text{Yes}}$. We lack the tools to prove this now, but here's an intuitive explanation. Because $\mathcal{M}$ and $\mathcal{P}$ are perpendicular and both pass through O, there is a line in $\mathcal{M}$ through O that is perpendicular to $\mathcal{P}$. (This is the statement we can't yet prove. It's intuitively obvious, but proving it rigorously is beyond the scope of this book.) Similarly, since $\mathcal{N}$ and $\mathcal{P}$ are perpendicular and both pass through O, there is a line in $\mathcal{N}$ through O that is perpendicular to $\mathcal{P}$. Since there is only one line through O perpendicular to $\mathcal{P}$, the aforementioned lines in $\mathcal{M}$ and $\mathcal{N}$ must be the same line. This line is in both $\mathcal{M}$ and $\mathcal{N}$, so it must be line m. Therefore, m is perpendicular to $\mathcal{P}$.

Exercises for Section 14.2

14.2.1 In a prism, any cross-section parallel to the bases is congruent to the bases.

14.2.2

(a) The volume is $2 \cdot 5 \cdot 3\sqrt{2} = \boxed{30\sqrt{2}}$.

(b) The total surface area is $2(2 \cdot 5 + 5 \cdot 3\sqrt{2} + 3\sqrt{2} \cdot 2) = \boxed{20 + 42\sqrt{2}}$.

(c) Each space diagonal is has length $\sqrt{2^2 + 5^2 + (3\sqrt{2})^2} = \boxed{\sqrt{47}}$.

(d) Since $2 < 3\sqrt{2} < 5$, the longest face diagonals are those in the faces with dimensions $3\sqrt{2}$ and 5. Therefore, the answer is $\sqrt{(3\sqrt{2})^2 + 5^2} = \boxed{\sqrt{43}}$.

14.2.3 Let the third side be a. Then we have $a^2 + 3^2 + 8^2 = 10^2$, so $a = 3\sqrt{3}$. Therefore, the volume is $3\sqrt{3} \cdot 3 \cdot 8 = \boxed{72\sqrt{3}}$.

14.2.4 Recall that in a cube of side length s, the space diagonal is $s\sqrt{3}$. Solving $s\sqrt{3} = 6$, we get $s = 2\sqrt{3}$. Therefore, the volume is $s^3 = (2\sqrt{3})^3 = \boxed{24\sqrt{3}}$.

14.2.5 A space diagonal always has a pair of two opposite vertices as its endpoints. Since a cube has eight vertices, there are four pairs of opposite vertices, making a total of $\boxed{\text{four}}$ space diagonals.

14.2.6 Let the original dimensions of the prism be l, w, and h, so the new dimensions are $2l$, $2w$, and $2h$.

The original surface area is $2(lw + lh + wh)$, and the new surface area is $2[(2l)(2w) + (2l)(2h) + (2w)(2h)] = 8(lw + lh + wh) = 4 \cdot 2(lw + lh + wh)$, so the surface area is multiplied by a factor of $\boxed{4}$.

The original volume is lwh, and the new volume is $2l \cdot 2w \cdot 2h = 8lwh$, so the volume is multiplied by a factor of $\boxed{8}$.

More generally, if each dimension is multiplied by a factor of k, then the surface area is multiplied by a factor of k^2, and the volume increases by a factor of k^3. Can you prove this?

14.2.7

(a) Segment $\overline{BD}$ is a diagonal of rectangle $ABCD$, so $BD = \sqrt{AB^2 + BC^2} = \sqrt{4^2 + 3^2} = \boxed{5}$.

(b) Since $ABGH$ is a square, $AB = BG$. Because $\overleftrightarrow{AB}$ is perpendicular to face $BCGF$, we have $\overline{AB} \perp \overline{BG}$. Therefore, we can apply the Pythagorean Theorem to $\triangle ABG$ to find $AG = \sqrt{AB^2 + BG^2} = \boxed{4\sqrt{2}}$.

We also could have solved the problem by finding the dimensions of the prism. Since BCG is a right triangle with right angle at C, we have $CG^2 = BG^2 - BC^2 = AB^2 - BC^2 = 4^2 - 3^2 = 7$. Therefore, the height between bases $ABCD$ and $EFGH$ of the rectangular prism is $\sqrt{7}$. Now that we have all three dimensions of the rectangular prism, we find that the length of space diagonal $\overline{AG}$ is $\sqrt{4^2 + 3^2 + (\sqrt{7})^2} = \boxed{4\sqrt{2}}$.

(c) Note that $\overline{FD}$ is just another space diagonal, so $FD = AG = \boxed{4\sqrt{2}}$.

(d) The volume is $(AB)(BC)(CG) = (4)(3)(\sqrt{7}) = \boxed{12\sqrt{7}}$.

14.2.8 The total surface area of the two boxes is $2(4 \cdot 6 + 6 \cdot 8 + 8 \cdot 4) + 2(2 \cdot 3 + 3 \cdot 5 + 5 \cdot 2) = 270$. Now we subtract twice the area of overlap to get the final surface area. This is because the area of overlap must be subtracted once from the big box and once from the small box. The largest area of overlap cannot exceed

$3 \cdot 5 = 15$, which is the largest area among all six faces of the smaller box. Bridget also will not paint the face that rests on the ground, so she places the first box with one of its two 6×8 faces on the ground. This reduces the amount she must paint by $(6)(8) = 48$. Therefore, she is left with $270 - 2 \cdot 15 - 48 = \boxed{192}$ to paint.

Exercises for Section 14.3

14.3.1

(a) Let F be the foot of the altitude from E to base $ABCD$. Because $ABCD$ is a square of side length 4, $AF = 2\sqrt{2}$. Since AFE is a right triangle with right angle at F, the length of altitude $\overline{EF}$ is equal to $\sqrt{AE^2 - AF^2} = \sqrt{8^2 - (2\sqrt{2})^2} = \boxed{2\sqrt{14}}$.

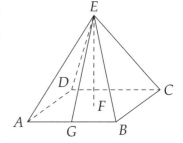

(b) Let G be the foot of the altitude from E to $\overline{AB}$ in isosceles triangle AEB. We know $AG = 2$, so the slant height EG is equal to $\sqrt{EA^2 - AG^2} = \sqrt{8^2 - 2^2} = \boxed{2\sqrt{15}}$. We also could have used right triangle $\triangle EFG$ to find EG.

(c) The volume of the pyramid is

$$\frac{1}{3} \cdot [ABCD] \cdot EF = \frac{1}{3} \cdot 4^2 \cdot 2\sqrt{14} = \boxed{\frac{32\sqrt{14}}{3}}.$$

(d) Because the pyramid is regular, the total surface area of the pyramid is

$$[ABCD] + 4[ABE] = 4^2 + 4\left(\frac{1}{2} \cdot 4 \cdot 2\sqrt{15}\right) = \boxed{16 + 16\sqrt{15}}.$$

14.3.2 If $BC = 8$, then the length of diagonal $\overline{CE}$ would be $8\sqrt{2}$, which would make $OC = 4\sqrt{2}$. Since triangle AOC has a right angle at O, the length of $\overline{AC}$ must be at least the length of OC. However, $AC = 5$, which is less than $4\sqrt{2}$, so the pyramid could not have $BC = 8$ and $AC = 5$.

14.3.3 The slant height is $\sqrt{3^2 + 4^2} = 5$, so the surface area is $6^2 + 4(6 \cdot 5/2) = 96$ in^2. The volume is $6^2 \cdot 4/3 = 48$. The ratio of the number of square inches in the surface area to the number of cubic inches in the volume is therefore $48/96 = \boxed{1/2}$.

14.3.4

(a) One of the faces of the shaded piece is a right triangle with legs 3 and 4. Looking at this triangle as the base, the height is then 2. The volume is then $\frac{1}{3}(\frac{1}{2} \cdot 3 \cdot 4)(2) = \boxed{4}$.

(b) The sum of the six edges is $2 + 3 + 4 + \sqrt{2^2 + 3^2} + \sqrt{3^2 + 4^2} + \sqrt{4^2 + 2^2} = \boxed{14 + \sqrt{13} + 2\sqrt{5}}$.

14.3.5 First, in each case, the solid is a tetrahedron with base STU and height 4. The area of triangle STU is $(ST)(TU)/2 = 8$, so the volume of each solid is $[STU](4)/3 = \boxed{32/3}$. Now we will find the surface areas.

(a) Triangles STU and STW are right triangles with legs 4 and 4, so their areas are $1/2 \cdot 4 \cdot 4 = 8$. Triangle WSU is a right triangle with legs $\overline{WS}$ and $\overline{SU}$. We know that $WS = 4$ and $\overline{SU}$ is a diagonal of square $STUV$, so $SU = 4\sqrt{2}$. Therefore, the area of triangle WSU is $4(4\sqrt{2})/2 = 8\sqrt{2}$. Triangle UTW is congruent to triangle WSU, so it also has area $8\sqrt{2}$. Therefore, the total surface area is
$$8 + 8 + 8\sqrt{2} + 8\sqrt{2} = \boxed{16 + 16\sqrt{2}}.$$

(b) By the same calculations as in part (a), triangles STU, STX, and TUX are right triangles with legs 4 and 4, so their areas are all 8. Triangle SUX has sides $\overline{SU}$, $\overline{SX}$, and $\overline{UX}$. Note that these are the diagonals of squares $STUV$, $STXW$, and $TUYX$, which are congruent, so they all have the same length, namely $4\sqrt{2}$. Therefore, triangle SUX is equilateral with side length $4\sqrt{2}$, so its area is $(4\sqrt{2})^2\sqrt{3}/4 = 8\sqrt{3}$. The total surface area is then $8 + 8 + 8 + 8\sqrt{3} = \boxed{24 + 8\sqrt{3}}$.

(c) By the same calculations as in parts (a) and (b), triangle STU has area 8, triangles STZ and TUZ have area $8\sqrt{2}$, and triangle SUZ has area $8\sqrt{3}$. Therefore, the total surface area is $8 + 8\sqrt{2} + 8\sqrt{2} + 8\sqrt{3} = \boxed{8 + 16\sqrt{2} + 8\sqrt{3}}$.

Exercises for Section 14.4

14.4.1 The quantity ends up being $\boxed{2}$ in all cases. This is not a coincidence. For a very tough challenge, see if you can prove that this quantity equals 2 for all polyhedra (such as hexagonal prisms or pentagonal pyramids, etc.)

14.4.2 Each vertex of the original octahedron is turned into a square face, which has 4 vertices. Therefore, the new number of vertices is $4 \cdot 6 = 24$. The number of faces is increased by 6, since each vertex is turned into a face. Portions of the original faces are still there, so there are now $8 + 6 = \boxed{14}$ faces. The number of edges is increased by $4 \cdot 6$ since the 6 new squares contribute 4 edges each, and a portion of each original edge still exists. The number of edges is then $12 + 24 = \boxed{36}$.

14.4.3

(a) Let M and N be the midpoints of $\overline{BC}$ and $\overline{CD}$, respectively. Then $\overline{AM}$ is a median of triangle ABC. Note that O is the centroid of triangle ABC, and we know that the centroid of a triangle divides the median into the ratio $AO/OM = 2/1$, so $AO/AM = 2/3$. Similarly, $AP/AN = 2/3$. Therefore, triangle AOP is similar to triangle AMN with ratio $2/3$ (by SAS Similarity). We conclude that $OP = 2MN/3$.

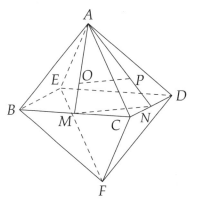

Since M is the midpoint of $\overline{BC}$ and N is the midpoint of $\overline{CD}$ in $\overline{BCD}$, we have $\triangle CMN \sim \triangle CBD$, so $MN = BD/2$. $\overline{BD}$ is a diagonal of square $BCDE$, so $BD = \sqrt{2}$, which gives us $MN = BD/2 = \sqrt{2}/2$. Therefore, $OP = 2MN/3 = \boxed{\sqrt{2}/3}$.

(b) *Solution 1:* The centers of the eight faces of a regular octahedron are the vertices of a cube. $\overline{OP}$ is an edge of this cube, and $\overline{OQ}$ is a space diagonal of it. Therefore, $OQ = OP\sqrt{3} = \boxed{\sqrt{6}/3}$.

Solution 2: Let X be the center of the octahedron. Consider tetrahedron $XABC$. Since X is the center of square $BCDE$, we have $BX = CX = \sqrt{2}/2$, since $\overline{BX}$ and $\overline{CX}$ are both half of a diagonal of $BCDE$. Similarly, $AX = \sqrt{2}/2$. Since $\overline{AX}$ is perpendicular to the plane of $BCDE$, the volume of $ABCX$ is $[BXC](AX)/3 = (BX \cdot CX/2)(AX)/3 = \sqrt{2}/24$.

Let Z be the foot of the altitude from X to face ABC. Since X is the center of the octahedron, we have $XA = XB = XC$. Therefore, $\triangle XZA \cong \triangle XZB \cong \triangle XZC$ by HL Congruence, so $ZA = ZB = ZC$. Hence, Z is point O, the center of $\triangle ABC$. Since $\triangle ABC$ is an equilateral triangle with side length 1, we have $[ABC] = 1^2\sqrt{3}/4 = \sqrt{3}/4$. Therefore, the volume of $XABC$ equals $[ABC](XO)/3 = (XO)(\sqrt{3}/12)$. Since we already know that the volume of $ABCX$ is $\sqrt{2}/24$, we have $(XO)(\sqrt{3}/12) = \sqrt{2}/24$, so $XO = \sqrt{6}/6$.

Since faces ABC and DEF are parallel (why?), the line through O perpendicular to ABC must also be perpendicular to DEF. Therefore, $\overrightarrow{OX}$ hits $\triangle DEF$ at the foot of the altitude from X to plane DEF. Just as the foot of the altitude from X to ABC is the circumcenter of $\triangle ABC$, so is the foot of the altitude from X to DEF the circumcenter of $\triangle DEF$. Thus, $\overrightarrow{OX}$ hits plane DEF at Q, and $XQ = XO = \sqrt{6}/6$. Finally, we have $OQ = 2OX = \boxed{\sqrt{6}/3}$.

(c) As in the first solution in the previous part, we note that the centers of the faces of the octahedron are vertices of a cube. $\overline{OP}$ is an edge of this cube, and $\overline{PQ}$ is a face diagonal of the cube. From our cube, we see that $\overline{OP} \perp \overline{PQ}$, so $[OPQ] = (OP)(PQ)/2 = (OP)(OP\sqrt{2})/2 = \boxed{\sqrt{2}/9}$. See if you can figure out how to prove $\overline{OP} \perp \overline{PQ}$ without noting that the centers of the faces of the octahedron are vertices of a cube!

Review Problems

14.13

(a) Each face of a cube is a square. Since the face diagonal is 4, the edge length is $4/\sqrt{2} = \boxed{2\sqrt{2}}$.

(b) In a cube, a space diagonal is $\sqrt{3}$ times as long as an edge. Since the edge is $2\sqrt{2}$, the space diagonal is $2\sqrt{2} \cdot \sqrt{3} = \boxed{2\sqrt{6}}$.

(c) Since the edge length is $2\sqrt{2}$, each of the six square faces has area $(2\sqrt{2})^2 = 8$. The total surface area is then $8 \cdot 6 = \boxed{48}$.

(d) The volume is $(2\sqrt{2})^3 = \boxed{16\sqrt{2}}$.

14.14 Since $\overline{TS}$ is perpendicular to plane $TUYX$, we know that $\overline{YT} \perp \overline{TS}$. Therefore, $\angle YTS$ is a right angle. Similarly, all other angles of $ZYTS$ are right. Thus, $ZYTS$ is a $\boxed{\text{rectangle}}$.

14.15

(a) Let the third side be a. Then $3^2 + 7^2 + a^2 = (3\sqrt{13})^2$, so $a = \boxed{\sqrt{59}}$.

(b) The volume is $3 \cdot 7 \cdot \sqrt{59} = \boxed{21\sqrt{59}}$.

(c) The total surface area is $2(3 \cdot 7 + 7 \cdot \sqrt{59} + \sqrt{59} \cdot 3) = \boxed{42 + 20\sqrt{59}}$.

14.16   Let cube $\mathcal{A}$ have side length s. Its space diagonals therefore have length $s\sqrt{3}$, so the edges of $\mathcal{B}$ have length $s\sqrt{3}$.

(a) The surface area of $\mathcal{B}$ is $6(s\sqrt{3})^2 = 18s^2$ and the surface area of $\mathcal{A}$ is $6s^2$. Therefore, the desired ratio is $\boxed{1/3}$. We also could note that the two cubes are similar. Since their corresponding sides have ratio $1/\sqrt{3}$, their corresponding surface areas have ratio $(1/\sqrt{3})^2 = 1/3$.

(b) The volume of $\mathcal{A}$ is s^3 and that of $\mathcal{B}$ is $(s\sqrt{3})^3 = 3s^3\sqrt{3}$. Therefore, the desired ratio is $1/(3\sqrt{3}) = \boxed{\sqrt{3}/9}$. We also could have noted that the cubes are similar, so the ratio of their volumes is $(1/\sqrt{3})^3 = 1/(3\sqrt{3}) = \sqrt{3}/9$.

14.17

(a) Let O be the center of base $WXYZ$. Since O is the midpoint of diagonal $\overline{YW}$ of the square, we have $YO = YW/2 = XY\sqrt{2}/2 = 5\sqrt{2}$. Since $\angle YOV = 90°$, $YO = 5\sqrt{2}$, and $YV = 13\sqrt{2}$, we know that $\triangle YOV$ is a 5-12-13 right triangle. Therefore, $OV = \boxed{12\sqrt{2}}$.

(b) Let M be the midpoint of $\overline{YZ}$. We then have $MO = 5$, so we have $MV = \sqrt{MO^2 + OV^2} = \boxed{\sqrt{313}}$.

(c) The total surface area of the pyramid is the sum of the areas of the four triangular faces and the area of the square base. Each of the triangular faces has base length 10 and height $\sqrt{313}$, so our total surface area is $4(10)(\sqrt{313})/2 + 10^2 = \boxed{100 + 20\sqrt{313}}$.

(d) The base has area 100, and the height of the pyramid is $12\sqrt{2}$, so our volume is $(100)(12\sqrt{2})/3 = \boxed{400\sqrt{2}}$.

14.18

(a) Note that triangle MBN is a right triangle with legs $BM = BA/2 = 4$ and $BN = BC/2 = 5$, so $MN = \sqrt{4^2 + 5^2} = \boxed{\sqrt{41}}$.

(b) Segment $\overline{MO}$ connects the midpoints of opposite sides of rectangle $ABCD$, so $MO = BC = \boxed{10}$.

(c) The solid can be viewed as a right prism with base $MNOP$ and height BF. Since $MNOP$ is formed by connecting the midpoints of rectangle $ABCD$, $MNOP$ has an area half that of $ABCD$. (Make sure you see why!) Therefore, the volume of $MNOPQRST$ is

$$[MNOP] \cdot BF = \frac{1}{2}[ABCD] \cdot BF = \frac{1}{2}AB \cdot BC \cdot BF = \boxed{200}.$$

14.19   In each case, we count the number of faces of the original figure. This gives us the number of vertices of the new figure, since there's one vertex of the new figure on each face of the old figure. A count of the vertices of the old figure gives us the number of faces of the new figure (make sure you see why). We've used symmetry a great deal in our solutions below – see if you can write proofs without using symmetry! (Specifically, prove that the faces we claim are regular polygons are in fact regular polygons.)

(a) There are 6 faces on a cube, so our new figure has 6 vertices. By symmetry, we can see that each face of our new figure is an equilateral triangle, so the new figure is a $\boxed{\text{regular octahedron}}$.

(b) There are 8 faces on an octahedron, so our new figure has 8 vertices. By symmetry, we can see that each of the faces of our new figure is a square, and we see that the new figure is a $\boxed{\text{cube}}$.

(c) There are 4 faces on an tetrahedron, so our new figure has 4 vertices. By symmetry, we can see that each of the faces of our new figure is an equilateral triangle, and we see that the new figure is a $\boxed{\text{regular tetrahedron}}$.

(d) There are 12 faces on a dodecahedron, so our new figure has 12 vertices. By symmetry, we can see that each of the faces of our new figure is an equilateral triangle, and we see that the new figure is a $\boxed{\text{regular icosahedron}}$.

(e) There are 20 faces on an icosahedron, so our new figure has 20 vertices. By symmetry, we can see that each of the faces of our new figure is a pentagon, and we see that the new figure is a $\boxed{\text{regular dodecahedron}}$.

14.20

(a) Note that $\overline{AC}$, $\overline{CF}$, and $\overline{FA}$ are face diagonals of the cube, so $AC = CF = FA = 8\sqrt{2}$. Hence, triangle ACF is an equilateral triangle, and its area is $(8\sqrt{2})^2\sqrt{3}/4 = \boxed{32\sqrt{3}}$.

(b) Triangle ACG has a right angle at C, so its area is $(AC)(CG)/2 = (8\sqrt{2})(8)/2 = \boxed{32\sqrt{2}}$.

14.21 There are six edges on each of the two bases and 6 edges connecting corresponding points on the bases. Therefore, there are $6 + 2(6) = \boxed{18}$ edges total.

14.22 Faces 1, 2, 3 and 4 share a vertex. Call these four faces together the 'top' of the octahedron, so the other four faces share the 'bottom' vertex. We can see that faces 2 and 4 share an edge with face 1 and face 3 does not, so we only have to find which of the bottom four faces shares an edge with face 1. Each of the bottom faces shares one edge with one of the top faces. Faces 6 and 3 are adjacent in our given diagram, as are faces 5 and 4. Just as face 4 folds to be adjacent to face 1, face 7 folds to be adjacent to face 2. Therefore, face 8 is left as the 'bottom' face that shares an edge with face 1. The the sum of the numbers on the faces adjacent to face 1 is $2 + 4 + 8 = \boxed{14}$

14.23

(a) Let H be the foot of the perpendicular from G to base $ABCDEF$. Because the pyramid is right, H is the center of the hexagon. Because $ABCDEF$ is a regular hexagon of side length 6, $AH = 6$. Since $\triangle GHA$ is a right triangle with hypotenuse $AG = 6\sqrt{3}$ and leg $AH = 6$, its other leg GH is equal to $\sqrt{AG^2 - AH^2} = \sqrt{(6\sqrt{3})^2 - 6^2} = \sqrt{108 - 36} = \sqrt{72} = 6\sqrt{2}$.

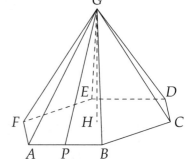

Now we need the base area. The regular hexagonal base can be divided into six equilateral triangles of side length 6, so the base area is $6(6^2\sqrt{3}/4) = 54\sqrt{3}$. Then the volume of the pyramid is

$$\frac{1}{3} \cdot [ABCDEF](GH) = \frac{1}{3} \cdot 54\sqrt{3} \cdot 6\sqrt{2} = \boxed{108\sqrt{6}}.$$

(b) Let P be the foot of the perpendicular from G to AB. Then P is also the midpoint of AB, so $AP = AB/2 = 3$. Furthermore, triangle APG has a right angle at P, so $GP^2 = AG^2 - AP^2 = (6\sqrt{3})^2 - 3^2 = 99$, so $GP = \sqrt{99} = 3\sqrt{11}$. The area of triangle ABG is then $(AB)(PG)/2 = 9\sqrt{11}$. Therefore, the total surface area is $54\sqrt{3} + 6 \cdot 9\sqrt{11} = \boxed{54\sqrt{3} + 54\sqrt{11}}$.

14.24 A regular tetrahedron has four equilateral triangles as its faces. Therefore, the total surface area of our regular tetrahedron is $4(9^2\sqrt{3}/4) = \boxed{81\sqrt{3}}$.

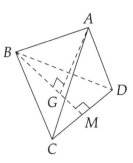

For the volume, we follow the steps we took in finding the volume of a tetrahedron with side length 6 in the text. Let $ABCD$ be our tetrahedron, M be the midpoint of $\overline{CD}$, and G be the foot of the altitude from A to face BCD. As described in the text, G is the centroid of $\triangle BCD$. Since $\triangle BCM$ is a 30-60-90 triangle, we have $BM = CM\sqrt{3} = CD\sqrt{3}/2 = 9\sqrt{3}/2$. Since G is the centroid of $\triangle BCD$, we have $BG = 2BM/3 = 3\sqrt{3}$. From right triangle $\triangle ABG$, we have $AG = \sqrt{AB^2 - BG^2} = \sqrt{81 - 27} = 3\sqrt{6}$. Therefore, the volume of $ABCD$ is $([BCD])(AG)/3 = \boxed{243\sqrt{2}/4}$.

14.25 *Solution 1:* Each face of the original cube has area 16 in^2; however, when we paint, 1 in^2 of each face is covered, so we use 15 in^2 on each face, for a total of $6(15) = 90$ in^2. We miss one of the six faces of each of the little cubes when we paint, so the six little cubes consume a total of $6(5) = 30$ in^2 of paint. Therefore, we use a total of $90 + 30 = \boxed{120 \text{ in}^2}$.

Solution 2: If we painted them before gluing, we would use $6(4^2) = 96$ in^2 for the large cube and $6(6) = 36$ in^2 total for the little cubes. However, when we glue faces, we cover up 2 in^2 that we don't have to paint for each little cube we glue on (1 square inch for the little cube, 1 for the big cube). Therefore, the total area we paint is $96 + 36 - 6(2) = \boxed{120 \text{ in}^2}$.

14.26

(a) Let G' be the foot of the perpendicular from W to base XYZ. Triangles $WG'X$, $WG'Y$, and $WG'Z$ are right triangles, so $G'X = \sqrt{WX^2 - G'W^2}$, $G'Y = \sqrt{WY^2 - G'W^2}$, and $G'Z = \sqrt{WZ^2 - G'W^2}$. We are given $WX = WY = WZ$, so $G'X = G'Y = G'Z$. Therefore, G' is the center of equilateral triangle XYZ. We are also given that G is the center of $\triangle XYZ$, so $G' = G$. In other words, G is the foot of the altitude from W to base XYZ.

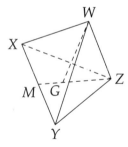

(b) Since $WX = WY$, median $\overline{WM}$ of isosceles $\triangle WXY$ is also an altitude. Therefore, $WM^2 = WX^2 - XM^2 = 18^2 - (9/2)^2 = 1215/4$, so $WM = \sqrt{1215/4} = \boxed{9\sqrt{15}/2}$.

(c) $\overline{ZM}$ is the median to base $\overline{XY}$ in equilateral triangle XYZ, so it is also an altitude. From 30-60-90 triangle XZM, we have $ZM = (\sqrt{3}/2)(XZ) = \boxed{9\sqrt{3}/2}$.

(d) Since G is the centroid of XYZ, G divides median $\overline{ZM}$ such that $ZG/GM = 2/1$. Hence, $GZ = 2ZM/3 = 3\sqrt{3}$. Triangle WGZ is a right triangle with hypotenuse $WZ = 18$ and one leg $GZ = 3\sqrt{3}$. Thus, $WG = \sqrt{WZ^2 - GZ^2} = \sqrt{18^2 - (3\sqrt{3})^2} = \sqrt{297} = 3\sqrt{33}$. The volume of $WXYZ$ is then

$$\frac{1}{3}[XYZ] \cdot WG = \frac{1}{3} \cdot \left(\frac{\sqrt{3}}{4} \cdot 9^2\right) \cdot (3\sqrt{33}) = \boxed{\frac{243\sqrt{11}}{4}}.$$

14.27 We can view such a solid as a triangular pyramid. The base is a right isosceles triangle with legs 6, and the height is also 6. Its volume is then $(6^2/2)(6)/3 = \boxed{36}$.

14.28

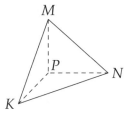

(a) From right triangle $\triangle KPN$, we have $PN = \sqrt{KN^2 - KP^2} = 6\sqrt{2}$. From right triangle $\triangle KPM$, we have $PM = \sqrt{KM^2 - KP^2} = 4\sqrt{2}$. Finally, from right triangle $\triangle PNM$, we have $MN = \sqrt{PN^2 + PM^2} = \boxed{2\sqrt{26}}$.

(b) Since $\triangle KPM$ is right, we have $[KPM] = (KP)(PM)/2 = 14\sqrt{2}$. Since $\overline{PN}$ is part of line n, it is perpendicular to both k and m. It is therefore perpendicular to plane KPM, so the height from N to face KPM of tetrahedron $KPMN$ has length $NP = 6\sqrt{2}$. Therefore, our volume is $([KPM])(NP)/3 = \boxed{56}$.

(c) Let the midpoint of $\overline{MN}$ be Q. Since $\triangle MPN$ is a right triangle, the length of the median to the hypotenuse equals half the length of the hypotenuse. Therefore, $PQ = MN/2 = \sqrt{26}$. Right triangle $\triangle KPQ$ gives $KQ = \sqrt{KP^2 + PQ^2} = \sqrt{49 + 26} = \boxed{5\sqrt{3}}$.

14.29 From the given side equalities, we have $\triangle MNO \cong \triangle OPM$ by SSS Congruence. Since $\angle MPO$ and $\angle MNO$ are corresponding angles of these triangles, we have $\angle MNO = \angle MPO$.

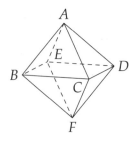

14.30 Let A and F be two opposite vertices of the octahedron, and let B, C, D, and E be the remaining vertices as shown. Note that square pyramid $ABCDE$ forms one half of the octahedron.

Let O be the center of the octahedron, which is also the center of square $BCDE$. Then the height AO of square pyramid $ABCDE$ is equal to half of AF. In turn, $\overline{AF}$ is a diagonal of square $ABFD$ that has side length s, so $AF = s\sqrt{2}$. Therefore, $AO = AF/2 = s\sqrt{2}/2$. The volume of square pyramid $ABCDE$ is then $[BCDE](AO)/3 = s^3\sqrt{2}/6$, so the volume of the octahedron is $2(s^3\sqrt{2}/6) = \boxed{s^3\sqrt{2}/3}$.

Challenge Problems

14.31

(a) Let P be the foot of the perpendicular from G to $\overline{EF}$. Then triangle GFP is a 30-60-90 triangle with hypotenuse $FG = 6$ and $\angle PFG = 60°$, so $PG = (\sqrt{3}/2)(FG) = 3\sqrt{3}$. Parallelogram $EFGH$ then has base $EF = 8$ and height $PG = 3\sqrt{3}$, so it has area $EF \cdot PG = 8 \cdot 3\sqrt{3} = \boxed{24\sqrt{3}}$.

(b) Because the solid is a right prism, the lateral faces are all rectangles. Therefore, the lateral surface area is equal to the perimeter of the base $EFGH$ times the height, which is $(8 + 6 + 8 + 6) \cdot 9 = 252$. The area of each base of the prism is $24\sqrt{3}$ (from part (a)), so the total surface area is $2 \cdot 24\sqrt{3} + 252 = \boxed{252 + 48\sqrt{3}}$.

(c) The volume is $[EFGH] \cdot 9 = \boxed{216\sqrt{3}}$.

14.32 Let the circumcenter of $\triangle WXY$ be O, so that $OW = OX = OY$. By LL Congruence, we have $\triangle VWO \cong \triangle VXO \cong \triangle VYO$, so $VW = VX = VY$.

14.33 We first find the total volume of the eight 'corners' that are part of the cube but not part of the desired figure. We can then subtract this result from the volume of the cube to get our answer. Let A be one of the vertices of the cube and let M, N, and O be midpoints of the three edges of the cube that have A as an endpoint. Then, we have $AM = AN = AO = 5$. Since $\overline{AM}$ is perpendicular to face ANO of tetrahedron $AMNO$, the volume of $AMNO$ is $[ANO](AM)/3 = ((AN)(AO)/2)(AM)/3 = 125/6$. Similarly, each of the eight corner tetrahedrons has volume $125/6$, so the desired volume is $10^3 - 8(125/6) = \boxed{2500/3}$.

14.34 Each of the 12 pentagons has 5 sides and each of the 20 hexagons has 6 sides, for a total of $(12)(5) + (20)(6) = 180$ edges. At each seam, two of these edges meet, so the total number of seams is $180/2 = \boxed{90}$.

14.35 Each of the squares has 4 edges and each of the triangles has 3. If we add up the total number of edges by adding the number of sides in each polygon, we count each edge of the folded-up solid twice, since two polygons meet at each edge. Therefore, there are $[5(4) + 10(3)]/2 = \boxed{25}$ edges on the folded up solid. The folded up solid is a regular pentagonal prism with a regular pentagonal pyramid glued onto each of the pyramids as shown at right. (To see this, start by ignoring the triangles. 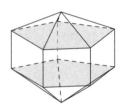 Each square face of the folded up solid will have two other squares adjacent to opposite edges. This gives us the sides of the pentagonal prism. There are 5 triangles that share a vertex on either side of the squares; these form the two pyramids.) The two pentagons contribute 5 vertices each, and the vertex points of the two pyramids contribute another two vertices, for a total of $2(5) + 2 = \boxed{12}$ vertices.

14.36 Let s be the edge length of the cube. Then the octahedron is a regular octahedron of edge length $s/\sqrt{2}$. (Make sure you see why!) We found in Problem 14.30 that the volume of a regular octahedron of edge length s is $s^3\sqrt{2}/3$, so the volume of our octahedron with edge length $s/\sqrt{2}$ is

$$\frac{\sqrt{2}}{3}\left(\frac{s}{\sqrt{2}}\right)^3 = \frac{s^3}{6}.$$

Since the volume of the cube is s^3, the desired ratio of volumes is $\boxed{6}$.

14.37 We unfold the tetrahedron as shown. The insect's starting point is M, the midpoint of $\overline{AB}$, and it walks to N, the midpoint of $\overline{CD}$. Our shortest path is the line segment $\overline{MN}$. Since $\triangle ABD$ and $\triangle ABC$ are equilateral, we have $\angle ADB = \angle DBC$ and $\angle ABD = \angle BDC$. Therefore, $\overline{AB} \parallel \overline{CD}$ and $\overline{AD} \parallel \overline{BC}$, so $ABCD$ is a parallelogram. Since $AM = AB/2 = CD/2 = ND$ and $\overline{AM} \parallel \overline{ND}$, $AMND$ is a parallelogram. Therefore, $MN = AD = \boxed{1}$.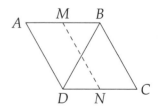

14.38

(a) Since $\overline{WU}$ is a space diagonal of the prism, we have $WU = \sqrt{3^2 + 4^2 + 12^2} = \boxed{13}$.

(b) Consider triangle WZU. We have $WZ = XY = 12$, and $ZU = \sqrt{UY^2 + ZY^2} = \sqrt{TX^2 + WX^2} = \sqrt{4^2 + 3^2} = 5$. Since $\overline{WZ}$ is perpendicular to plane $VUZY$, we hvae $\overline{WZ} \perp \overline{ZU}$. Therefore, WZU is a right triangle with legs 5 and 12. Therefore, $[WZU] = (5)(12)/2 = 30$. The distance h from Z to

$\overline{WU}$ is the altitude from Z to base $\overline{WU}$ in triangle WZU. Since the area of $\triangle WZU$ is 30, we have $13h/2 = 30$, so $h = \boxed{60/13}$.

(c) Solid $TWXY$ can be viewed as a triangular pyramid with base WXY and height TX. Its volume is therefore

$$\frac{1}{3} \cdot [WXY] \cdot TX = \frac{1}{3} \cdot \frac{3 \cdot 12}{2} \cdot 4 = \boxed{24}.$$

(d) *Solution 1:* Pyramid $VSUYW$ has base $SUYW$ and apex V. We can calculate the volume of this pyramid by taking the volume of the triangular prism $SUVWYZ$ and subtracting the volume of the triangular pyramid $WZVY$. The triangular prism $SUVWYZ$ is half of the rectangular prism $STUVWXYZ$, so its volume is $(3)(4)(12) = 72$. Triangular pyramid $WZVY$ has base WZY and height VZ, so its volume is $[WZY](VZ)/3 = [(3)(12)/2](4)/3 = 24$. Therefore, the volume of pyramid $VSUYW$ is $72 - 24 = \boxed{48}$.

Solution 2: The height from V to $SUYW$ equals the altitude from V to $\overline{SU}$. Letting this height be h, we have $[SUV] = (SV)(VU)/2 = (SU)(h)/2$, so $h = (SV)(VU)/SU$. Therefore, our volume is $[SUYW](h/3) = (SW)(SU)[(SV)(VU)/(SU)]/3 = (SW)(SV)(SU)/3 = \boxed{48}$.

(e) $WXZU$ is a triangular pyramid with base WXZ and apex U. The area of the base is $[WXZ] = 12 \cdot 3/2 = 18$. The height is the distance from U to face WXZ, which is equal to UY, and $UY = TX = 4$. Therefore, the volume is $[WXZ](UY)/3 = (18)(4)/3 = \boxed{24}$.

14.39 *Solution 1:* All three of these planes pass through the line $x = y = z$, so this question is equivalent to asking into how many regions is space divided by three planes that pass through the same line. Clearly the first two planes divide space into four regions. The next plane does not pass through two of these regions, and divides each of the other two regions in two. Therefore, there are $\boxed{6}$ regions.

Solution 2: Each order of x, y, z corresponds to a different region of the cheese after the cuts. (Make sure you see why! For example, a point with $x < y$ will be on the opposite side of the plane $x = y$ from a point with $y < x$.) Since there are $3 \cdot 2 \cdot 1 = 6$ ways to order x, y, and z (3 ways to choose the highest, then 2 ways to choose the next, and only 1 way to choose the last), there are $\boxed{6}$ regions.

14.40 Let P be the plane. There are several cases.

Plane P contains three or more edges of the cube. In this case, P must be one of the faces. There are six planes for this case.

Plane P contains exactly two edges of the cube. The only way for this to happen occurs when P contains two opposite edges of the cube. There are twelve edges in a cube, making six pairs of opposite edges. There are six planes for this case.

Plane P contains exactly one edge of the cube. There are no planes that pass through three or more vertices of a cube, but only pass through exactly one edge.

Plane P contains no edge of the cube. In this case, P must contain three vertices adjacent to some vertex. Since a cube has eight vertices, there are eight planes for this case.

The answer is then $6 + 6 + 8 = \boxed{20}$.

14.41 We find the volume of triangular pyramid $BACF$ in two ways. Looking at the solid as a triangular pyramid with base ABC and height BF, we get $[BACF] = [ABC] \cdot BF/3 = (12^2/2)(12)/3 = 288$.

Another way of looking at *BACF* is using triangle *ACF* as a base. The height h from B to base *ACF* is the distance we are seek. Triangle *ACF* is equilateral with side length $12\sqrt{2}$, so its area is $(12\sqrt{2})^2\sqrt{3}/4 = 72\sqrt{3}$. Therefore, the volume of *BACF* is $[ACF]h/3 = 24h\sqrt{3}$. Since we know the volume of *BACF* is 288, we have $24h\sqrt{3} = 288$, so $h = \boxed{4\sqrt{3}}$.

14.42 Let the three dimensions of the box be a, b, and c. We are given $a + b + c = 140/4 = 35$ and $a^2 + b^2 + c^2 = 21^2$. Then the surface area is $2(ab + bc + ca) = (a + b + c)^2 - (a^2 + b^2 + c^2) = 35^2 - 21^2 = \boxed{784}$.

14.43 The remainder of wedge A has the same triangular base as wedge A, but a height between the triangular bases of $12 - 6/4 = 21/2$. Hence, the ratio of the volume of the remainder of A to the original wedge is the ratio of the heights between their triangular bases. This ratio is $(21/2)/12 = 7/8$.

The height of the wedge removed from B is $(6/4)/6 = 1/4$ of the height of wedge B. The horizontal base of the wedge removed from B is a rectangle with one side equal in length to the corresponding side of the rectangle that is the horizontal base of wedge B. The other side of the horizontal base of the removed wedge has $1/4$ the length of the corresponding side of the horizontal base of wedge B (for the same reason the height of the removed wedge is $1/4$ the height of B). Therefore, the volume of the removed piece is $(1/4)^2 = 1/16$ the volume of B, which means the ratio of the remainder of B to the original wedge is $15/16$.

Since wedges A and B originally were identical, our desired ratio is $(7/8)/(15/16) = \boxed{14/15}$.

14.44 Intuitively, it is obvious that *MNOPQRST* is a prism with square bases *MNOP* and *QRST*. We'll prove this explicitly before finding the volume. Since M and N are the midpoints of $\overline{AB}$ and $\overline{AC}$, we have $AM/AB = AN/AC = 1/2$ and $\angle MAN = \angle BAC$, so $\triangle AMN \sim \triangle ABC$. Therefore, $\overline{MN} \parallel \overline{BC}$ and $MN = BC/2 = 1/2$. Similarly, $\overline{NO} \parallel \overline{CD}$ and $NO = CD/2 = 1/2$. Since $\overline{BC} \perp \overline{CD}$, we have $\overline{MN} \perp \overline{NO}$. Similarly, we can show that all four angles of *MNOP* are $90°$, and each has side of *MNOP* has side length 1. Similarly, *QRST* is a square with side length 1.

Since M and Q are midpoints of $\overline{BA}$ and $\overline{BF}$ of $\triangle ABF$, we have $\overline{MQ} \parallel \overline{AF}$ and $MQ = AF/2 = \sqrt{2}/2$ ($AF = \sqrt{2}$ because it is a diagonal of square *ACFE*.) Since $\overline{MQ} \parallel \overline{AF}$ and $\overline{AF}$ is perpendicular to plane *BCDE*, we know $\overline{MQ}$ is perpendicular to plane *BCDE*. Since *BCDE* is parallel to *MNOP*, we know that $\overline{MQ}$ is perpendicular to face *MNOP* of prism *MNOPQRST*. Similarly, each of $\overline{MQ}$, $\overline{NR}$, $\overline{OS}$ and $\overline{PT}$ is perpendicular to both *MNOP* and *QRST*. Therefore, *MNOPQRST* is a rectangular prism.

The volume of rectangular prism *MNOPQRST* is $[MNOP](MQ) = (1/2)^2(\sqrt{2}/2) = \boxed{\sqrt{2}/8}$.

14.45 To calculate the lengths of space diagonals $\overline{FL}$ and $\overline{EK}$, we first need to calculate the lengths of diagonals $\overline{FH}$ and $\overline{EG}$ of parallelogram *EFGH*.

Let P be foot of the perpendicular from G to $\overline{EF}$. Triangle *GFP* is a 30-60-90 triangle, so $FP = FG/2 = 3$ and $PG = (\sqrt{3}/2)(FG) = 3\sqrt{3}$. Then $EP = EF - FP = 8 - 3 = 5$. Triangle *EPG* has a right angle at P, so $EG^2 = EP^2 + PG^2 = 5^2 + (3\sqrt{3})^2 = 52$, so $EG = 2\sqrt{13}$.

Similarly, let Q be foot of the perpendicular from H to $\overline{EF}$, extended. Then by SAS, triangle *HEQ* is congruent to triangle *GFP*, so it has the same dimensions; in particular, $EQ = FP = 3$, so $FQ = FE + EQ = 11$. Triangle *FQH* has a right angle at Q, so $FH^2 = FQ^2 + QH^2 = 11^2 + (3\sqrt{3})^2 = 148$, so $FH = 2\sqrt{37}$.

Now, to calculate the length of $\overline{FL}$, consider triangle FHL. Triangle FHL has a right angle at H, so $FL^2 = FH^2 + HL^2 = (2\sqrt{37})^2 + 9^2 = 229$, so $FL = \boxed{\sqrt{229}}$. Similarly, to calculate the length of $\overline{EK}$, consider triangle EGK. Triangle EGK has a right angle at G, so $EK^2 = EG^2 + GK^2 = (2\sqrt{13})^2 + 9^2 = 133$, so $EK = \boxed{\sqrt{133}}$.

14.46

(a) *Solution 1:* We consider the cross-section containing the centers of the squares that form the bases of the frustum, the apex of the original pyramid, and the midpoints of opposite sides of the bases. Since $BCDE$ is a square, $MN = BE = 6$. Since altitude AO of triangle AMN is 9 and the smaller base is 3 units from the larger, the altitude from A to base $\overline{PQ}$ is $9 - 3 = 6$. Since $\triangle AMN \sim \triangle AQP$ (because $\overline{PQ} \parallel \overline{MN}$), the ratio of corresponding sides PQ/MN equals the ratio of the altitudes to these sides. Therefore, $PQ = (MN)(2/3) = 4$. The volume of the small upper pyramid then is $(4^2)(6)/3 = 32$, and the volume of the original pyramid is $(6^2)(9)/3 = 108$, so the volume of the frustum is $108 - 32 = \boxed{76}$.

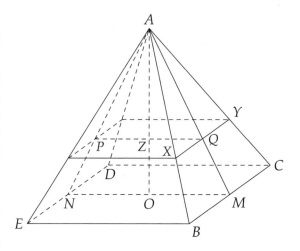

Solution 2: The part we cut off is another right square pyramid that is similar to the original right square pyramid. Hence, the ratio of their volumes is the cube of the ratio of corresponding lengths. The ratio of their heights is $6/9 = 2/3$, so the ratio of their volumes is $(2/3)^3 = 8/27$. This means that the frustum has a volume of $1 - 8/27 = 19/27$ times the volume of the original pyramid, so its volume is $(19/27)[BCDE](AO)/3 = \boxed{76}$.

(b)   Triangle AOM has a right angle at O, $OM = 6/2 = 3$, and we are given $AO = 9$, so $AM^2 = AO^2 + OM^2 = 3^2 + 9^2 = 90$. Therefore, $AM = \sqrt{90} = 3\sqrt{10}$. Let X and Y be the points where the plane meets $\overline{AB}$ and $\overline{AC}$, respectively. From $\triangle APQ \sim \triangle ANM$, we have $AQ/AM = AZ/AO = 2/3$. Therefore, $AQ = (2/3)AM = 2\sqrt{10}$ and $QM = AM - AQ = \sqrt{10}$.

    Since $\overline{XY} \parallel \overline{BC}$, we have $\angle AXY = \angle ABC$ and $\angle AYX = \angle ACB$, so $\triangle AXY \sim \triangle ABC$. $\overline{AQ}$ and $\overline{AM}$ are corresponding altitudes of these triangles, so we have $XY/BC = AQ/AM = 2/3$. Hence, $XY = (2/3)BC = 4$. We can now find the area of each slanted face of the frustum. Each such face is a trapezoid with bases of length 4 and 6, and height $\sqrt{10}$. Therefore, each such face has area $(4 + 6)(\sqrt{10})/2 = 5\sqrt{10}$. The two other faces of the frustum are squares with areas $4^2 = 16$ and $6^2 = 36$. Therefore, the total surface area of the frustum is $16 + 36 + 4(5\sqrt{10}) = \boxed{52 + 20\sqrt{10}}$.

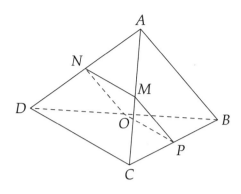

14.47 The two new solids are identical. We could find the surface area of each face of solid *MNOPCD* at left, but we have a slicker solution. Because the two new solids are identical, the surface area of each equals half the surface area of *ABCD* plus the area of the 'new' face, *MNOP*. By symmetry, *MNOP* is a square. Since *M* and *N* are the midpoints of sides $\overline{AC}$ and $\overline{AD}$ of $\triangle ADC$, we have $\overline{MN} \parallel \overline{CD}$ and $MN = CD/2 = 1$. Therefore, $[MNOP] = 1$. Since the surface area of *ABCD* is $4[ABC]$ and $\triangle ABC$ is an equilateral triangle of side length 2, the surface area of *ABCD* is $4(2^2 \sqrt{3}/4) = 4\sqrt{3}$. As noted earlier, the surface area of *MNOPCD* is $[MNOP]$ plus half the surface area of *ABCD*. Therefore, the desired surface area is $1 + (4\sqrt{3})/2 = \boxed{1 + 2\sqrt{3}}$.

CHAPTER 15

_____Curved Surfaces

Exercises for Section 15.1

15.1.1

(a) The volume of the cylinder is given by $\pi r^2 h = \pi \cdot 8^2 \cdot 4 = \boxed{256\pi}$.

(b) The lateral surface area of the cylinder is given by $2\pi rh = 2\pi \cdot 8 \cdot 4 = \boxed{64\pi}$.

(c) The total surface area of the cylinder is given by $2\pi rh + 2\pi r^2 = 2\pi \cdot 8 \cdot 4 + 2\pi \cdot 8^2 = 64\pi + 128\pi = \boxed{192\pi}$.

15.1.2 Let r be the radius of the cylinder. Then the height of the cylinder is $2r$, so the total surface area of the cylinder is $2\pi rh + 2\pi r^2 = 2\pi r \cdot 2r + 2\pi r^2 = 6\pi r^2 = 150\pi$. Therefore, $r = 5$. The volume of the cylinder is $\pi r^2 h = \pi \cdot 5^2 \cdot 10 = \boxed{250\pi}$.

15.1.3

(a) Since $\overline{AC}$ is perpendicular to the bases, it is the height of the cylinder. By the Pythagorean Theorem,
$$AC = \sqrt{AB^2 - BC^2} = \sqrt{8^2 - 4^2} = \boxed{4\sqrt{3}}.$$

(b) The area of the base is $\pi \cdot BC^2 = 16\pi$, so the volume of the cylinder is $16\pi \cdot 4\sqrt{3} = \boxed{64\sqrt{3}\pi}$.

15.1.4 Let r be the radius, and h be the height. The lateral surface area is $2\pi rh$, and the sum of the areas of the bases is $2\pi r^2$, so $2\pi rh = 2\pi r^2$. So, $r = h$ and the ratio of the radius to the height is $\boxed{1}$.

15.1.5 Since $AB = 4$, the radius r of the cylinder is 2. Since $AE = 8$ is the distance between the bases of the cylinder, the height h of the cylinder is 8. The volume of the cylinder is $\pi r^2 h = \pi \cdot 2^2 \cdot 8 = \boxed{32\pi}$.

15.1.6 $\boxed{\text{Yes}}$, we can arrange 7 cylinders such that each touches the other 6. See diagram below.

Exercises for Section 15.2

15.2.1 The volume of the cone is given by $\frac{1}{3}\pi r^2 h = \frac{1}{3}\pi \cdot 2^2 \cdot 5 = \boxed{\dfrac{20\pi}{3}}$. The height, radius, and slant height make a right triangle, with the slant height as the hypotenuse. Since the height is 5 and the radius is 2, the slant height is $\sqrt{2^2 + 5^2} = \sqrt{29}$. Therefore, the total surface area is $\pi r^2 + \pi r l = \pi \cdot 2^2 + \pi \cdot 2 \cdot \sqrt{29} = \boxed{(4 + 2\sqrt{29})\pi}$.

15.2.2 Let l be the slant height. Since $6\pi l = 54\pi$, we have $l = \boxed{9}$. The height, radius, and slant height make a right triangle, with the slant height as the hypotenuse. Since the radius is 6 and the slant height is 9, the height is $\sqrt{9^2 - 6^2} = \boxed{3\sqrt{5}}$. The volume of the cone is $\frac{1}{3}\pi r^2 h = \frac{1}{3}\pi \cdot 6^2 \cdot 3\sqrt{5} = \boxed{36\pi\sqrt{5}}$.

15.2.3

(a) A circle with radius 4 has circumference $2\pi \cdot 4 = 8\pi$, and the circumference of the base of the cone is the arc length of one quarter of this circle, or $\frac{1}{4} \cdot 8\pi = 2\pi$. Therefore, the radius r of the base is 1. The slant height l is the radius of the quarter-circle, so $l = 4$. Hence, the lateral surface area is $\pi r l = \boxed{4\pi}$. We also could have noted that the area of the sector is the lateral surface area of the cone, which is $(1/4)(\pi)(4^2) = 4\pi$.

(b) The height, radius, and slant height make a right triangle, with the slant height as the hypotenuse. Since the radius $r = 1$ and the slant height $l = 4$, the height h is equal to $\sqrt{l^2 - r^2} = \sqrt{15}$. Therefore, the volume of the cone is $\frac{1}{3}\pi r^2 h = \frac{1}{3}\pi \cdot 1^2 \cdot \sqrt{15} = \boxed{\dfrac{\pi\sqrt{15}}{3}}$.

15.2.4 Let the radius and height of cone $\mathcal{B}$ be r and h, respectively. Then cone $\mathcal{A}$ has radius $2r$ and height $h/2$, so its volume is $\frac{1}{3}\pi(2r)^2\left(\frac{h}{2}\right) = \frac{2}{3}\pi r^2 h$, and the volume of cone $\mathcal{B}$ is $\frac{1}{3}\pi r^2 h$. Hence, the ratio is $\boxed{2}$.

15.2.5 Let the vertex of the cone be V, the center of the base be O, the radius of the base of the cone be r and let X be a point on the circumference of the base. Since the cone is right, we have $\overline{OV} \perp \overline{OX}$. Therefore, the Pythagorean Theorem gives $VX = \sqrt{VO^2 + OX^2} = \sqrt{VO^2 + r^2}$. There's nothing special about X; V is similarly $\sqrt{VO^2 + r^2}$ away from every point on the circumference of the base.

15.2.6 Let r, h, and l be the radius, height, and slant height of the cone, respectively. These three lengths make a right triangle, with the slant height as the hypotenuse, so $l^2 = r^2 + h^2 \geq r^2$, so $l \geq r$. Thus, it is not possible for the slant height l to be smaller than the radius r.

15.2.7 The smaller cone and the bigger cone are similar. This means that the ratio of their volumes equals the cube of the ratio of their corresponding lengths. The ratio of their corresponding lengths can be found by looking at the ratio of their heights. This ratio is $6/9 = 2/3$, so the ratio of their volumes is $(2/3)^3 = 8/27$. We then multiply this by the volume of the bigger cone to find the volume of the smaller cone.

15.2.8

(a) Because $\overline{AB} \parallel \overline{YZ}$, triangle VAB is similar to triangle VZY with ratio $AB/ZY = r_1/r_2$.

(b) Let d be the distance from V to $\mathcal{B}_1$. Since triangle VAB is similar to triangle VZY with ratio r_1/r_2, their heights are also in this ratio. Hence, $\frac{d}{d+h} = \frac{r_1}{r_2}$. Solving for d gives

$$d = \boxed{\frac{r_1 h}{r_2 - r_1}}.$$

(c) The volume of the cone with vertex V and base $\mathcal{B}_1$ is

$$\frac{1}{3}\pi r_1^2 d = \frac{1}{3}\pi r_1^2 \cdot \frac{r_1 h}{r_2 - r_1} = \boxed{\frac{1}{3}\pi \cdot \frac{r_1^3 h}{r_2 - r_1}}.$$

(d) The volume of the cone with vertex V and base $\mathcal{B}_2$ is

$$\frac{1}{3}\pi r_2^2 (d + h) = \frac{1}{3}\pi r_2^2 \left(\frac{r_1 h}{r_2 - r_1} + h \right) = \frac{1}{3}\pi r_2^2 \left(\frac{r_1 h}{r_2 - r_1} + \frac{h(r_2 - r_1)}{r_2 - r_1} \right) \boxed{\frac{1}{3}\pi \cdot \frac{r_2^3 h}{r_2 - r_1}}.$$

(e) The volume of the frustum is the difference between the answers in parts (c) and (d), which is

$$\frac{1}{3}\pi \cdot \frac{r_2^3}{r_2 - r_1} h - \frac{1}{3}\pi \cdot \frac{r_1^3}{r_2 - r_1} h = \frac{1}{3}\pi \cdot \frac{r_2^3 - r_1^3}{r_2 - r_1} h = \frac{1}{3}\pi \cdot \frac{(r_2 - r_1)(r_2^2 + r_1 r_2 + r_1^2)}{r_2 - r_1} h = \boxed{\frac{1}{3}\pi (r_1^2 + r_1 r_2 + r_2^2) h}.$$

Exercises for Section 15.3

15.3.1 Let r be the radius of the sphere. We are given $\frac{4}{3}\pi r^3 = 36\pi$, so $r = 3$. Hence, the surface area of the sphere is $4\pi r^2 = \boxed{36\pi}$.

15.3.2

(a) Let r be the radius of the sphere. Then the radius and the height of the cylinder are r and $2r$, respectively. Hence, the volume of the sphere is $\frac{4}{3}\pi r^3$, and the volume of the cylinder is $\pi r^2 \cdot 2r = 2\pi r^3$. The ratio of the volume of the sphere to the volume of the cylinder is $\boxed{2/3}$.

(b) The surface area of sphere is $4\pi r^2$. The lateral surface area of the cylinder is $2\pi r \cdot 2r = 4\pi r^2$. Therefore, the ratio of the surface area of the sphere to the lateral surface area of the cylinder is $\boxed{1}$.

15.3.3 The volume of the cone is $\frac{1}{3}\pi r^2 h = \frac{1}{3}\pi \cdot 1^2 \cdot 4 = 4\pi/3$. Let r be the radius of the sphere. Because the volume of the sphere equals that of the cone, we have $\frac{4}{3}\pi r^3 = \frac{4}{3}\pi$, so $r = \boxed{1}$.

15.3.4 In the solution, we prove that every point where the plane meets the sphere is the same distance, namely XA, from X, where X is the foot of the perpendicular from the center of the sphere to the plane. All this proves is that every point in the intersection of plane $\mathcal{P}$ and the sphere is on the circle in $\mathcal{P}$ with center X and radius XA. It does not show that the intersection is the entire circle – it could just be an arc of the circle. To show that every point on the circle is in the intersection, consider point B on the circle in $\mathcal{P}$ with center X and radius XA. Then, since $XB = XA$, $XO = XO$, and $\angle OXA = \angle OXB$, we have $\triangle OXA \cong \triangle OXB$. Thus, $OB = OA$ since they are corresponding lengths of these triangles. Since OA is the radius of the sphere and $OB = OA$, we know that B is on the sphere as well. Hence, B is on the intersection of $\mathcal{P}$ and the sphere. This proof holds for any point on the circle in $\mathcal{P}$ with center X and radius XA, so the intersection is the entire circle.

15.3.5 Let r be the radius of the sphere. We build a right triangle with a radius of the sphere as hypotenuse, a radius of the circle left in the surface of the ice as a leg, and the segment connecting the center of the ball to the center of this circle as the other leg. Since the hole is 8 cm deep, the center of the ball was $r - 8$ cm from the plane containing the surface of the lake. Since the hole is 24 cm across, its radius is 12 cm. Applying the Pythagorean Theorem gives $(r-8)^2 + 12^2 = r^2$. Solving gives $r = \boxed{13}$.

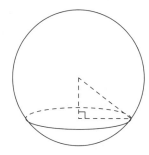

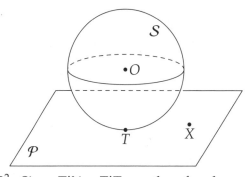

15.3.6

(a) We first show that it is impossible for $\overline{OT}$ to not be perpendicular to $\overline{TX}$. If $\overline{OT}$ is not perpendicular to $\overline{TX}$, then T is not the foot of the perpendicular from O to $\mathcal{P}$. Let T' be the foot of perpendicular from O to plane $\mathcal{P}$. Then TOT' is a right triangle with hypotenuse $\overline{OT}$, so $OT' < OT$. Thus, the point T' on plane $\mathcal{P}$ is in the interior of sphere $\mathcal{S}$. Consider point Y on the circle in plane $\mathcal{P}$ with center T' and radius $T'T$. Since $T'Y = T'T$ and $\overline{OT'} \perp \overline{T'Y}$ (because $\overline{OT'} \perp \mathcal{P}$), we have $OY^2 = T'Y^2 + T'O^2$ from right triangle $\triangle T'OY$. Right triangle $\triangle T'OT$ gives us $OT^2 = T'T^2 + T'O^2$. Since $T'Y = T'T$, we therefore have $OY^2 = OT^2$, so $OY = OT$. This means Y is on $\mathcal{S}$. Therefore, the whole circle in $\mathcal{P}$ with center T' and radius $T'T$ is on the surface of $\mathcal{S}$. This means that if we do not have $\overline{OT} \perp \mathcal{P}$, then plane $\mathcal{P}$ must meet the sphere at more than one point. Therefore, our assumption that $\overline{OT}$ is not perpendicular to $\overline{TX}$ cannot be true if $\mathcal{P}$ is tangent to $\mathcal{S}$.

We now show that if $\overline{OT} \perp \mathcal{P}$, then $\mathcal{P}$ is tangent to $\mathcal{S}$. Let Z be a point in $\mathcal{P}$ besides T. Since $\overline{OT} \perp \mathcal{P}$, we know that $\triangle OTZ$ is a right triangle. Therefore, $OZ^2 = OT^2 + TZ^2$, so $OZ > OT$. Hence, Z is outside $\mathcal{S}$, so all points in $\mathcal{P}$ besides T are outside $\mathcal{S}$. We conclude that $\mathcal{P}$ is tangent to $\mathcal{S}$.

(b) Let r_1, r_2 be the radii of $\mathcal{S}_1, \mathcal{S}_2$ respectively. Then $O_1T = r_1$ and $O_2T = r_2$. There are three cases.

Case 1: $O_1T + O_2T < O_1O_2$. This case is impossible because it violates the Triangle Inequality.

Case 2: $O_1T + O_2T = O_1O_2$. This case implies O_1, T, O_2 are collinear. In other words, $\overleftrightarrow{O_1O_2}$ passes through T.

Case 3: $O_1T + O_2T > O_1O_2$. Since $O_1T + O_2T > O_1O_2$, we know that T is not on $\overline{O_1O_2}$. Therefore, we can find some point R such that $\overline{O_1O_2}$ is the perpendicular bisector of $\overline{RT}$. To find point R, we draw altitude $\overline{TX}$ from T to $\overline{O_1O_2}$, then we draw $\overline{RX}$ such that $RX = TX$ and $\overline{RX} \perp \overline{O_1O_2}$. Since $O_1X = O_1X$, $\angle TXO_1 = \angle RXO_1$, and $RX = TX$, we have $\triangle TO_1X \cong \triangle RO_1X$. Therefore, $TO_1 = RO_1$, so R is on sphere $\mathcal{S}_1$. Similarly, we have $\triangle O_2TX \cong \triangle O_2RX$, so $O_2T = O_2R$ and R is on $\mathcal{S}_2$. Therefore, if $O_1T + TO_2 > O_1O_2$, it is impossible for

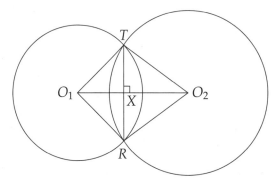

the spheres to be tangent because there is a second point at which the spheres meet. (In fact, there's a whole circle of points at which the spheres meet.)

Since only Case 2 is valid, we conclude that $O_1T + O_2T = O_1O_2$, so $\overline{O_1O_2}$ passes through T.

(c) Let r_1, r_2 be the radii of $\mathcal{S}_1, \mathcal{S}_2$ respectively, with $r_1 < r_2$. Then $O_1T = r_1, O_2T = r_2$. There are three cases.

Case 1: $O_1O_2 + O_2T < O_1T$. This case is impossible because it violates the triangle inequality.

Case 2: $O_1O_2 + O_2T = O_1T$. This case implies O_1, O_2, T are collinear. In other words, $\overleftrightarrow{O_1O_2}$ passes through T.

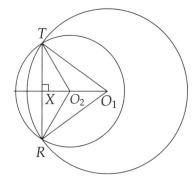

Case 3: $O_1O_2 + O_2T > O_1T$. As with Case 3 in part (b), we can show that there is another point on both spheres. Specifically, since $O_1O_2 + O_2T > O_1T$, we know that T is not on $\overleftrightarrow{O_1O_2}$. As before, we take the point R such that $\overleftrightarrow{O_1O_2}$ is the perpendicular bisector of $\overline{TR}$. To find point R, we draw altitude $\overline{TX}$ from T to $\overleftrightarrow{O_1O_2}$, then we draw $\overline{RX}$ such that $RX = TX$ and $\overline{RX} \perp \overleftrightarrow{O_1O_2}$. As before, we can show $\triangle TO_1X \cong \triangle RO_1X$, so R is on sphere $\mathcal{S}_1$. Similarly, we have $\triangle O_2TX \cong \triangle O_2RX$, so $O_2T = O_2R$ and R is on $\mathcal{S}_2$. Therefore, if $O_1O_2 + O_2T > O_1T$, it is impossible for the spheres to be tangent because there is a second point at which the spheres meet. (In fact, there's a whole circle of points at which the spheres meet.)

Since only Case 2 is valid, we conclude that $O_1O_2 + O_2T = O_1T$, so $\overleftrightarrow{O_1O_2}$ passes through T.

15.3.7 Let the cube be $ABCDEFGH$ as shown, let O be the midpoint of diagonal $\overline{AG}$, and let N be the midpoint of face diagonal $\overline{EG}$. Since $OG/AG = NG/GE$ and $\angle AGE = \angle OGN$, we have $\triangle AGE \sim \triangle OGN$ by SAS Similarity. Therefore, $\overline{ON} \parallel \overline{AE}$, and $ON = AE/2$. Since $\overline{AE}$ is perpendicular to face $EFGH$ and $\overline{ON} \parallel \overline{AE}$, we know that $\overline{ON}$ is perpendicular to $EFGH$. Therefore, we conclude O is midway between faces $ABCD$ and $EFGH$. Similarly, we can show that O is midway between each pair of opposite faces of the cube. Since $ADGF$, $ACGE$, and $ABGH$ are rectangles, the midpoint of $\overline{AG}$ is also the midpoint of space diagonals $\overline{BH}$, $\overline{CE}$, and $\overline{DF}$.

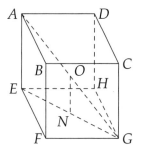

Since O is the midpoint of all four of the space diagonals, and all of these diagonals have the same length, we conclude that the sphere with center O and diameter equal to the length of the diagonal of the cube must pass through all 8 vertices of the cube. Therefore, each space diagonal is a diameter of the sphere that passes through all 8 vertices of the cube.

We must also show that all other points on the surface of the cube are inside this sphere. Point N is the midpoint of $\overline{EG}$, which is the hypotenuse of both $\triangle EHG$ and $\triangle EFG$. Because these triangles are right triangles, the circle with this hypotenuse as diameter is the circumcircle of both triangles. Therefore, square $EFGH$ is entirely on or inside the circle with center N and diameter $\overline{EG}$. This circle is a cross-section of our sphere, so $EFGH$ is entirely on or inside the sphere. Similarly, no point on any of the other faces can be outside the sphere, so the cube is inscribed in the sphere.

(We could also use a symmetry approach – the solution above is intended to show how to solve the problem with the fundamental concepts we have learned in the book.)

Exercises for Section 15.4

15.4.1 Let a, b, and c be the dimensions of the rectangular prism. Then

$$bc = 24,$$
$$ca = 32,$$
$$ab = 36.$$

Multiplying all three equations gives $a^2b^2c^2 = 24 \cdot 32 \cdot 36$, so $abc = \sqrt{27648} = 96\sqrt{3}$. Dividing this by $bc = 24$ gives $a = (abc)/(bc) = (96\sqrt{3})/24 = 4\sqrt{3}$. Similarly, dividing $abc = 96\sqrt{3}$ by $ca = 32$ gives $b = 3\sqrt{3}$, and dividing it by $ab = 36$ gives $c = 8\sqrt{3}/3$. Therefore, the dimensions of the rectangular prism are $\boxed{8\sqrt{3}/3, 3\sqrt{3}, \text{ and } 4\sqrt{3}}$.

15.4.2 Let E be the midpoint of $\overline{BC}$. In $\triangle ABC$, median $\overline{AE}$ passes through centroid G. In $\triangle BCD$, median $\overline{DE}$ passes through centroid H. Since the centroid of a triangle divides its medians into a ratio of $2:1$ as described in the text, we have $EG/EA = EH/ED = 1/3$. Hence, by SAS Similarity we know $\triangle GEH$ is similar to $\triangle AED$ with ratio $1/3$. Therefore, $GH = AD/3 = \boxed{8/3}$.

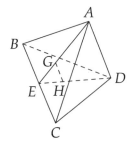

Figure 15.1: Diagram for Problem 15.4.2   Figure 15.2: Diagram for Problem 15.4.3

15.4.3

(a) Note that $\overline{WN}$ is a median in equilateral triangle WYZ, so it is also an altitude. Since $\angle WYN = 60°$, $\triangle WYN$ is a 30-60-90 triangle, so $WN = WY(\sqrt{3}/2) = 2\sqrt{3}$. By an analogous argument on triangle XYZ, we have $XN = 2\sqrt{3}$. $\overline{MN}$ is therefore the median to base $\overline{WX}$ in isosceles triangle WNX. Therefore, $\overline{MN} \perp \overline{WX}$.

(b) $\overline{MN}$ is the median to base $\overline{WX}$ in isosceles triangle WNX. Therefore, it is also an altitude. Since $WN = 2\sqrt{3}$ and $WM = WX/2 = 2$, we use the Pythagorean Theorem to find $MN = \sqrt{WN^2 - WM^2} = \boxed{2\sqrt{2}}$.

15.4.4 The radius of the can is the same as the radius of a ball. (Make sure you see why!) By taking a cross-section of the can including the axis of the can, we see that the height of the can equals the sum of the diameters of the three balls, or $3(4) = 12$. Therefore, the volume of the can is $(12)(2^2)\pi = \boxed{48\pi}$.

15.4.5 We saw our general technique in the text when Adam was crawling on the surface of a cube. Here, however, we have six 'unfolding' options for Annie to consider. Fortunately, they come in equivalent pairs, so we only have to consider three of them. First, she could walk across face $ABCD$, then across face *BCGF*. (This is equivalent to crossing *ADHE*, then *EHGF*.) Second, she could walk across face *ABFE*, then across face *BCGF* (equivalent to *ADHE-DCGH*). Third, she could cross *ABCD*, then *DCGH* (equivalent to *ABFE-EFGH*). We could also view each in terms of the edge Annie crosses on her journey. The resulting unfoldings for these three options are shown below.

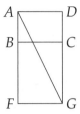

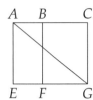

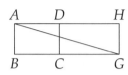

Figure 15.3: *ABCD-BCGF*   Figure 15.4: *ABFE-BCGF*   Figure 15.5: *ABCD-DHGC*

On the left above, Annie's optimal path is diagonal $\overline{AG}$ of rectangle *AFGD*. Since $AF = AB + BF = 12$ and $FG = 6$, we have $AG = \sqrt{AF^2 + FG^2} = 6\sqrt{5}$. In the center above, Annie's best path is diagonal $\overline{AG}$ of rectangle *ACGE*. Since $AC = 10$ and $CG = AE = 8$, we have $AG = \sqrt{AC^2 + CG^2} = 2\sqrt{41}$. On the right above, Annie's best path is diagonal $\overline{AG}$ of rectangle *ABGH*. Since $BG = BC + CG = 14$ and $AB = 4$, we have $AG = \sqrt{AB^2 + BG^2} = 2\sqrt{53}$. This is greater than $2\sqrt{41}$, so this is not the shortest path. Since $2\sqrt{41} < 6\sqrt{5}$ (which we can see by squaring both), Annie's shortest path has length $\boxed{2\sqrt{41}}$.

15.4.6

(a) We can show that the two spheres intersect by showing that there is a point on both spheres. We know that there exists a right triangle with sides of length 6, 8, and 10. We place this triangle so that $\overline{OP}$ is its hypotenuse, and point A is the vertex of the right angle such that $AO = 6$ and $PA = 8$. Since point A is 6 units from O, it is on $\mathcal{S}$. Similarly, A is 8 units from P, so it is on $\mathcal{T}$. Therefore, A is on both spheres, so the spheres intersect.

(b) Since X is on $\mathcal{S}$, $\overline{XO}$ is a radius of $\mathcal{S}$, so $XO = \boxed{6}$. Similarly, $\overline{XP}$ is a radius of $\mathcal{T}$, so $XP = \boxed{8}$.

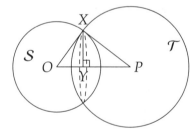

(c) $\triangle XOP$ is a $\boxed{\text{right triangle}}$ because its sides are 6, 8, and 10, and $6^2 + 8^2 = 10^2$.

(d) The area of triangle XOP equals both $(XO)(XP)/2 = (6)(8)/2 = 24$ and $(OP)(XY)/2 = 5XY$. Therefore, $XY = \boxed{24/5}$.

(e) Since $\angle OXP = \angle OYX = 90°$ and $\angle YOX = 90° - \angle OXY = \angle OXP - \angle OXY = \angle YXP$ we have $\triangle OXY \sim \triangle OPX$ by AA Similarity. Therefore, we have $OY/OX = OX/OP$, so $OY = OX^2/OP = \boxed{18/5}$.

(f) There's nothing special about X except that it is on both spheres. The arguments above hold for any point that is on both spheres. Specifically, the foot of the altitude from Z to $\overline{OP}$ is the point on $\overline{OP}$ that is 18/5 from point O. This is point Y. Also as above, we have $ZY = XY = 24/5$.

(g) By part (f), any point Z that lies on both spheres must satisfy $ZY = 24/5$ and $\overline{ZY} \perp \overline{YO}$. Since Y is fixed, this means that Z lies on a circle centered at Y with radius 24/5 that is in the plane through Y perpendicular to $\overline{OP}$. We show that this entire circle is the intersection of $\mathcal{S}$ and $\mathcal{T}$ by letting W be a point on the circle and showing that W is on both spheres. Since W is on the circle and $\overline{OP}$ is perpendicular to the plane containing the circle, we have $\overline{WY} \perp \overline{OP}$. Therefore,

$WO = \sqrt{WY^2 + OY^2}$ and $WP = \sqrt{WY^2 + YP^2}$. Since WY equals the radius of the circle, we have $WY = XY$, so $WO = \sqrt{XY^2 + OY^2} = XO$. Since $WO = XO$, W is on sphere $\mathcal{S}$. Similarly, $WP = \sqrt{XY^2 + YP^2} = XP$, so W is on $\mathcal{T}$.

(h) The radius is 24/5, so the area of the circle is $\pi(24/5)^2 = \boxed{576\pi/25}$.

15.4.7

(a) Since M and N are midpoints of sides $\overline{FG}$ and $\overline{HG}$ of $\triangle FGH$, we have $MN = FH/2 = \boxed{3\sqrt{2}}$.

(b) $\overline{AM}$ is a space diagonal of a rectangular prism with dimensions FB, FE, and FM. Therefore, $AM = \sqrt{FB^2 + FE^2 + FM^2} = \sqrt{6^2 + 6^2 + 3^2} = \boxed{9}$.

(c) We know $AM = 9$. By a similar argument, $AN = 9$. Hence, $\triangle AMN$ is an isosceles triangle. We also know $MN = 3\sqrt{2}$. Let P be the midpoint of $\overline{MN}$. Then triangle APM has a right angle at P, and $MP = MN/2 = 3\sqrt{2}/2$, so $AP^2 = AM^2 - MP^2 = 9^2 - (3\sqrt{2}/2)^2 = 153/2$, so $AP = \sqrt{153/2} = 3\sqrt{17/2}$. Therefore, the area of triangle AMN is

$$\frac{1}{2}MN \cdot AP = \frac{1}{2} \cdot 3\sqrt{2} \cdot 3\sqrt{\frac{17}{2}} = \boxed{\frac{9\sqrt{17}}{2}}.$$

(d) The solid $AEMN$ is a triangular pyramid with base MEN and height AE. Since $[MEN] = [EFGH] - [EFM] - [MGN] - [EHN] = (EH)^2 - (EF)(FM)/2 - (MG)(GN)/2 - (EH)(HN)/2 = 27/2$, the volume of $AEMN$ is $[MEN](AE)/3 = \boxed{27}$.

15.4.8 If we unroll the shortest possible rope, we get the hypotenuse of a right triangle whose legs are the height of the cylinder and four times the circumference of the cylinder (since we must 'spin the cylinder' 4 times to unroll the rope). Hence, the length of the rope is at least $\sqrt{12^2 + (4 \cdot 2)^2} = \sqrt{208} = \boxed{4\sqrt{13}}$.

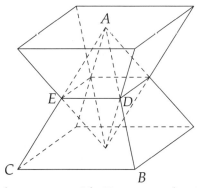

15.4.9 As shown in the diagram, the edges from the base to vertex of each pyramid meet and form a square midway between the two bases, because the two pyramids are congruent and all corresponding edges are parallel. Since they meet halfway between the bases, the plane containing the square where the sides intersect divides in half each of the slanted edges from a base to vertex, such as $\overline{AB}$, of the original pyramids. Therefore, the slanted edges of the little pyramids with the square of intersection as a base have length 6. Each triangular face of one of these little pyramids is similar to a triangular face of the large pyramid, with sides of length 1/2 those of the triangular face of the large pyramid. For example, $AE/AC = AD/AB = 1/2$ and $\angle EAD = \angle CAB$ gives us $\triangle ADE \sim \triangle ABC$, so $ED = BC/2 = 6$. Thus, we find that all eight faces of our desired intersection are equilateral triangles, so the region is a regular octahedron of side length $s = 6$. As we have shown in Problem 14.30, the volume of an octahedron with side length s is $s^3\sqrt{2}/3$. Therefore, our desired volume is $\frac{\sqrt{2}}{3} \cdot 6^3 = \boxed{72\sqrt{2}}$.

Review Problems

15.20 Let the radius of the sphere be r. The volume of the sphere is $4\pi r^3/3$ and the surface area is $4\pi r^2$, so we have $4\pi r^3/3 = 4\pi r^2$. Solving, we find $r = \boxed{3}$.

15.21

(a) The height, radius, and slant height make a right triangle, with the slant height as the hypotenuse. Since the height is 5 and the slant height is 7, the radius is $\sqrt{7^2 - 5^2} = \sqrt{24} = 2\sqrt{6}$. Therefore, the volume of the cone is $\frac{1}{3}\pi r^2 h = \frac{1}{3}\pi \cdot (2\sqrt{6})^2 \cdot 5 = \boxed{40\pi}$.

(b) The total surface area of the cone is $\pi rl + \pi r^2 = \pi \cdot 2\sqrt{6} \cdot 7 + \pi(2\sqrt{6})^2 = \boxed{(24 + 14\sqrt{6})\pi}$.

15.22 The sphere has radius $r = 4$. Therefore, the surface area of the sphere is $4\pi r^2 = \boxed{64\pi}$, and the volume of the sphere is $\frac{4}{3}\pi r^3 = \boxed{\dfrac{256\pi}{3}}$.

15.23

(a) The volume of the cylinder is $\pi r^2 h = \pi \cdot 3^2 \cdot 6 = \boxed{54\pi}$.

(b) The lateral surface area of the cylinder is $2\pi rh = 2\pi \cdot 3 \cdot 6 = \boxed{36\pi}$.

(c) The total surface area of the cylinder is $2\pi rh + 2\pi r^2 = 36\pi + 2\pi \cdot 3^2 = \boxed{54\pi}$.

15.24 Since the edges of the cube have length 8, the a space diagonal has length $8\sqrt{3}$. Therefore, the radius of our sphere is $4\sqrt{3}$, so its volume is $(4/3)\pi(4\sqrt{3})^3 = \boxed{256\pi\sqrt{3}}$.

15.25 Let the radius of cone $\mathcal{B}$ be r and let h be the height of the cone, so its volume is $\pi r^2 h/3$. The radius of cylinder $\mathcal{A}$ is $3r$ and its height is $h/2$, so its volume is $\pi(3r)^2(h/2) = 9\pi r^2 h/2$. The ratio of the volume of $\mathcal{A}$ to the volume of cone $\mathcal{B}$ is $[9\pi r^2 h/2]/[\pi r^2 h/3] = \boxed{27/2}$.

15.26 Diagonal $\overline{BD}$ of $ABCD$ is a diameter of a base of the cylinder. Hence, the radius of the cylinder is $BD/2 = 9\sqrt{2}/2$. The height of the cylinder is the edge length of the cube, which is 9. Therefore, the volume of the cylinder is $9\pi \left(9\sqrt{2}/2\right)^2 = \boxed{\dfrac{729\pi}{2}}$.

15.27 *Solution 1:* Since the sector has area $(40°/360°)(9^2\pi) = 9\pi$, the lateral surface area of the cone formed when the sector is rolled up will be 9π. Therefore, if r is the radius of the cone and l its slant height, we have $\pi rl = 9\pi$. The radius of our sector will be the slant height of the cone when the sector is rolled up (since the vertex of the cone is equidistant from all the points on the circumference of the base). Therefore, the slant height of our cone is 9, which means our radius is $r = (9\pi)/(l\pi) = 1$. Since the height (h), slant height (l), and radius (r) of a cone together form the sides of a right triangle with the slant height as hypotenuse, we have $h = \sqrt{l^2 - r^2} = 4\sqrt{5}$. Therefore, the volume of the cone is $\pi r^2 h/3 = \boxed{4\pi\sqrt{5}/3}$.

Solution 2: As before, the slant height of our cone is 9. After rolling up the sector, the arc of the sector will be the circumference of the base of the cone. Therefore, the circumference of the cone is $(40°/360°)(18\pi) = 2\pi$, so the radius is 1. We then find the height and volume as in the previous solution.

15.28 Let O be the center of the sphere, X be the center of the circle, and Y be a point on the circle that is the intersection of the plane and the sphere. We are given that $OX = 8$ and $XY = 6$ (since $\overline{XY}$ is a radius of a circle with area 36π). As shown in the text, the segment connecting the center of a sphere to the center of a circular cross-section of the sphere is perpendicular to the plane of the cross-section. Therefore, we have $OY = \sqrt{OX^2 + XY^2} = 10$, so the volume of the sphere is $4\pi(OY^3)/3 = \boxed{4000\pi/3}$.

15.29 Since each sphere is tangent to the 'top' and the 'bottom' of the box, the height of the box equals the length of the diameter of one of the spheres, 8. We take a cross-section of the box that includes the centers of the spheres, the point of tangency of the spheres, and three of the points on each ball where the ball touches the box. The result is shown at right. We see that another dimension of the box (besides the height 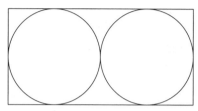 we already found) equals a diameter of a sphere, and the final dimension equals two times a diameter of the sphere. Therefore, the dimensions of the box are 8, 8 and 16, so its volume is $(8)(8)(16) = \boxed{1024}$.

15.30 Because the axis of the cylinder is parallel to our cross-section, two of the sides of our square are heights of the cylinder. Therefore, the height of the cylinder is 6, since the side length of the square is $\sqrt{36} = 6$. Let $\overline{AB}$ be a chord on one of the bases of the cylinder such that $\overline{AB}$ is one of the sides of our square. Therefore, $AB = 6$. Since the axis is 4 units from the plane of the cross-section, the center of this base, point O in the diagram, is 4 units from $\overline{AB}$. We draw the perpendicular segment from O to $\overline{AB}$, meeting $\overline{AB}$ at M. We thus have $OM = 4$. Since $\overline{OM}$ is part of a radius that is perpendicular to chord $\overline{AB}$, it must bisect $\overline{AB}$. Therefore, $AM = 3$ and 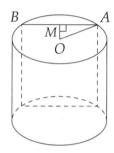 $OA = \sqrt{OM^2 + AM^2} = 5$. Our cylinder has radius 5 and height 6, so it has volume $\pi(5^2)(6) = \boxed{150\pi}$.

15.31 The two pieces are a frustum and a smaller cone. We can find the dimensions of the smaller cone as described in the similar problem in the text, or we can note that the smaller cone is similar to the original cone. The height of the original cone is given as 12 and the height of the smaller cone is $12 - 8 = 4$. Therefore, the ratio of a length of the small cone to the corresponding length in the original cone is $1/3$. Since the radius of the original cone is 9, the radius of the smaller cone is 3. Since the height of the small cone is 4 and its radius is 3, its slant height is $\sqrt{4^2 + 3^2} = 5$. Therefore, the lateral area of the small cone is $\pi(3)(5) = 15\pi$ and the area of its base is $3^2\pi = 9\pi$, so the surface area of the small cone is $\boxed{24\pi}$.

To find the area of the other piece, we note that it has bases of radii 3 and 9, which have areas 9π and 81π, respectively. These two together have area 90π. To find the area of the curved surface of the other piece, we first find the area of the curved surface of the original cone. Since the slant height of the smaller cone is 5, the slant height of the original cone is $3(5) = 15$. Therefore, the lateral surface area of the original cone is $\pi(9)(15) = 135\pi$. Since 15π of this area is 'lost' to the smaller cone, the remaining curved surface has area $135\pi - 15\pi = 120\pi$. We add this to the areas of the bases for a total surface area of $120\pi + 90\pi = \boxed{210\pi}$. (We also could have noted that the curved area of the original cone must be $3^2 = 9$ times that of the smaller cone because the cones are similar.)

15.32 To see that the center of the cube is the center of the sphere, we take a cross-section parallel to one pair of opposite faces of the cube that passes through the center of the cube. Since this plane is mid-way between opposite faces of the cube, it bisects the four edges of the cube that connect these two faces. Therefore, the four vertices of the cross-section are four of the points through which the sphere in the problem passes. This cross-section of the cube is a square whose sides equal the side length of the cube, and the center of the square is the center of the cube. Each diagonal of the square has length $8\sqrt{2}$, so the center of this

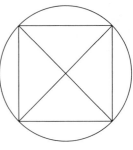

square is $4\sqrt{2}$ from the vertices of the square, which in turn are four of the midpoints of the edges of the cube through which S must pass. Similarly, the center of the cube is $4\sqrt{2}$ from all 12 of the midpoints of the edges, so the sphere has radius $4\sqrt{2}$ and therefore has volume $4\pi r^3/3 = \boxed{512\pi\sqrt{2}/3}$.

15.33 We start with a cross-section including the axis of the cone. The resulting cross-section of the cone is an isosceles triangle. The midpoint of the base of this triangle is the center of the cone. The sphere is tangent to the curved surface all the way around the cone, and to the base at the center of the base. So, the resulting cross-section of the sphere is a circle that is tangent to all three sides of the triangle. Letting O be the center of the base, we have $OB = 9$ because it is a radius of the cone, and $AO = 12$ because it is

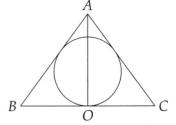

the height of the cone. Therefore, we have $AB = AC = \sqrt{9^2 + 12^2} = 15$, and the semiperimeter of $\triangle ABC$ is $(15 + 15 + 18)/2 = 24$. The area of $\triangle ABC$ is $(BC)(AO)/2 = 108$. Therefore, the inradius of $\triangle ABC$ is $[ABC]/(24) = \boxed{9/2}$. (Note: it is a coincidence that the radius of the sphere is half the radius of the cone. This will not always be the case!)

We could also have drawn a radius of the sphere to the point of tangency with $\overline{AB}$ and used similar triangles to find the radius. See if you can work out that approach on your own.

15.34

(a) Let H be the foot of altitude from D to ABC. Since $DA = DB = DC$, triangles DHA, DHB, and DHC are congruent right triangles by HL. Hence, $HA = HB = HC$, so H is the circumcenter of equilateral triangle ABC. The circumcenter of ABC is also the centroid and orthocenter of ABC. We can find AH by noting that it is $2/3$ the length of median $\overline{AM}$. Since $\overline{AM}$ is also an altitude and $\angle ACB = 60°$, we know that $\triangle ACM$ is a 30-60-90 triangle. Therefore, $AM = AC(\sqrt{3}/2) = 3\sqrt{3}$ and $AH = (2/3)(AM) = 2\sqrt{3}$. Finally,

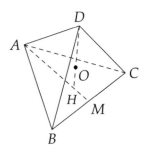

right triangle $\triangle ADH$ gives us $DH = \sqrt{AD^2 - AH^2} = \boxed{2\sqrt{6}}$.

(b) Equilateral triangle ABC has area $6^2\sqrt{3}/4 = 9\sqrt{3}$. Therefore, the volume of tetrahedron $ABCD$ is $(DH)([ABC])/3 = \boxed{18\sqrt{2}}$.

(c) The sphere is tangent to all four faces, so each perpendicular from O to a face of the tetrahedron is a radius of the sphere. Therefore, O is r away from each face of the tetrahedron.

(d) We can view $OABC$ as a triangular pyramid with base $\triangle ABC$. The height of this pyramid is the distance from O to $\triangle ABC$, which is r. Therefore, the volume of tetrahedron $OABC$ is $(r)([ABC])/3 = \boxed{3r\sqrt{3}}$.

(e) Since O is equidistant to all four faces and all four faces have equal areas, the volumes of $OABD$, $OACD$, and $OBCD$ are equal to that of $OABC$, $\boxed{3r\sqrt{3}}$.

(f) The four solids have a combined volume equal to that of $ABCD$, so $18\sqrt{2} = 4 \cdot 3\sqrt{3}r$. Solving for r gives $r = \boxed{\sqrt{6}/2}$.

15.35 Let O be the center of the base, so O is the midpoint of $\overline{XY}$. Therefore, height $\overline{AO}$ of the cone is also a median of $\triangle XAY$. Since we are given that $AO = XY/2$, we know that the length of median $\overline{AO}$ equals half the side to which it is drawn. Therefore, $\triangle XAY$ must be a right triangle with right angle at $\angle XAY$.

Challenge Problems

15.36

(a) Let X be the point where S is tangent to the base of C. Since S is tangent to the base of C, $\overline{VX}$ must be perpendicular to the base. Because the cone is right, the center of the base, point O, is the foot of the perpendicular from the cone's vertex, V, to the base. Therefore, X is O, and our sphere is tangent to the base at the center of the base. Therefore, the radius of the sphere equals the height of the cone, $\boxed{6}$.

(b) Let P be the foot of the perpendicular from X to $\overline{OV}$ and let Z be the point where $\overrightarrow{VX}$ hits the base. We are given $VO = VZ = 6$, so right triangle VOZ is a 45-45-90 triangle. Since $\overline{XP}$ and $\overline{ZO}$ are both perpendicular to $\overline{OV}$, we have $\overline{XP} \parallel \overline{ZO}$. Therefore, $\triangle VXP$ is a 45-45-90 triangle and $VX = 6$ (since it is a radius of the sphere), and we have $VP = XP = VX/\sqrt{2} = \boxed{3\sqrt{2}}$.

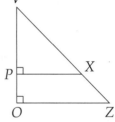

(c) From part (b), every point at which the sphere hits the curved surface of the cone is $3\sqrt{2}$ away from P, the point on $\overline{OV}$ such that $VP = 3\sqrt{2}$. Conversely, we can show that every point on the cross-section of the cone that is the intersection of the curved surface of C and the plane through P perpendicular to $\overline{OV}$ is 6 units from V (and therefore on S). Therefore, the intersection of S and the curved surface of C is a circle.

(d) The radius of this circle is $3\sqrt{2}$, so the area of the circle is $\pi(3\sqrt{2})^2 = \boxed{18\pi}$.

15.37 Since the heights of the cylinder and the prism are the same, the ratio of their volumes equals the ratio of the areas of their bases. Each base of our prism is a regular hexagon. We dissect the regular hexagon into equilateral triangles. As shown in the diagram, the radius of the circle equals the height of one of these triangles. If we let the side length of the regular hexagon be s, the radius of the circle is $s\sqrt{3}/2$. The area of the hexagon then is $3s^2\sqrt{3}/2$ and the area of the circle is $\pi(s\sqrt{3}/2)^2 = 3\pi s^2/4$. The ratio of the area of the circle to the area of the hexagon is $(3\pi s^2/4)/(3s^2\sqrt{3}/2) = \pi/(2\sqrt{3}) = \boxed{\pi\sqrt{3}/6}$.

15.38 Let O be the center of the ice cream scoop as shown at left below. Also, let A be the vertex of the cone, and let B be a point on the circumference of the base of the cone. Let C be the foot of the perpendicular from B to $\overline{OA}$. Since $OB = 2$, $BC = \sqrt{3}$, and $\angle BCO = 90°$, we know that $\triangle BCO$ is a 30-60-90 triangle, so $\angle BOA = 60°$. Since $\overline{BA}$ is tangent to the scoop, we have $\overline{BA} \perp \overline{OB}$. Therefore, $\triangle BOA$ and $\triangle ABC$ are a 30-60-90 triangles, as well. Therefore, $AC = \sqrt{3}BC = 3$. Hence, the volume of the cone is $\frac{1}{3}\pi \cdot (\sqrt{3})^2 \cdot 3 = 3\pi$. The volume of the ice cream is $\frac{4}{3}\pi \cdot 2^3 = \frac{32\pi}{3}$. Therefore, Dennis ate $32\pi/3 - 3\pi = \boxed{23\pi/3 \text{ cm}^3}$ of ice cream.

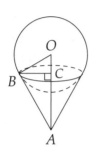
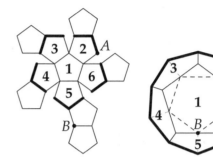

15.39 When we form our dodecahedron, we get the picture shown at right above. Shown are a couple strategies to figure out where A and B are on the dodecahedron. One way is to track an 'equator' of the dodecahedron, as shown in bold. Another is to number some of the faces, where face 1 is the 'front' face of the dodecahedron, and point A is a vertex of faces 2 and 6. We can now more easily see that the shortest path along the edges our ant can take is $\boxed{3}$ edges long.

15.40 Let the planes be $\mathcal{P}_1$ and $\mathcal{P}_2$ and the centers of the circles be C_1 and C_2, respectively. Let the center of the sphere be O. We have $\overline{OC_1} \perp \mathcal{P}_1$ and $\mathcal{P}_1 \parallel \mathcal{P}_2$, so $\overleftrightarrow{OC_1} \perp \mathcal{P}_2$ as well. The line through O that is perpendicular to $\mathcal{P}_2$ meets $\mathcal{P}_2$ at C_2, the center of the intersection of $\mathcal{P}_2$ and the sphere. Therefore, $\overleftrightarrow{OC_1}$ passes through C_2.

Let X be a point on the circumference of $\odot C_1$ and Y be a point on the circumference of $\odot C_2$. Since $OX = OY$, $C_1X = C_2Y$, and $\angle XC_1O = \angle YC_2O = 90°$, we have $\triangle XC_1O \cong \triangle YC_2O$ by HL Congruence. Therefore, $OC_1 = OC_2$, so O is the midpoint of $\overline{C_1C_2}$.

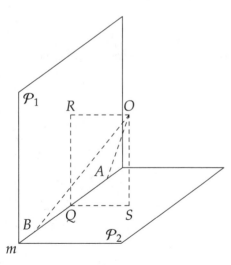

15.41 Let Q, R, S be the feet of altitudes from O to $m, \mathcal{P}_1, \mathcal{P}_2$, respectively. Then $\overline{OQ}$ is a diagonal of rectangle $ORQS$, and $OQ = \sqrt{OR^2 + OS^2} = \sqrt{4^2 + 6^2} = 2\sqrt{13}$. On the other hand, $\overline{OQ}$ is the median from O to $\overline{AB}$ in isosceles triangle AOB. Since $OA = OB = 9$ and $OQ = 2\sqrt{13}$, $AB = 2\sqrt{OB^2 - OQ^2} = \boxed{2\sqrt{29}}$.

15.42 Let Z be the point such that $\overline{ZX}$ is a diameter of the cylinder. Then $\triangle XYZ$ is a right triangle, with a right angle at Z. Since $XY = 12$ and $ZX = 8$, $YZ = \sqrt{XY^2 - XZ^2} = \sqrt{12^2 - 8^2} = \sqrt{80} = 4\sqrt{5}$. Hence, the volume of the cylinder is $\pi r^2 h = \pi \cdot 4^2 \cdot 4\sqrt{5} = \boxed{64\pi\sqrt{5}}$.

15.43 We unroll the inside and outside of the cylinder separately, and place these unrolled pieces so the sides representing the 'top rim' of the cylinder coincide, as shown at right. Arnav starts at Z and must walk to X. We can find the length of this path by constructing right triangle $\triangle XYZ$. Because X and Z are originally diametrically opposite, YZ must be half the circumference of the glass, or 2π. Since X and Z are halfway up the glass, they must each be $8/2 = 4$ from the rim, so $XT = TY = 4$. Therefore, we have $XZ = \sqrt{XY^2 + YZ^2} = \boxed{\sqrt{64 + 4\pi^2}}$.

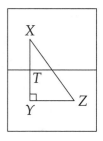

15.44 If we unroll the outer handrail, we get the hypotenuse of a right triangle whose legs are the height of the staircase and the length of the 270° arc. These have lengths 10 and $(3/4) \cdot 2\pi \cdot 3 = 9\pi/2$, respectively. Then the length of the handrail is

$$\sqrt{10^2 + \left(\frac{9\pi}{2}\right)^2} = \boxed{\sqrt{100 + \frac{81\pi^2}{4}}}.$$

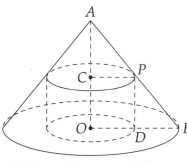

15.45 Let r be the radius of the cylinder, so the height of the cylinder is $2r$. Let O be the center of the base of the cylinder (and cone), and let A be the vertex of the cone. Let B be a point on the circumference of the base of the cone, and let P be the point where $\overline{AB}$ touches the cylinder. Finally, let C and D be the feet of the altitudes from P to $\overline{AO}$ and $\overline{OB}$, respectively.

Then $OD = r$, so $BD = 5 - r$. Also, $OC = 2r$ and $PDOC$ is a rectangle, so $PD = OC = 2r$. Since $\overline{PD} \parallel \overline{AO}$, we have $\angle BPD = \angle BAO$ and $\angle BDP = \angle BOA$, so $\triangle BDP \sim \triangle BOA$. From this similarity, we have $BD/DP = BO/OA$, so $(5 - r)/2r = 5/12$. Solving for r yields $r = \boxed{30/11}$.

15.46 *Solution 1:* Let s be the side length of the tetrahedron, M be the midpoint of $\overline{PQ}$, N be the midpoint of $\overline{RS}$, and G be the centroid of $\triangle PQR$. Since $MN = MN$, $RN = SN$, and $RM = SM$ (corresponding medians of congruent faces of the tetrahedron), we have $\triangle MNR \cong \triangle MNS$. Therefore, $\angle MNR = \angle MNS$. Since $\angle MNR + \angle MNS = 180°$, we have $\angle MNR = \angle MNS = 90°$. Similarly, $\angle QMR = 90°$. From 30-60-90 triangle $\triangle MQR$, we have $MR = MQ\sqrt{3} = (QR/2)\sqrt{3} = s\sqrt{3}/2$. We also have $NR = RS/2 = s/2$, and from right triangle $\triangle MNR$, we have $MN^2 + NR^2 = MR^2$. Therefore, we have $36 + s^2/4 = 3s^2/4$, so $s^2 = 72$ and $s = 6\sqrt{2}$. We can now find the volume of the tetrahedron in one of several ways. Here are two of them.

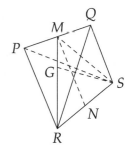

Method 1: Connect S to the centroid of face $\triangle PQR$. As described in the text, $\overline{SG}$ is an altitude of the tetrahedron. Since G is the centroid of $\triangle PQR$, we have $GR = (2/3)MR = s\sqrt{3}/3 = 2\sqrt{6}$. From right triangle RSG, we have $SG = \sqrt{SR^2 - GR^2} = \sqrt{72 - 24} = 4\sqrt{3}$. Therefore, the volume of $PQRS$ is $(SG)([PQR])/3 = (4\sqrt{3})[(6\sqrt{2})^2 \sqrt{3}/4]/3 = \boxed{72}$.

Method 2: In the text, we found that the volume of a regular tetrahedron with side length 6 has volume $18\sqrt{2}$. All regular tetrahedra are similar. Since the tetrahedron of this problem has a side length that is $(6\sqrt{2})/6 = \sqrt{2}$ longer, its volume is $(\sqrt{2})^3 = 2\sqrt{2}$ the volume of our side length 6 tetrahedron.

Therefore, our volume is $(18\sqrt{2})(2\sqrt{2}) = \boxed{72}$. Can you use this method to develop a formula for the volume of a regular tetrahedron with side length s?

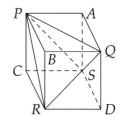

Solution 2: Let the six edges of the tetrahedron be the face diagonals of a cube, as shown. $\overline{MN}$ then connects the centers of opposite faces, so each edge of the cube has length $MN = 6$. Therefore, the cube has volume $6^3 = 216$. We find the volume of $PQRS$ by subtracting the volumes of the corners we have to cut off the cube to make $PQRS$. There are four such corners, and each has volume equal to that of pyramid $PAQS$. Since $PA = AQ = AS = 6$, the volume of $PAQS$ is $[PAQ](AS)/3 = 6^3/6 = 36$. Therefore, the volume of $PQRS$ is $216 - 4(36) = \boxed{72}$.

15.47 Let P and Q be the centers of the quarter-circles $\overset{\frown}{AB}$ and $\overset{\frown}{BC}$ as shown in the diagram, and let O be the center of the sphere that fits snugly in the wire frame. As described in Section 15.3, $\overline{OP}$ is perpendicular to the plane containing quarter-circle $\overset{\frown}{AB}$. Similarly, $\overline{OQ}$ is perpendicular to the plane containing quarter-circle $\overset{\frown}{BC}$. Therefore, $\overline{OP} \perp \overline{BP}$ and $\overline{OQ} \perp \overline{BQ}$. Furthermore, $\overline{BP}$ and $\overline{BQ}$ are each perpendicular to $\overleftrightarrow{BD}$, and hence form the dihedral angle between planes ADB and BDC. Therefore, $\overline{BP} \perp \overline{BQ}$ and we see that $OPBQ$ is a rectangle. Since $BP = BQ$, the rectangle is a square with side length $BP = 3$. Since B is on the surface of the sphere, $\overline{OB}$ is a radius of the sphere. Therefore, our desired radius is $OB = BP\sqrt{2} = \boxed{3\sqrt{2}}$.

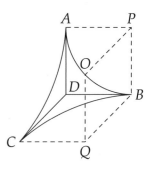

15.48 We start with a cross-section containing the axis of the cones, as shown. Our region of intersection is a pair of cones that share a base. This base has radius $\overline{BE}$. Since $\overline{BE}$, $\overline{AD}$, and $\overline{CF}$ are all perpendicular to $\overline{DF}$, they are all parallel. Therefore, we have $\triangle ADF \sim \triangle BEF$ and $\triangle CFD \sim \triangle BED$, so $BE/AD = EF/DF$ and $BE/CF = DE/DF$. Adding these gives $BE/AD + BE/CF = (EF + DE)/DF = 1$, so $BE = 1/(1/AD+1/CF)$. (You might remember this from Problem 5.17 in the similar triangles chapter.) We are given $AD = 4$ and $CF = 6$, so $BE = 12/5$. The total volume of the two cones then is $(BE^2)(DE)\pi/3 + (BE^2)(EF)\pi/3 = [(BE^2)(\pi)/3](DE + EF) = [(BE^2)(\pi)/3](DF) = \boxed{576\pi/25}$.

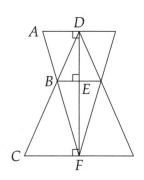

15.49

(a) Let I denote the center of the sphere. Consider tetrahedron $[IABC]$. The height IA of this tetrahedron is the distance from I to base ABC, which is just r. The area of the base ABC is K_4, so

$$[IABC] = \frac{1}{3}K_4 r.$$

Similarly,

$$[IBCD] = \frac{1}{3}K_1 r, \qquad [IACD] = \frac{1}{3}K_2 r, \qquad [IABD] = \frac{1}{3}K_3 r.$$

Hence, $V = [ABCD] = [IABC] + [IBCD] + [IACD] + [IABD] = \frac{1}{3}K_4 r + \frac{1}{3}K_3 r + \frac{1}{3}K_2 r + \frac{1}{3}K_1 r = \frac{1}{3}(K_1 + K_2 + K_3 + K_4)r$. Therefore, $r = \frac{3V}{K_1+K_2+K_3+K_4}$.

(b) In Problem 14.26, we found the volume of the tetrahedron to be $243\sqrt{11}/4$. Since triangle XYZ is equilateral, we have $[XYZ] = 9^2\sqrt{3}/4 = 81\sqrt{3}/4$. Let M be the midpoint of $\overline{XY}$. Then $XM = 9/2$, $WX = 18$, and $\angle WMX = 90°$. By the Pythagorean Theorem, $WM = \sqrt{18^2 - (9/2)^2} = \sqrt{1215/4} = 9\sqrt{15}/2$. Therefore,

$$[WXY] = \frac{1}{2}XY \cdot WM = \frac{1}{2} \cdot 9 \cdot \frac{9\sqrt{15}}{2} = \frac{81\sqrt{15}}{4}.$$

Since $\triangle WYZ \cong \triangle WXZ \cong \triangle WXY$ by SSS, we have $[WYZ] = [WXZ] = [WXY] = 81\sqrt{15}/4$. Then by part (a), the radius of the sphere inscribed in tetrahedron $WXYZ$ is

$$r = \frac{3 \cdot 243\sqrt{11}/4}{81\sqrt{3}/4 + 3 \cdot 81\sqrt{15}/4} = \frac{9\sqrt{11}}{\sqrt{3} + 3\sqrt{15}} = \frac{9\sqrt{11}}{3\sqrt{15} + \sqrt{3}} \cdot \frac{3\sqrt{15} - \sqrt{3}}{3\sqrt{15} - \sqrt{3}} = \boxed{\frac{9\sqrt{165} - 3\sqrt{33}}{44}}.$$

15.50 Assume that $r_1 < r_2$. We follow the same strategy as in Problem 15.2.8, namely extending the curved surface of the frustum to a point V, and consider the frustum as the portion left over when a small cone C_1 is removed from a large cone C_2. Let B and D be the centers of the bases and let A and C be points on the bases, as shown, such that $\overline{BA}$ and $\overline{DC}$ are radii of the bases and $\overline{AC}$ is part of the slant height of the large cone. Let the slant height of the large cone be l_2 and the slant height of the small cone be l_1. Therefore, $l_2 - l_1 = l$.

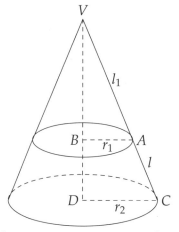

The area of the curved surface of the frustum is the difference of the areas of the curved surfaces of the cones, or $\pi r_2 l_2 - \pi r_1 l_1$. Since $\overline{AB}$ and $\overline{CD}$ are both perpendicular to $\overline{VD}$, we have $\overline{AB} \parallel \overline{CD}$. Therefore, $\angle ABV = \angle CDV$ and $\angle BAV = \angle DCV$, so $\triangle ABV \sim \triangle CDV$. These similar triangles give us $AB/AV = CD/CV$, so $r_1/l_1 = r_2/l_2$. Therefore, $r_1 = l_1 r_2/l_2$. Substituting this into our expression for the area of the curved surface of the frustum gives $\pi r_2 l_2 - \pi r_1 l_1 = \pi(r_2 l_2 - l_1^2 r_2/l_2) = \pi r_2(l_2^2/l_2 - l_1^2/l_2) = \pi(r_2/l_2)(l_2^2 - l_1^2) = \pi(r_2/l_2)(l_2 - l_1)(l_2 + l_1) = \pi r_2 l[(l_2 + l_1)/l_2] = \pi r_2 l(1 + l_1/l_2) = \pi l(r_2 + r_2 l_1/l_2) = \pi l(r_2 + r_1)$.

To find the total surface area of the frustum, we must include the areas of the bases, so our desired area is $\boxed{\pi(r_1 + r_2)l + \pi r_1^2 + \pi r_2^2}$.

15.51 Let O be the center of the larger ball, and let P, Q, and R be the feet of the perpendiculars from O to the two walls and the floor, respectively. Let O' be the center of the smaller ball, and let P', Q', and R' be the feet of the perpendiculars from O' to the two walls and the floor, respectively. Finally, let T be the point of tangency between the two balls, and let X be the corner.

Let r be the radius of the smaller ball. Note that $\overline{OX}$ is a diagonal of a cube with side length 1 (whose vertices also include P, Q, and R), so $OX = \sqrt{3}$. Similarly, $\overline{O'X}$ is a diagonal of a cube with side length r (whose vertices also include P', Q', and R'), so $O'X = \sqrt{3}r$.

Since T is the point of tangency between the two balls, points O, T, and O' are collinear, so $OO' = OT + TO' = 1 + r$. But $OO' = OX - O'X = \sqrt{3} - \sqrt{3}r$, so $1 + r = \sqrt{3} - \sqrt{3}r$. Solving for r gives $r = (\sqrt{3} - 1)/(\sqrt{3} + 1) = \boxed{2 - \sqrt{3}}$. (Can you prove that O, O', and X are collinear?)

CHAPTER 16

The More Things Change...

Exercises for Section 16.1

16.1.1 *ABCDEF* in the diagram is a regular hexagon. Since ∠*CDE* = ∠*FAB*, *CD* = *FA*, and *DE* = *AB*, we have △*CDE* ≅ △*FAB*. Therefore, *FB* = *CE*. Because ∠*FAB* = 120°, we have ∠*FBA* = ∠*BFA* = (180° − ∠*FAB*)/2 = 30°. Therefore, ∠*CBF* = ∠*CBA* − ∠*FBA* = 90°. Similarly, ∠*BCE* = 90°. Since ∠*BCE* + ∠*CBF* = 180°, we have $\overline{BF} \parallel \overline{CE}$. Because *BF* = *CE* and $\overline{BF} \parallel \overline{CE}$, the same translation that maps *E* to *C* maps *F* to *B*.

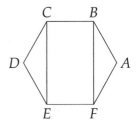

Exercises for Section 16.2

16.2.1

(a) The octagon is regular ∠*ABC* = 135° and *AB* = *BC*, so a $\boxed{135°}$ clockwise rotation about *B* maps *A* to *C*.

(b) Since △*BCD* is isosceles with *BC* = *CD*, we have ∠*CBD* = (180° − ∠*BCD*)/2 = 22.5°. Similarly, ∠*ABH* = 22.5°. Therefore, ∠*HBD* = ∠*ABC* − ∠*ABH* − ∠*CBD* = 90°. Isosceles triangles △*HAB* and △*BCD* are congruent by SAS, so we have *HB* = *BD*. Similarly, we can show that all the angles of *BDFH* are 90° and all the sides are equal in length. Therefore, *BDFH* is a square, so a $\boxed{90°}$ rotation about *B* maps *H* to *D*.

16.2.2 Since *B'* is the image upon rotating *B* 90° around *A*, we have *AB'* = *AB* = 5 and ∠*BAB'* = 90°. Therefore, $BB' = \sqrt{BA^2 + B'A^2} = \boxed{5\sqrt{2}}$.

16.2.3 Because ∠*AOC* = 180° (since *O* is on $\overline{AC}$), it is impossible for the image of *A* upon a 90° rotation about *O* to be point *C*. Therefore the image of *A* must be either *B* or *D*. Let the image upon rotation be *B*. (The proof when the image is *D* is the same.) Because the image of *A* upon 90° rotation about *O* is *B*, we have *OA* = *OB* and $\overline{OA} \perp \overline{OB}$. Furthermore, the image of *B* under this rotation is given as *C* (it cannot be *D* because ∠*BOD* = 180°), the image of *C* is *D* and the image of *D* is *A*. Therefore, we have *OA* = *OB* = *OC* = *OD* and $\overline{AC} \perp \overline{BD}$. We thus find that △*AOB*, △*BOC*, △*COD*, and △*DOA* are all congruent isosceles right triangles. Therefore, *AB* = *BC* = *CD* = *DA* and each angle of *ABCD* is the sum of two 45° angles, or 90°. Hence, *ABCD* must be a square.

16.2.4

(a) Because plane *BCD* is perpendicular to the line about which we are rotating, its image is in the same plane as *BCD*. Therefore, to determine the image of rotating *BCD* about the altitude from *A*, we look at the result of rotating *BCD* 60° about the foot of the altitude from *A*. As we have seen earlier in the text, the foot of the altitude from *A* to *BCD* is the center of △*BCD* because *ABCD* is a regular tetrahedron. Let this be point *O* and the image of *BCD* be *B'C'D'*. Therefore, $\angle B'OB = \angle C'OC = \angle D'OD = 60°$. Since we also have $\angle BOC = \angle BOD = \angle DOC = 120°$, we have

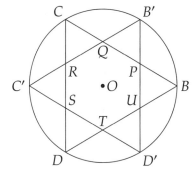

$\angle COB' = \angle DOC' = \angle BOD' = 60°$. Therefore, we see that *BB'CC'DD'* is a regular hexagon, since its vertices are equally spaced about a circle with center *O* and radius *OB*.

The intersection of *BCD* and *B'C'D'* is also a regular hexagon. We can see this quickly by using the aforementioned circle. Let *PQRSTU* be the hexagon formed by the intersection points of the two triangles. We have $\angle CQR = (\overarc{CC'} + \overarc{BB'})/2 = 60°$. Similarly, we can show that our outer six little triangles are congruent equilateral triangles. We can also use our arcs to show that each angle of *PQRSTU* is 120°: $\angle PQR = (\overarc{BDC'} + \overarc{B'C})/2 = 120°$. Therefore, the sides of *PQRSTU* are all equal and the angles of *PQRSTU* are all equal, so *PQRSTU* is a regular hexagon. Hence, our region of intersection is a regular hexagonal pyramid.

(b) As we found in Problem 14.10 of the textbook, the altitude of the tetrahedron is $2\sqrt{6}$. Since the small outer triangles of the previous part are all congruent, each side of the hexagonal base is 1/3 the side length of the tetrahedron. Therefore, the base is a regular hexagon with side length 2, and thus has area $3(2)^2\sqrt{3}/2 = 6\sqrt{3}$. Therefore, the volume of the intersection region is $(2\sqrt{6})(6\sqrt{3})/3 = \boxed{12\sqrt{2}}$.

Exercises for Section 16.3

16.3.1 For each of the following parts, we search for lines of symmetry using the fact that a reflection over a line of symmetry of a polygon must map each vertex of the polygon to a vertex of the polygon.

(a) Let *ABCD* be the rectangle. The line connecting the midpoints of $\overline{AB}$ and $\overline{CD}$ is a line of symmetry (mapping *A* to *B*), as is the line connecting the midpoints of $\overline{AD}$ and $\overline{BC}$ (mapping *A* to *D*). There are no other lines of symmetry (we can't map *A* to itself or to *C* and still successfully map the whole rectangle to itself). Therefore, a rectangle that is not a square has $\boxed{2}$ lines of symmetry.

(b) Each of the diagonals of a rhombus is part of a line of symmetry of the rhombus. If the rhombus is not a square, then it has no other lines of symmetry. Therefore, such a rhombus has $\boxed{2}$ lines of symmetry.

(c) A parallelogram that is not a rectangle or a rhombus has $\boxed{0}$ lines of symmetry. We can see this by considering reflections that map *A* to each vertex of parallelogram of *ABCD*. In each case, the image of *ABCD* is not *ABCD*.

(d) There are $\boxed{5}$ lines of symmetry of a regular pentagon: one through each vertex and the midpoint of the side opposite the vertex.

(e) There are $\boxed{12}$ lines of symmetry of a regular dodecagon, one through each of the six pairs of opposite vertices and one through each of the six pairs of midpoints of opposite sides.

(f) There are $\boxed{n}$ lines of symmetry of a regular n-gon. Let the polygon be $A_1A_2 \cdots A_n$. For each i from 2 to n, the perpendicular bisector of A_iA_1 is a line of symmetry. In addition to these $n - 1$ lines of symmetry, there is also a line of symmetry through A_1 and either the opposite vertex (if n is even) or the opposite side (if n is odd).

16.3.2 Since $\overline{DA'}$ is the image of $\overline{DA}$, and $\overline{AD} \perp \overline{CD}$, we have $\overline{A'D} \perp \overline{CD}$ as well, so $\angle ADA' = 180°$ and $AA' = AD + A'D = 2AD = \boxed{8}$. $\triangle AA'B$ is a right triangle with legs $\overline{AA'}$ and $\overline{AB}$. Therefore, $A'B = \sqrt{A'A^2 + AB^2} = \boxed{4\sqrt{5}}$.

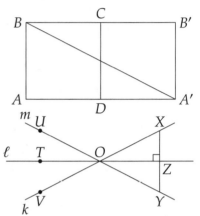

16.3.3 Let X be on k and Y be on m such that Y is the image of X upon reflection over ℓ. Therefore, $\overline{XY}$ is perpendicular to ℓ. Let Z be the point where $\overline{XY}$ meets ℓ. Since Y is the image of X upon reflection over ℓ, we have $XZ = YZ$. We also have $\angle XZO = \angle YZO$ (where O is the point where k and ℓ meet) and $OZ = OZ$, so $\triangle XZO \cong \triangle YZO$ by SAS. Therefore, $\angle XOZ = \angle YOZ$, so ℓ bisects $\angle XOY$. Because $\angle TOV = \angle XOZ$ and $\angle YOZ = \angle UOT$, we have $\angle TOV = \angle UOT$, so ℓ bisects $\angle UOV$ as well. Therefore, ℓ bisects a pair of the angles between k and m.

16.3.4

(a) $\boxed{\text{Yes}}$. Suppose $\ell \perp m$ as shown at left below, and X, Y, and Z are as described in the problem. Y' is the image of Y upon reflection over m. Since $\ell \perp m$, $\overline{XY} \perp \ell$, $\overline{XZ} \perp m$, and $\overline{YY'} \perp m$, we have $\overline{XZ} \parallel \ell \parallel \overline{YY'}$ and $\overline{XY} \parallel m$. Since $XSQY$ is a rectangle, we have $XS = YQ$. Since $ZS = XS$ and $Y'Q = QY$, we have $ZX = Y'Y$. Since $XZ = YY'$ and $\overline{XZ} \parallel \overline{YY'}$, we know that $XZY'Y$ is a parallelogram. Since $\angle X = 90°$, $XZY'Y$ is a rectangle. Therefore, $ZY' = XY$ and $\overline{ZY'} \parallel \overline{XY}$ and $ZRPX$ and $RY'YP$ are both rectangles. Hence, $\overline{ZY'} \perp \ell$ and $ZR = XP = PY = RY'$, so Y' is the image of Z upon reflection over ℓ in addition to being the image of Y over m.

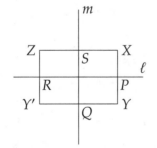

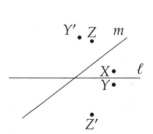

(b) $\boxed{\text{No}}$. If ℓ and m are not perpendicular, the the reflection of Y over m cannot be the same point as the reflection of Z over ℓ. An example is shown at right above.

16.3.5

(a) The image of the square base must be itself, and the image of the apex of the pyramid must also be itself. Therefore, the planes of symmetry are those that pass through the apex of the pyramid and a line of symmetry of the base. As we saw earlier, there are $\boxed{4}$ lines of symmetry of a square, and each gives us a different plane of symmetry of the pyramid.

(b) There is a plane of symmetry for each pair of faces of the prism: the plane mid-way between those faces. There are no other planes of symmetry of a rectangular prism that does not have a square as a face (see if you can prove why!). Therefore, there are $\boxed{3}$ planes of symmetry of such a prism.

(c) First, we note that it is impossible for a plane of symmetry of a tetrahedron to pass through 1 or 3 vertices, since then there will be an odd number of vertices left to reflect through the plane. We couldn't then pair off the remaining vertices to be images of each other.

   Clearly, no plane can go through all 4 vertices. Letting our tetrahedron be $ABCD$, if a reflection through a plane maps A to B, then that plane must be perpendicular to $\overline{AB}$ and pass through the midpoint of $\overline{AB}$. Therefore, this plane consists of all points equidistant from A and B, so it must contain C and D. Hence, all planes of symmetry pass through two of the vertices of the tetrahedron and the midpoint of the segment connecting the other two vertices. There are $\boxed{6}$ ways to choose the edge that is bisected by the plane, so there are 6 planes of symmetry of a regular tetrahedron.

(d) There are two types of planes of symmetry of a cube. First, there are planes that are parallel to one pair bases and cut the cube in half. There are 3 such planes – one for each pair of opposite faces of the cube. Second, there are planes that contain opposite edges of a cube (and therefore intersect on pair of opposite faces along diagonals of those faces). There is one of these planes for each pair of opposite edges. Since there are 6 such pairs of edges (there are 12 edges on a cube), there are 6 such planes. Therefore, we have a total of $3 + 6 = \boxed{9}$ planes of symmetry of a cube.

(e) The image of the vertex of the cone must be itself, and the image of the base of the cone must be itself. Therefore, the plane of symmetry must pass through the vertex and be perpendicular to the base, so it must pass through the center of the base as well as the vertex of the cone. Since the plane contains both the vertex of the cone and the center of the cone's base, it must contain the axis of the cone.

Exercises for Section 16.4

16.4.1

(a) Because $\overline{X'Y'}$ is the image of $\overline{XY}$ under a dilation with scale factor 4, we have $X'Y' = 4XY = \boxed{12}$.

(b) Triangle XYZ is a 3-4-5 right triangle with legs $\overline{XY}$ and $\overline{YZ}$, so $\triangle X'Y'Z'$ is a right triangle with legs $\overline{X'Y'}$ and $\overline{Y'Z'}$. We already found that $X'Y' = 12$, and we have $Y'Z' = 4YZ = 16$, so $[X'Y'Z'] = (X'Y')(Y'Z')/2 = (12)(16)/2 = \boxed{96}$.

(c) $\boxed{\text{No}}$. To see why, let O be the center of the dilation. Then, we have $OX' = 4OX$, and X is on $\overline{OX'}$. Therefore, we have $XX' = OX' - OX = 3OX$, but we don't know anything about OX! So, we can't determine XX'.

16.4.2 Because Q is the image of P under a dilation with center O and scale factor 5, we have $OQ = 5OP$, and point P is on $\overline{OQ}$. Therefore, we have $OP + PQ = OQ = 5OP$. Since $PQ = 20$, we have $OP + 20 = 5OP$, from which we find $OP = \boxed{5}$.

16.4.3 We found in the text that if A' and B' are the images of A and B, respectively, under a dilation with scale factor k, then we have $A'B'/AB = k$. Therefore, since A', B', and C' are the images of A, B, and C, under a dilation with scale factor k, we have

$$\frac{A'B'}{AB} = \frac{A'C'}{AC} = \frac{B'C'}{BC} = k.$$

Therefore, we have $\triangle ABC \sim \triangle A'B'C'$ by SSS Similarity.

16.4.4

(a) Because A' and X' are the images of A and X, respectively, under the same dilation, we have $\overline{A'X'} \parallel \overline{AX}$. Since O is the center of this dilation, points O, X, and X' are collinear. Combining this with the fact that $\overline{AX} \parallel \overline{A'X'}$, we find that $\angle OXA = \angle OX'A'$.

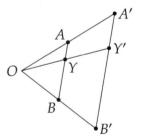

In the same way, we find that $\overline{BX} \parallel \overline{B'X'}$, so $\angle OXB = \angle OX'B'$.

(b) Because X is on $\overline{AB}$, we have $\angle AXB = 180°$. Using the results of part (a), we also have

$$\angle AXB = \angle AXO + \angle OXB = \angle A'X'O + \angle OX'B' = \angle A'X'B'.$$

Therefore, we have $\angle A'X'B' = \angle AXB = 180°$, which means that X' is on $\overline{A'B'}$. So, the image of each point on $\overline{AB}$ under the dilation is a point on $\overline{A'B'}$.

(c) All we have shown so far is that the image of each point on $\overline{AB}$ is a point on $\overline{A'B'}$. We haven't shown that the image upon dilating $\overline{AB}$ is the entire segment $\overline{A'B'}$. It could just be part of the segment. To show that it is the whole segment, we have to show that each point on $\overline{A'B'}$ is the image of some point on $\overline{AB}$ under the dilation. To do so, we essentially reverse our steps from above.

Let Y' be a point on $\overline{A'B'}$. Then, let Y be the point on $\overrightarrow{OY'}$ such that $OY = kOY'$. In other words, Y' is the image of Y under the dilation. We wish to show that Y is on $\overline{AB}$. Since A' and Y' are the images of A and Y, respectively, under the dilation, we have $\overline{AY} \parallel \overline{A'Y'}$. Therefore, we have $\angle A'Y'O = \angle AYO$. Similarly, we find $\angle B'Y'O = \angle BYO$. Because Y' is on $\overline{A'B'}$, we have $\angle A'Y'B' = 180°$. Moreover, we have

$$180° = \angle A'Y'B' = \angle A'Y'O + \angle B'Y'O = \angle AYO + \angle BYO = \angle AYB.$$

Since $\angle AYB = 180°$, point Y is on $\overline{AB}$. Therefore, every point on $\overline{A'B'}$ is the image of some point on $\overline{AB}$ under the dilation. Combining this with our previous observation that the image of each point on $\overline{AB}$ is a point on $\overline{A'B'}$, we find that $\overline{A'B'}$ is the image of $\overline{AB}$ under the dilation.

Exercises for Section 16.5

16.5.1 The center of the circle is also the center of the rectangle, since the midpoint of each diagonal is also the midpoint of the hypotenuse of a right triangle inscribed in the circle. We connect this center to each of the four vertices of the rhombus, thus dividing our original rectangles into four smaller rectangles. Each side of the rhombus is a diagonal of one of these rectangles. The other diagonal of each of these rectangles is a radius of the circle. Since the diagonals of a rectangle are equal in length, the sides of the rhombus must therefore have length equal to the radius of the circle. Thus, the perimeter of the rhombus is $4(4) = \boxed{16}$ feet.

16.5.2 Because $ABCD$ is a rectangle, we have $\overline{AB} \parallel \overline{DC}$, so $\angle EDA = \angle FAB$ and $\angle DEF = \angle FBA$. We also have $AB = CD = DE$, so $\triangle FDE \cong \triangle FAB$ by SAS Congruence. Therefore, $FA = FD$, so $DA = 2FD$. The area of the rectangle is $(CD)(AD) = (DE)(2FD) = 2(DE)(DF)$. We are given that $(DE)(DF)/2 = 4$ because the area of right triangle $\triangle FDE$ is 4, so we have $[ABCD] = 2(DE)(DF) = \boxed{16}$.

Notice that we could have quickly arrived at this numerical answer by letting $DE = 4$ and $FD = 2$; then, we have $CD = 4$, $FA = 2$, $DA = 4$, so $[ABCD] = 16$. This doesn't prove that the area is 16 for all possible arrangements, but it gives us a quick check.

16.5.3 Since the arcs have radius 5, we have $BD = 10$. We draw $\overline{BD}$, then 'complete' rectangle $BDEF$ by moving sectors BOC and COD into sectors BFA and DEA. Our rearranged figure is a rectangle with dimensions $BD = 10$ and $DE = 5$, and its area is the same as the original area, so our answer is $(BD)(DE) = \boxed{50}$.

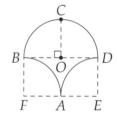

16.5.4 Let O be the center of the square. O is the midpoint of diagonal $\overline{BD}$, and the midpoint of $\overline{MN}$, which connects the midpoints of the opposite sides of the square. Therefore, the image of $\overline{BN}$ upon $180°$ rotation about O is $\overline{DM}$. Furthermore, the image of $\overline{AC}$ upon this rotation is $\overline{CA}$, so the image of the intersection of $\overline{BN}$ and $\overline{AC}$ (point Y) is the intersection of $\overline{DM}$ and $\overline{CA}$ (these are the images of $\overline{BN}$ and $\overline{AC}$), which is point X. Therefore, $DNYX$ is the image of $BMXY$ upon $180°$ rotation about the center of the square.

16.5.5 We can at least get a quick answer by letting $ABCD$ be a rectangle. Then, $BC = AD = 7$, and $AB = CD$. Since $BC = AB + CD$, we have $AB = CD = BC/2 = 3.5$, so $(AB)(CD) = \boxed{12.25}$.

To prove this is the case for all trapezoids that fit the problem, we let $\overline{AB}$ be the shorter base of the trapezoid and draw altitude $\overline{BX}$ to $\overline{CD}$. Then, $ADXB$ is a rectangle, so $BX = 7$ and $XD = AB$. Therefore, $CX = CD - AB$. From right triangle $\triangle BXC$ we have $BX^2 + CX^2 = BC^2$, so $(AB + CD)^2 = (CD - AB)^2 + 49$. Expanding and rearranging, we have $4(AB)(CD) = 49$, so $(AB)(CD) = \boxed{12.25}$.

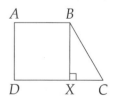

Exercises for Section 16.6

16.6.1 Since A' and Q' are the images of A and Q, respectively, under the same translation, we have $AA' = QQ'$ and $\overline{AA'} \parallel \overline{QQ'}$, so $AA'Q'Q$ is a parallelogram. To construct Q', we construct a line through Q parallel to $\overline{AA'}$, then draw a line through A' parallel to $\overline{AQ}$. We take the point where these lines intersect to be our Q'. Since $\overline{QQ'} \parallel \overline{AA'}$ and $\overline{Q'A'} \parallel \overline{QA}$, $AA'Q'Q$ is a parallelogram. Therefore $\overline{QQ'} \parallel \overline{AA'}$ and $QQ' = AA'$, so Q' is the image of Q under the same translation that maps A to A'.

16.6.2 The image of P upon reflection over ℓ is the point P' on the other side of ℓ such that ℓ is perpendicular to $\overline{PP'}$ and bisects the segment. We first construct the line through P that is perpendicular to ℓ. Let this line meet ℓ at X. We then draw a circle with center X and radius XP. Let the other point (besides P) where this circle hits $\overleftrightarrow{XP}$ be P'. Since we have $P'X = XP$ and $\overline{PP'} \perp \ell$, we have found our desired image.

After following this procedure with enough points, you may start to suspect that the result of a composition of two reflections is a rotation. Your suspicion is correct. Can you prove it?

Review Problems

16.22 The octagon is shown at left below. Since $HCDG$ is a rectangle (why?), we have $\overline{HC} \parallel \overline{GD}$ and $HC = GD$. The same translation that maps G to D also maps H to $\boxed{C}$.

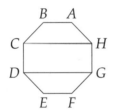 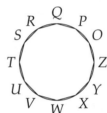

16.23 Our dodecagon and its circumcircle are shown at right above. We include the circumcircle because in each part we seek an angle that is formed by connecting the vertices of the polygon. Each such angle is inscribed in an arc of the circumcircle, and we can easily find the measure of each of these arcs because each arc between successive vertices of the dodecagon is $360°/12 = 30°$.

(a) The angle $\angle VUT$ is an angle of the regular dodecagon, so it has measure $\angle VUT = 180° - (360°/12) = \boxed{150°}$.

(b) The angle $\angle XUR$ cuts off 6 small arcs of the circumcircle, so its measure is $(6 \cdot 30°)/2 = \boxed{90°}$. We might also have noted that this is one of the angles of square $XURO$.

(c) $\angle WUS$ cuts off 8 small arcs of the circumcircle, so it has measure $(8 \cdot 30°)/2 = \boxed{120°}$. We might also have noted that this is one of the angles of regular hexagon $WUSQOY$.

(d) $\angle YUQ$ cuts off 4 small arcs of the circumcircle, so it has measure $(4 \cdot 30°)/2 = \boxed{60°}$. We might also have noticed that this is one of the angles of equilateral triangle $\triangle UQY$.

(e) $\angle ZUP$ cuts off 2 small arcs of the circumcircle, so it has measure $(2 \cdot 30°)/2 = \boxed{30°}$.

16.24 The only lines of symmetry are those lines that pass through a pair of midpoints of opposite sides. There are four pairs of opposite sides, so there are $\boxed{4}$ lines of symmetry. Make sure you see both why these are lines of symmetry, and why there are no other lines of symmetry.

16.25

(a) Yes. $\triangle GHI$ is isosceles with $GH = GI$. Therefore, the altitude from G to $\overline{HI}$ is also the perpendicular bisector of $\overline{HI}$, so the reflection of H over this altitude is I. Since G is on this altitude, the image of G upon reflection over this altitude is itself. Hence, the image of $\overline{GH}$ upon reflection over the altitude is $\overline{GI}$.

(b) No. Suppose $XYBA$ is a parallelogram that is not a rectangle. Then, we have $XY = AB$. However, the reflection that maps X to A (the reflection over the perpendicular bisector of $\overline{AX}$) does not map Y to B. Nor does the reflection that maps X to B map Y to A.

16.26

(a) Yes. Since $AD = BC$, $\angle C = \angle D$, and $CM = DM$, we have $\triangle ADM \cong \triangle BCM$ by SAS Congruence. Therefore, we have $AM = BM$, so a rotation of $\angle AMB$ (in the appropriate direction) maps A to B.

(b) No. Since $AM = \sqrt{AD^2 + DM^2} > \sqrt{DM^2} = DM = CM$, we know that $AM > CM$. The image of A upon any rotation about M must be the same distance from M as A is, so C cannot be the image of A upon some rotation about M.

(c) Yes. The reflection over the perpendicular bisector of $\overline{AM}$ maps M to A.

16.27 The two unshaded pieces of the central square can be combined to form a square that is 1/4 the central square. There are four shaded pieces that are outside the central square. Each of these is congruent to one of the unshaded pieces of the central square. Hence, two of these pieces can be combined to form one of the aforementioned squares that is 1/4 the central square, while the other two of these pieces can be used to 'complete' the central square. The area of the central square is 1, so our shaded area is $1 + 1/4 = \boxed{5/4}$.

16.28

(a) If a point is reflected over a line that passes through that point, the image is the point itself. Therefore, $W' = W$ and $E' = E$.

(b) Since $\angle QWE = \angle Q'WE = 120°$, we have $\angle QWQ' = 360° - \angle QWE - \angle Q'WE = \boxed{120°}$.

(c) We build a right triangle by drawing $\overline{TT'}$. We find ET by drawing the altitude from R to X, and noting that $\triangle RTX$ is a 30-60-90 triangle. Therefore, $TX = RT\sqrt{3}/2 = 3\sqrt{3}$, so $TT' = 4XT = 12\sqrt{3}$. Finally, we have $YT' = \sqrt{T'T^2 + YT^2} = \boxed{6\sqrt{13}}$.

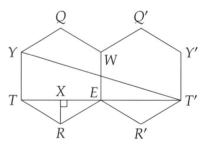

16.29

(a) If line ℓ is a line of symmetry of $\triangle ABC$, then there must be the same number of vertices on each side of the line. Therefore, there must be an even number of vertices of $\triangle ABC$ that are not on the line. Since the triangle has an odd number of vertices, this means that one of the vertices must be on ℓ if ℓ is a line of symmetry.

(b) Let the triangle be $\triangle ABC$. As we saw in the previous part, the line of symmetry must go through a vertex of the triangle. Call the vertex that the line of symmetry passes through vertex A. Then,

$\overline{AB}$ is the image of $\overline{AC}$ upon reflection over ℓ. Hence, $AB = AC$, so it is impossible for $\triangle ABC$ to be scalene.

16.30

(a) Because R is the image of S under a dilation with center T and scale factor 3, we have $TR = 3TS$. Therefore, we have $TS = TR/3 = \boxed{6}$.

(b) Because E is the image of D under a dilation with center O and scale factor -3, we have $EO = 3DO$. However, because the scale factor is negative, point O is on $\overline{ED}$, so $EO + OD = ED$. Since $ED = 9$ and $EO = 3OD$, we have $3OD + OD = 9$, from which we find $OD = \boxed{9/4}$.

(c) Because X is the image of Y under a dilation with center A and scale factor 0.25, we have $AX = 0.25AY$. Furthermore, because the scale factor is between 0 and 1, point X is on $\overline{AY}$. Therefore, we have $AX + XY = AY$. Since $XY = 20$ and $AX = 0.25AY$, we have $0.25AY + 20 = AY$, from which we find $AY = \boxed{80/3}$.

16.31 Because $\odot O'$ is the image of $\odot O$ under a dilation with scale factor 4, the radius of $\odot O'$ is 4 times the radius of $\odot O$. Therefore, the area of $\odot O'$ is $4^2 = 16$ times the area of $\odot O$. So, the area of $\odot O$ is $(48\pi)/16 = \boxed{3\pi}$.

16.32 The shortest path from A to the wall then to B will have the same length as the path from A to the wall, then to the image of B upon reflection over the wall. Let B' be the image of B upon reflection over the wall. The shortest path from A to B' is a straight line. To find AB', we build right triangle $\triangle AXB'$ as shown. We have $B'A = \sqrt{B'X^2 + AX^2} = \boxed{400\sqrt{13} \text{ m}}$.

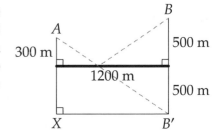

16.33 Because B is the image of A upon reflection over ℓ, we have $\overline{AB} \perp \ell$. Since we also have $\ell \perp m$, we have $\overline{AB} \parallel m$. Because C is the image of B upon reflection over m, we have $\overline{BC} \perp m$. Since $\overline{AB} \parallel m$ and $\overline{BC} \perp m$, we have $\overline{BC} \perp \overline{AB}$, so $\angle ABC = 90°$.

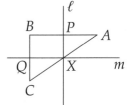

Let $\overline{AB}$ meet ℓ at P and $\overline{BC}$ meet m at Q. Because B is the image of A upon reflection over ℓ, we have $AP = BP$. Similarly, $BQ = QC$. We have $\triangle XPA \cong \triangle XPB$ by SAS Congruence, so $XA = XB$. Similarly, we have $\triangle XQB \cong \triangle XQC$. Therefore, $XB = XC$, so $XA = XB = XC$.

From our triangle congruences, we have $\angle BXQ = \angle QXC$ and $\angle AXP = \angle BXP$. Therefore, we have $\angle AXC = \angle AXP + \angle BXP + \angle BXQ + \angle CXQ = 2(\angle BXP + \angle BXQ) = 2(90°) = 180°$, so $\angle AXC$ is a straight angle.

We could also have argued that because $\triangle ABC \sim \triangle XQC$ by SAS Similarity ($BC/QC = AB/XQ = 2$ and $\angle B = \angle XQC$), we have $AC/XC = AB/XQ = 2$. Therefore, $AC = 2XC = XC + XA$, so by the Triangle Inequality, X must be on $\overline{AC}$.

We also could have solved this problem by noting that $\overline{BX}$ is the image of $\overline{AX}$ upon reflection over ℓ, so $\angle AXP = \angle BXP$. Similarly, $\angle BXQ = \angle CXQ$. Since $\angle BXP + \angle BXQ = 90°$, we have $\angle AXC = \angle AXB + \angle BXC = 2(\angle BXP + \angle BXQ) = 180°$. Therefore, $\angle AXC$ is a straight angle.

16.34 Because the shaded area equals the remaining uncovered area of the paper and the shaded area equals the area where that shaded region used to be, the shaded area must be 1/3 the area of the whole square. The shaded area plus the original location of the shaded triangle is therefore a square that has area 2/3 the original square, or 2(6)/3 = 4. Therefore, the side length of this square is 2, so $AA' = \boxed{2\sqrt{2}}$.

16.35 Because a rotation about A maps X to Y, we have $AX = AY$. Since a rotation about X maps A to Y, we have $XA = XY$. Therefore, $AY = AX = XY$, so $\triangle AXY$ is equilateral.

16.36

(a) $\boxed{\text{Yes}}$. The image of A must be a point A' such that $\angle AOA' = 180°$. Since O is inside the quadrilateral (and therefore not on a side), the image of A cannot be B or D. Therefore, the image of A is C, and, similarly, the image of B is D. Hence, we have $AO = OC$ and $BO = OD$, and AOC and BOD are straight lines. Therefore, $\triangle AOB \cong \triangle COD$ by SAS Congruence, so $\angle OAB = \angle OCD$, which gives us $\overline{AB} \parallel \overline{CD}$. Similarly, we can show that $\overline{AD} \parallel \overline{BC}$, so $ABCD$ is a parallelogram.

Alternatively, we could note that the diagonals of $ABCD$ bisect each other, so $ABCD$ is a parallelogram.

(b) $\boxed{\text{No.}}$ The image of any parallelogram upon a rotation of 180° is itself, because the diagonals of a parallelogram bisect each other. Therefore, $ABCD$ need not be a rectangle.

(c) $\boxed{\text{No.}}$ Since $ABCD$ need not be a rectangle, it need not be a square either.

16.37 Our given information tells us that $\angle ABC = 90°$, $BA = BC$, $\angle ADC = 90°$, and $AD = DC$. We also know from the directions of the rotations than D and A are not the same point. Therefore, $\triangle ADC$ and $\triangle BAC$ are both isosceles right triangles. Since they have the same hypotenuse, they are congruent by ASA. Therefore, all the sides of $ABCD$ are equal, so $ABCD$ is a rhombus. Since one of the angles of this rhombus is right, $ABCD$ is a square.

16.38 Since $\angle A = \angle C = 60°$, we know $\overline{BD}$ divides the rhombus into two equilateral triangles. Therefore, the image of A upon a 60° clockwise rotation about B is D, and the image of D upon this rotation is C. Thus, the image of $\overline{AD}$ is $\overline{DC}$. Since $\angle PBQ = 60°$, the image of P must be on $\overrightarrow{BQ}$. Since P is on $\overline{AD}$, the image of P must be on $\overline{CD}$. Therefore, the image of P is Q, so $PB = BQ$. Since $\angle PBQ = 60°$ and $PB = BQ$, we know that $\triangle PBQ$ is equilateral, so $\angle BPQ = \angle PQB = \boxed{60°}$.

16.39 Since $AB^2 + BC^2 = AC^2$, $\triangle ABC$ is a right triangle with hypotenuse $\overline{AC}$.

(a) This cone has radius $BC = 4$ and height $AB = 3$, so its volume is $\frac{1}{3}\pi \cdot 4^2 \cdot 3 = \boxed{16\pi}$.

(b) This cone has radius $AB = 3$ and height $BC = 4$, so its volume is $\frac{1}{3}\pi \cdot 3^2 \cdot 4 = \boxed{12\pi}$.

(c) This solid is a combination of two cones. Both have radius equal to BX, the distance from B to $\overline{AC}$. Since $[ABC] = (AB)(BC)/2 = (AC)(BX)/2$, we have $BX = (AB)(BC)/AC = 12/5$.

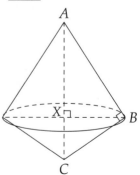

Let h_1, h_2 be the heights of the two cones. Then $h_1 + h_2 = AC = 5$, so the volume of this solid is

$$\frac{1}{3}\pi BX^2 h_1 + \frac{1}{3}\pi BX^2 h_2 = \frac{1}{3}\pi BX^2(h_1 + h_2) = \frac{1}{3}\pi\left(\frac{12}{5}\right)^2 \cdot 5 = \boxed{\frac{48\pi}{5}}.$$

(We could also have used similar triangles or the Pythagorean Theorem to find the heights of each cone.)

16.40 Because X' and Y' are the images of X and Y, respectively, under a dilation about O with scale factor k, we have

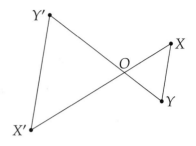

$$\frac{OX'}{OX} = \frac{OY'}{OY} = |k|.$$

(Note that we use $|k|$ instead of just k so our proof covers both dilations with negative scale factors and dilations with positive scale factors.) Since $\overline{XX'}$ and $\overline{YY'}$ both pass through O, we have $\angle XOY = \angle X'OY'$. Therefore, we have $\triangle XOY \sim \triangle X'OY'$ by SAS Similarity, which gives us $X'Y'/XY = OX'/OX = |k|$. Moreover, we have $\angle OXY = \angle OX'Y'$, so $\overline{XY} \parallel \overline{X'Y'}$.

Challenge Problems

16.41 $\boxed{\text{Yes}}$. Because the image of $\overline{AB}$ upon rotation is $\overline{CD}$, we have $AB = CD$, and the image of A must be either C or D. Let the image of A be C and the center of rotation be O. Since A is the image of C upon rotation about O, we have $AO = OC$, so O must be on the perpendicular bisector of $\overline{AC}$. Similarly, B maps to D under this rotation, so O is on the perpendicular bisector of $\overline{BD}$. There is only one intersection of these two perpendicular bisectors. If $\overline{AB}$ and $\overline{CD}$ share midpoint M, then M is on both of these perpendicular bisectors. Therefore, O must be M.

16.42 The image of $\angle AOC$ upon reflection over $\overleftrightarrow{XY}$ is $\angle BOD$. Since $\angle BAD = 90°$, $\overparen{BD} = 90°$. Therefore, $\angle BOD$ is inscribed in a $360° - 90° = 270°$ arc (if we draw the whole circle of which $\overparen{BD}$ is a part), so its measure is $(270°)/2 = 135°$. So, our answer is $\angle AOC = \angle BOD = \boxed{135°}$.

16.43

(a) Let the perpendicular bisectors of $\overline{HI}$ and $\overline{GH}$ meet at O. Therefore, $OG = OH = OI$. Since $IJ = GH$, $IO = GO$, and $\angle OIJ = \angle HIJ - \angle HIO = \angle IHG - \angle IHO = \angle OHG = \angle OGH$, we have $\triangle OIJ \cong \triangle OGH$ by SAS Congruence. Therefore, $OJ = OH = OI = OG$. Similarly, we can go around the octagon, showing that each vertex of the octagon is the same distance from O.

(b) From quadrilateral $OIHG$, we have $\angle IOG = 360° - \angle OIH - \angle IHG - \angle OGH = 360° - \angle OHI - 135° - \angle OHG = 360° - 135° - (\angle OHI + \angle OHG) = 360° - 2(135°) = 90°$. Similarly, we can show that $\angle JOH = 90°$, $\angle KOI = 90°$, etc. Combined with the fact that O is equidistant from the vertices of the octagon, this tells us that $GHIJKLMN$ maps to itself under a $90°$ rotation about its center.

16.44 Point A is 6 units in the negative direction from the line $x = 8$. Its image upon reflection over $x = 8$ is therefore 6 units in the positive direction from $x = 8$, or $(14, 4)$. Similarly, the image of B is $(10, 4)$ and the image of C is $(12, 10)$. The sum of the x-coordinates of these images is $\boxed{36}$. Note that the original sum of the x-coordinates is 12 and $(12 + 36)/2 = 24 = 3 \cdot 8$. (Remember, we reflected the three points over $x = 8$.) Is this a coincidence?

16.45 The center of the incircle of an equilateral triangle is also the centroid and the center of the circumcircle. Because the circumcircle is the image of the incircle, the scale factor equals the ratio of the circumradius to the inradius. Therefore, we seek the ratio AO/OY in the diagram at right. Because $\overline{AY}$ and $\overline{CX}$ are also medians, point O is the centroid of $\triangle ABC$. So, we have $AO/OY = \boxed{2}$.

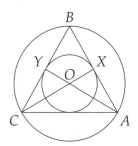

16.46 *Solution 1:* We draw perpendiculars from B to $\overline{AO}$ and $\overline{CO}$, thus forming square $BXOY$. Let Z be the foot of the perpendicular from B to $\overline{AC}$. Since $AB = AB$, $\angle BZA = \angle BXA$, and $\angle ZBA = \angle XAB$ (each is half one of the interior angles of the octagon), we have $\triangle ABX \cong \triangle BAZ$ by AAS Congruence. Since $[BXOY] = [ABCO] - 2[ABX]$, $[AOC] = [ABCO] - 2[ABZ]$, and $[ABX] = [ABZ]$, we have $[BXOY] = [AOC]$. Therefore, $OX^2 = (OA)(OC)/2 = OA^2/2$. Hence, we have $OA = OX\sqrt{2}$. Since $[BOA]/[BOX] = OA/OX = \sqrt{2}$, we have $[ABCO]/[BXOY] = (2[BOA])/(2[BOX]) = \sqrt{2}$. We already showed that $[BXOY] = [AOC]$, so we have $[ABCO] = [AOC]\sqrt{2}$. Since $[ABCO] = [ABCDEFGH]/4$ and $[AOC] = [ACEG]/4$, we have the desired $[ABCDEFGH] = [ACEG]\sqrt{2}$.

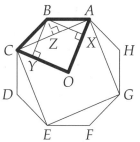

Solution 2: Let Z be the intersection of $\overline{AC}$ and $\overline{BF}$. Since $\overline{BF}$ is an axis of symmetry of octagon $ABCDEFG$, $\overline{AC}$ is perpendicular to $\overline{BF}$, and Z is the midpoint of $\overline{AC}$. Then $[BOC] = (OB)(CZ)/2 = (OB)(AC)/4$. Triangle AOC is 45-45-90, so $AC = \sqrt{2}OA$. Since $OB = OA = AC/\sqrt{2}$, we have $(OB)(AC)/4 = (\sqrt{2}/8)AC^2$. Therefore, $[ABCDEFG] = 8[BOC] = AC^2\sqrt{2} = [ACEG]\sqrt{2}$.

16.47 The two shortest paths possible are shown in the diagram. The path shown by the solid lines covers a total horizontal distance of $1 + 6 + 1 = 8$ and vertical distance of $5 + 5 = 10$, for a total path length of $\sqrt{8^2 + 10^2} = \sqrt{164} = 2\sqrt{41}$. The dashed path covers a horizontal distance of 4 and a vertical distance of $5 + 2 + 2 + 5 = 14$, for a total path length of $\sqrt{4^2 + 14^2} = \sqrt{212} = 2\sqrt{53}$. Therefore, the shortest path has length $\boxed{2\sqrt{41}}$.

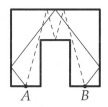

16.48

(a) Because $A'B' = |k|AB$, where $k \neq 1$, we do not have $A'B' = AB$, so $A'B'BA$ is not a parallelogram. (Note that we use $|k|$ instead of just k so our proof covers both dilations with negative scale factors and dilations with positive scale factors.) However, we do have $\overline{AB} \parallel \overline{A'B'}$, so we cannot have $\overleftrightarrow{AA'} \parallel \overleftrightarrow{BB'}$ as well. Therefore, $\overleftrightarrow{AA'}$ and $\overleftrightarrow{BB'}$ intersect

(b) Because $\overline{AB} \parallel \overline{A'B'}$, we have $\angle PAB = \angle PA'B'$ and $\angle PBA = \angle PB'A'$, so $\triangle PAB \sim \triangle PA'B'$ by AA Similarity. Notice that our proof so far addresses both possibilities shown below.

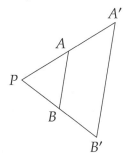
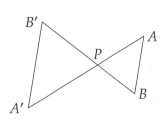

(c)   Because $\triangle PAB \sim \triangle PA'B'$, we have

$$\frac{A'B'}{AB} = \frac{PA'}{PA} = \frac{PB'}{PB} = |k|.$$

In the case at left above, we have A' on $\overrightarrow{PA}$ and B' on $\overrightarrow{PB}$ along with $PA'/PA = PB'/PB$, so there is a dilation about P that maps A to A' and B to B'. In the case at right above, P is on both $\overline{AA'}$ and $\overline{BB'}$, and we have $PA'/PA = PB'/PB$, so there is a dilation about P with a negative scale factor that maps A to A' and B to B'.

16.49   Since N and D are the images of M and C, respectively upon the same $90°$ rotation about the center of the square, $\overline{DN}$ is the image of $\overline{CM}$ under a $90°$ rotation about the center of the square. We showed in the text that if lines m and m' are such that one is the image of a rotation of θ degrees of the other, then the angle between the two is θ. Therefore, the angle between $\overline{DN}$ and $\overline{CM}$ is $90°$, so $\overline{DN} \perp \overline{CM}$.

16.50   As with the reflection problems we have already solved, we want to aim the cue ball at the reflection of B. However, we have a little wrinkle here: the center of the ball cannot reach the rail of the table. When the ball bounces off the rail, the center of the ball will be 1 inch away from the rail. Therefore, if we shoot the ball such that the center of the ball passes over B, the path of the center of the ball goes from A to a point 1 inch from the rail, then to B. Hence, to find the shortest path, we reflect B over the line that is one inch closer to B than the rail is. Let B' be the image of this reflection.

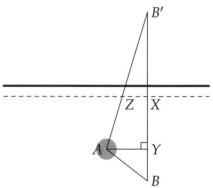

Let Y be the foot of the perpendicular from A to $\overline{BB'}$. Since $\overline{BB'}$ is perpendicular to both the rail and to $\overline{AY}$, the rail and $\overline{AY}$ are parallel. Since B is 3 units farther from the rail than A is, we have $YB = 3$. Since $AB = 5$, we have $AY = 4$ from right triangle $\triangle AYB$. Since $B'X = BX = 9 - 1 = 8$, we have $B'Y = BX + XY = 13$. Because $\overline{ZX} \parallel \overline{AY}$, we have $\triangle B'XZ \sim \triangle B'YA$, so $ZX/AY = B'X/B'Y = 8/13$. Therefore, $ZX = (AY)(8/13) = \boxed{32/13}$. ZX is our desired distance, since the point where the ball touches the rail when the ball bounces is just as far from $\overline{BB'}$ as the center of the ball is at that time. (Make sure you see why!)

16.51   We let A be $(0, 4)$, B be the point on our path on the y-axis, C be the point on the line $y = 6$, and D be $(18, 4)$. Let D' be the reflection of D over the line $y = 6$. Wherever B is along the line $y = 0$, the shortest path that goes from B to the line $y = 6$, then to point D, must meet the line $y = 6$ at the point where $\overline{BD'}$ hits the line $y = 6$. Let D'' be the image of D' upon reflection over $y = 0$. Wherever B is, the shortest path from A to the line

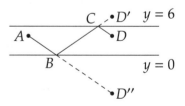

$y = 0$, then to D' (which equals the path from A to B to C to D), must meet the line $y = 0$ at the point where $\overline{AD''}$ meets the line $y = 0$.

The reflection of D over $y = 6$ is $(18, 8)$ and the reflection of this point over $y = 0$ is $(18, -8)$. Therefore, point D'' is 18 units to the right of A and 12 units below it, so our shortest path has length $\sqrt{18^2 + 12^2} = \boxed{6\sqrt{13}}$.

16.52 Let my path hit $\overrightarrow{BC}$ at E and $\overrightarrow{BD}$ at F. Wherever I hit $\overrightarrow{BC}$, I should then walk towards the image of A upon reflection over $\overleftrightarrow{BD}$. We let this image be A', and we see that wherever E is, the point F that minimizes my path length is the intersection of $\overline{EA'}$ and $\overrightarrow{BD}$, since $FA = FA'$. Now we must minimize $EA + EA'$. We know how to do this: we reflect A' over $\overrightarrow{BC}$ to get point A''. Since $EA' = EA''$, the Triangle Inequality gives us $AA'' \le AE + EA'' = AE + EA' = AE + EF + FA' = AE + EF + FA$, with $AA'' = AE + EF + FA$ if and only if E is on $\overline{AA''}$. Therefore, $AE + EF + FA \ge AA''$, so the length of my shortest path is AA'', which occurs when E is on $\overline{AA''}$. So, where $\overline{AA''}$ hits $\overrightarrow{BC}$ is our desired point E. We can thus construct my shortest path by constructing the reflection of A over $\overleftrightarrow{BD}$, then constructing the image of that image point upon reflection over $\overleftrightarrow{BC}$.

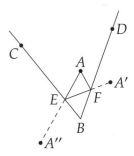

16.53 We can divide the region inside the octahedron into six regions based on which vertex of the octahedron the points are closest to. We want the volume of the region that corresponds to A. By symmetry, all six of these regions must have the same volume, so the volume of the region that is closest to A is $360/6 = \boxed{60}$.

16.54 A $60°$ clockwise rotation about R maps A to Q and P to B. Therefore, this rotation maps $\overline{AP}$ to $\overline{QB}$, so $AP = QB$. Similarly, a $60°$ rotation counterclockwise about Q maps $\overline{AP}$ to $\overline{RN}$, so $AP = RN$. (We could also have proved $\triangle ARP \cong \triangle QRB$ and $\triangle PQA \cong \triangle NQR$ to show that $AP = BQ = RN$.)

16.55 Let O be the center of the circle, P be the center of rotation, r be the radius of the circle, A be a point on $\odot O$, and O' and A' be the images of O and A, respectively, under our rotation about P. We therefore have $\angle OPO' = \angle APA'$, $OP = O'P$, and $AP = A'P$, so $\triangle OPA \cong \triangle O'PA'$ by SAS Congruence. Therefore, $O'A' = OA = r$, so the image of A must be on a circle with center O' and radius r.

Going backwards, let B' be a point on $\odot O$ with radius r. Let B the point such that B' is the image of B under our rotation. Therefore, $\angle BPB' = \angle OPO'$, $BP = B'P$, and $OP = O'P$, so $\triangle OPB \cong \triangle O'PB'$. We thus have $OB = O'B' = r$, so for each point B' on $\odot O'$, there is a point on $\odot O$ such that the image of that point under our rotation is B'. Therefore, the image of a circle under rotation is a circle.

CHAPTER 17

Analytic Geometry

Exercises for Section 17.1

17.1.1 The slope of the line is $(9 - 5)/[-5 - (-2)] = -4/3$, so a point-slope equation of the line is $y - 5 = (-4/3)[x - (-2)]$. Multiplying both sides by 3 then expanding both sides gives $3y - 15 = -4x - 8$. Rearranging this gives us $\boxed{4x + 3y = 7}$.

17.1.2 First, we find the equation of the line. Using the given slope and point, a point-slope equation of the line is $y - 2 = 4(x + 7)$. Rearranging this gives $4x - y = -30$. Letting $x = 0$ in this equation gives $y = 30$, so $\boxed{(0, 30)}$ is the y-intercept. Letting $y = 0$ gives $x = -15/2$, so $\boxed{\left(-7\frac{1}{2}, 0\right)}$ is the x-intercept.

17.1.3 Applying the distance formula gives us

$$\sqrt{(a + 4)^2 + (0 - 3)^2} = 5.$$

Squaring both sides and expanding the left side gives

$$a^2 + 8a + 25 = 25.$$

Rearranging this equation gives $a^2 + 8a = 0$. Factoring gives us $a(a + 8) = 0$, so $\boxed{a = 0 \text{ or } a = -8}$.

17.1.4 $\boxed{\text{No}}$. Because $(x_2 - x_1)^2 = (x_1 - x_2)^2$ and $(y_2 - y_1)^2 = (y_1 - y_2)^2$, it doesn't matter in which order we label the points (x_1, y_1) and (x_2, y_2).

17.1.5

(a) We have $PQ = \sqrt{(-8 + 5)^2 + (8 - 2)^2} = \sqrt{9 + 36} = \boxed{3\sqrt{5}}$.

(b) The midpoint of $\overline{PQ}$ is $\left(\frac{-5-8}{2}, \frac{2+8}{2}\right) = \left(-\frac{13}{2}, 5\right) = \boxed{\left(-6\frac{1}{2}, 5\right)}$.

(c) Let T be (a, b). Then, the midpoint of $\overline{PT}$ is $\left(\frac{a-5}{2}, \frac{b+2}{2}\right)$. Since this must be the same as Q, which is $(-8, 8)$, we have $(a - 5)/2 = -8$ and $(b + 2)/2 = 8$. Solving these equations for a and b gives us $(a, b) = \boxed{(-11, 14)}$.

17.1.6 Putting $4x - 3y + 14 = 0$ in slope-intercept form gives us $y = \frac{4}{3}x + \frac{14}{3}$. The slope of the graph of this equation is $\frac{4}{3}$. Therefore, the slope of k is $-\frac{3}{4}$. Because k has slope $-\frac{3}{4}$ and passes through $(-3, 2)$, a point-slope form of the equation whose graph is k is $y - 2 = -\frac{3}{4}(x + 3)$. Multiplying both sides by 4 then rearranging gives $\boxed{3x + 4y = -1}$.

17.1.7 Suppose (x_1, y_1) and (x_2, y_2) are on the graph of $y = mx + b$. Then, we have $y_1 = mx_1 + b$ and $y_2 = mx_2 + b$. So, the slope of the line through (x_1, y_1) and (x_2, y_2) is

$$\frac{y_2 - y_1}{x_2 - x_1} = \frac{mx_2 + b - (mx_1 + b)}{x_2 - x_1} = \frac{m(x_2 - x_1)}{x_2 - x_1} = m,$$

as desired.

17.1.8 Suppose the slopes of k and ℓ in the diagram at right have product -1. Since the product of these slopes is defined, neither of the lines is vertical. We draw two right triangles with legs parallel to the axes, as shown. Then, the slope of k is $-DA/DP$ and the slope of ℓ is BE/PE. Because the product of the slopes of k and ℓ is -1, we have

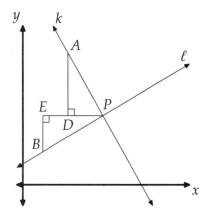

$$\left(\frac{BE}{PE}\right)\left(-\frac{DA}{DP}\right) = -1.$$

Rearranging this equation gives $BE/PE = DP/DA$. Since we also have $\angle PDA = \angle BEP$, we have $\triangle PDA \sim \triangle BEP$ by SAS Similarity. Therefore, we have $\angle EPB = \angle DAP$. Since $\angle DAP + \angle DPA = 90°$ and $\angle EPB = \angle DAP$, we have $\angle EPB + \angle DPA = 90°$. So, $\angle APB = 90°$, which means $k \perp \ell$.

17.1.9 Solving for y to put the equation in slope-intercept form gives

$$y = -\frac{A}{B}x - \frac{C}{B}.$$

Therefore, the slope of the graph of the equation is $-A/B$.

Exercises for Section 17.2

17.2.1 We divide both sides of the equation by 3 to make the coefficients of x^2 and y^2 equal to 1. This gives us
$$x^2 - 4x + y^2 + 2y = 5.$$

We add 4 and 1 to both sides to complete the squares on the left side. This gives us

$$x^2 - 4x + 4 + y^2 + 2y + 1 = 5 + 4 + 1,$$

which means

$$(x - 2)^2 + (y + 1)^2 = 10.$$

The graph of this equation is a circle with center $\boxed{(2, -1)}$ and radius $\boxed{\sqrt{10}}$.

17.2.2 The distance between the center and the point given on the circle is

$$\sqrt{[-5 - (-2)]^2 + (9 - 7)^2} = \sqrt{13},$$

so the radius of the circle is $\sqrt{13}$. Therefore, our equation is $\boxed{(x + 2)^2 + (y - 7)^2 = 13}$.

17.2.3 The longest chords of a circle have length equal to the diameter of the circle, so we must find the diameter of the circle. We add 1 and 9 to both sides to complete the squares on the left, and we get

$$(x + 1)^2 + (y - 3)^2 = 16.$$

Therefore, the radius is $\sqrt{16} = 4$, so the diameter is $2 \cdot 4 = \boxed{8}$.

17.2.4 The points at which the graphs intersect are the points (x, y) that satisfy both equations. Solving $2x - y = 7$ for y gives $y = 2x - 7$. Substituting this into the other equation gives

$$x^2 - 10x + (2x - 7)^2 + 4(2x - 7) = -4.$$

Simplifying the left side of this equation gives us

$$x^2 - 10x + (2x - 7)^2 + 4(2x - 7) = x^2 - 10x + 4x^2 - 28x + 49 + 8x - 28$$
$$= 5x^2 - 30x + 21,$$

so our equation is $5x^2 - 30x + 21 = -4$. Simplifying this equation gives $x^2 - 6x + 5 = 0$, so we have $(x - 1)(x - 5) = 0$. Therefore, we have $x = 1$ or $x = 5$. When $x = 1$, we have $y = 2x - 7 = -5$, and when $x = 5$, we have $y = 2x - 7 = 3$, so the two graphs intersect at $\boxed{(1, -5) \text{ and } (5, 3)}$.

Exercises for Section 17.3

17.3.1 We can use the Pythagorean Theorem to show that a triangle is a right triangle. Using the distance formula, we find that $AB = \sqrt{13}$, $BC = \sqrt{65}$, and $AC = \sqrt{52}$. So, we have $AB^2 + AC^2 = BC^2$, which means that $\triangle ABC$ is a right triangle with hypotenuse $\overline{BC}$.

17.3.2 The graphs of $y = x$ and $y = -x$ are perpendicular, so the triangle is a right triangle. The graphs of $y = x$ and $y = -x$ at $(0,0)$, and they meet the line $y = 6$ at $(6,6)$ and $(-6,6)$, respectively. This means that the vertices of the triangle are $(0,0)$, $(6,6)$, and $(-6,6)$, so the triangle is 45-45-90 triangle with legs of length $6\sqrt{2}$. Therefore, its area is $(6\sqrt{2})^2/2 = \boxed{36}$.

17.3.3 Completing the squares in both equations gives us the two equations

$$(x + 3)^2 + (y - 4)^2 = 25 \qquad \text{and} \qquad (x - 7)^2 + (y + 5)^2 = 9.$$

The graph of the first is a circle with center $(-3, 4)$ and radius 5. The graph of the second is a circle with center $(7, -5)$ and radius 3.

Using the distance formula, the distance between the two centers is $\sqrt{100 + 81} = \sqrt{181}$. Since $(\sqrt{181})^2$ is larger than $(3 + 5)^2$, we see that the distance between the centers is larger than the sum of the radii of the two circles. This means the circles do not intersect. The pair of points such that one point is on each circle and the two points are as close as possible consists of the two points where the segment connecting the circle's centers hits the two circles. The segment connecting the two centers has length $\sqrt{181}$. The portion of this segment inside one or the other of the circles has length $3 + 5 = 8$, so the remaining portion of this segment has length $\boxed{\sqrt{181} - 8}$.

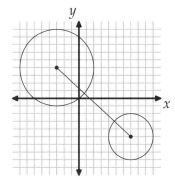

17.3.4 Let B be the origin and A be $(8, 0)$, so that $AB = 8$. We can find a point C by building a 30-60-90 right triangle as shown in the diagram at right. From 30-60-90 triangle BCX, we find $CX = 6$ and $BX = 6\sqrt{3}$, so point C is $(6\sqrt{3}, 6)$. Therefore, we can use the distance formula to find

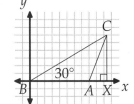

$$\begin{aligned} AC &= \sqrt{(6\sqrt{3} - 8)^2 + (6 - 0)^2} \\ &= \sqrt{108 - 96\sqrt{3} + 64 + 36} \\ &= \sqrt{208 - 96\sqrt{3}} \\ &= \boxed{4\sqrt{13 - 6\sqrt{3}}}. \end{aligned}$$

17.3.5 Because the circle is tangent to both axes, its center must be on a line that bisects the angle between the axes. Because the circle passes through a point for which x and y are both positive, this line must be the graph of $y = x$. Therefore, the center is (a, a) for some positive value of a. Because the circle is tangent to both axes, the radius of the circle must be a. So the equation of the circle is

$$(x - a)^2 + (y - a)^2 = a^2.$$

Since the circle passes through $(9, 2)$, we must have

$$(9 - a)^2 + (2 - a)^2 = a^2.$$

Expanding the left side gives $a^2 - 18a + 81 + a^2 - 4a + 4 = a^2$, so we have $a^2 - 22a + 85 = 0$. Factoring the left side gives $(a - 5)(a - 17) = 0$, so we must have $a = 5$ or $a = 17$. Therefore, the two possible radii are $\boxed{5 \text{ and } 17}$.

Exercises for Section 17.4

17.4.1

(a) The line connecting the first two vertices is horizontal while the line through each of the last two vertices is vertical. Therefore, these four vertices represent a quadrilateral in which the opposite sides are on perpendicular lines, not parallel lines. So, these vertices cannot be used as the vertices of a general parallelogram. In fact, they can't be used to represent any parallelogram.

(b) All four sides of the quadrilateral formed by connecting these points in the order given have length $\sqrt{a^2 + b^2}$. So, these points can be used to represent a rhombus to prove facts about rhombi, but they cannot be used to prove facts about a parallelogram, because not every parallelogram is a rhombus. For example, these points cannot be the vertices of a parallelogram with adjacent sides of different lengths.

(c) Here, we have two horizontal sides, and these opposite sides both have length $|a|$. So, the quadrilateral formed by connecting these points in the given order is indeed a parallelogram. However, the two non-horizontal sides each have slope 1, so every parallelogram represented by these points as vertices must have a 45° angle. But not every parallelogram has a 45° angle (for example, rectangles only have 90° angles), so we cannot use these points as vertices of a parallelogram to prove a fact for all parallelograms.

(d) Finally, we have one that works. Two sides are horizontal with length $|a|$, so the quadrilateral is a parallelogram. There are no restrictions on the slope or length of the other sides, except that the other two sides have the same slope and the same length. So, these points can be used to represent the vertices of a parallelogram to prove a fact about all parallelograms.

17.4.2

(a) Let the parallelogram be $ABCD$, where $A = (0,0)$, $B = (2b,0)$, $C = (2c,2e)$, and $D = (2c-2b,2e)$, where $b > 0$. Note that $\overline{AB} \parallel \overline{CD}$ (both are horizontal) and $AB = CD = 2b$. The diagonals are $\overline{AC}$ and $\overline{BD}$. If the two diagonals have the same midpoint, they must bisect each other. The midpoint of $\overline{AC}$ is (c,e) and the midpoint of $\overline{BD}$ is $((2b+2c-2b)/2, 2e/2) = (c,e)$. Since $\overline{AC}$ and $\overline{BD}$ have the same midpoint, they bisect each other.

(b) Let $ABCD$ be a quadrilateral with $\overline{AC}$ and $\overline{BD}$ as its diagonals. Let $A = (0,0)$, $B = (2b,0)$, $C = (2c,2e)$ and $D = (2d,2f)$. Since diagonals $\overline{AC}$ and $\overline{BD}$ bisect each other, they must have the same midpoint. The midpoint of $\overline{AC}$ is (c,e) and the midpoint of $\overline{BD}$ is $(b+d,f)$. Since these midpoints must be the same, we must have $c = b+d$ and $e = f$. So, $C = (2c,2e) = (2b+2d,2f)$ and $D = (2d,2f)$. Therefore, $\overline{AB} \parallel \overline{CD}$ (both are horizontal) and $AB = CD = 2b$, so $ABCD$ is a parallelogram.

17.4.3 The diagonals of any rhombus must be perpendicular. We take advantage of this by letting the intersection of the diagonals be the origin. Because the diagonals are perpendicular, we can choose the axes such that the diagonals lie along the axes. Suppose the diagonals are $\overline{AC}$ and $\overline{BD}$. We let $A = (a,0)$. The origin is the midpoint of $\overline{AC}$, because the diagonals of a rhombus bisect each other. So, point C is $(-a,0)$. Similarly, we let $B = (0,b)$, so point D is $(0,-b)$ because the origin is the midpoint of $\overline{BD}$. Therefore, the distance formula gives us

$$AB = BC = CD = DA = \sqrt{a^2 + b^2},$$

so $ABCD$ is a rhombus.

17.4.4 Any ordered pair that satisfies two equations must also satisfy the equation that results from adding those two equations. The point (x,y) that is the intersection of $\overleftrightarrow{AD}$ and $\overleftrightarrow{BE}$ must satisfy the equations for both lines. Since the sum of these two equations gives us the equation whose graph is $\overleftrightarrow{CF}$, we know that the point where $\overleftrightarrow{AD}$ and $\overleftrightarrow{BE}$ intersect also satisfies the equation whose graph is $\overleftrightarrow{CF}$. Therefore, the point where $\overleftrightarrow{AD}$ and $\overleftrightarrow{BE}$ intersect is on $\overleftrightarrow{CF}$, as well.

Exercises for Section 17.5

17.5.1 Applying the formula we derived in the text, the desired distance is

$$\frac{|4(3) - 3(4) + 7|}{\sqrt{4^2 + (-3)^2}} = \boxed{\frac{7}{5}}.$$

17.5.2 Let $(a, 10)$ be any point on the graph of $y = 10$. We rewrite the equation of the line as $x + 3y - 9 = 0$ so we can apply our formula for the distance between a point and a line. The distance from $(a, 10)$ to this line is

$$\frac{|a + 30 - 9|}{\sqrt{1^2 + 3^2}} = \frac{|a + 21|}{\sqrt{10}}.$$

Since this distance is 5, we must have $\frac{|a+21|}{\sqrt{10}} = 5$, or

$$|a + 21| = 5\sqrt{10}.$$

Therefore, either $a + 21 = 5\sqrt{10}$ or $a + 21 = -5\sqrt{10}$. So, we must have $a = -21 + 5\sqrt{10}$ or $a = -21 - 5\sqrt{10}$. This means the two desired points are $\boxed{(-21 + 5\sqrt{10}, 10) \text{ and } (-21 - 5\sqrt{10}, 10)}$.

17.5.3 *Horizontal lines.* A horizontal line on the Cartesian plane is the graph of an equation $y = k$, for some constant k. The point (x_0, y_0) is $k - y_0$ below (if $k > y_0$) or $y_0 - k$ above (if $y_0 \geq k$) the graph of $y = k$. Either way, we can write this distance as $|y_0 - k|$. Rearranging the equation $y = k$ gives $y - k = 0$. The formula for the distance between the graph of this equation and the point (x_0, y_0) gives us

$$\frac{|0 \cdot x_0 + 1 \cdot y_0 - k|}{\sqrt{0^2 + 1^2}} = |y_0 - k|.$$

Therefore, the formula does give the correct distance.

Vertical lines. A vertical line on the Cartesian plane is the graph of an equation $x = h$, for some constant h. The point (x_0, y_0) is $h - x_0$ to the left (if $h > x_0$) or $x_0 - h$ to the right (if $x_0 \geq h$) of the graph of $x = h$. Either way, we can write this distance as $|x_0 - h|$. Rearranging the equation $x = h$ gives $x - h = 0$. The formula for this distance between the graph of this equation and the point (x_0, y_0) gives us

$$\frac{|1 \cdot x_0 + 0 \cdot y_0 - h|}{\sqrt{1^2 + 0^2}} = |x_0 - h|.$$

Therefore, the formula does give the correct distance.

17.5.4 These two lines both have slope $3/4$, so they are parallel. We can find the distance between these parallel lines by finding a point on one line, then using our formula for the distance between a point and line. The point $(0, -2)$ is on the graph of the equation $3x = 4y + 8$. Rearranging $3x = 4y + 17$ gives $3x - 4y - 17 = 0$, so the distance between $(0, -2)$ and the graph of this equation is

$$\frac{|3(0) - 4(-2) - 17|}{\sqrt{3^2 + 4^2}} = \boxed{\frac{9}{5}}.$$

17.5.5 The graph of $(x-3)^2 + (y+2)^2 = 100$ is a circle with center $(3, -2)$ and radius $\sqrt{100} = 10$. We compute how far the center is from the line $2x - 3y = 48$ to see at how many points the line and circle intersect. We rewrite the equation $2x - 3y = 48$ as $2x - 3y - 48 = 0$ to use our formula for the distance between a point and a line. The distance between $(3, -2)$ and this line is

$$\frac{|2(3) - 3(-2) - 48|}{\sqrt{2^2 + (-3)^2}} = \frac{36}{\sqrt{13}}.$$

We can compare this to 10 by noting that $(36/\sqrt{13})^2 = 1296/13 < 100$, so $36/\sqrt{13} < 10$. Because the center is less than 10 from the line, there is a point on the line that is inside the circle. Therefore, the line must intersect the circle at $\boxed{2}$ points.

Exercises for Section 17.6

17.6.1

(a) Let the desired point be X. The midpoint of X and $(5, 6)$ is $(0, 0)$, so X must be $\boxed{(-5, -6)}$.

(b) The translation that maps $(2, -3)$ to $(0, 6)$ moves each point left 2 units and up 9 units. So, it maps $(5, 6)$ to $(5 - 2, 6 + 9) = \boxed{(3, 15)}$.

(c) The point $(5, 6)$ is 8 units to the right of $x = -3$, so its image is 8 units to the left of $x = -3$. The y-coordinate of the image is the same as that of $(5, 6)$, so the image is $(-3 - 8, 6) = \boxed{(-11, 6)}$.

(d) Let C be the center of rotation $(-3, 2)$, let A be $(5, 6)$, and let X be the image of A upon a $90°$ counterclockwise rotation about C. Point A is on the line $y = 6$, which is a horizontal line 4 units above C. The image of this line under the rotation is the vertical line 4 units to the left of C, which is $x = -7$. So, the image of A is on the line $x = -7$. Similarly, A is on $x = 5$, which is a vertical line 8 units to the right of C. The image of this line under the rotation is a horizontal line 8 above C, which is $y = 10$. So, the image of A is on the line $y = 10$. Therefore, the image of A under the rotation is $\boxed{(-7, 10)}$.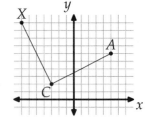

(e) When we graph $x = y$ and $(5, 6)$, it looks like $(6, 5)$ is the image of $(5, 6)$ upon reflection over the graph of $x = y$. To see that this is the case, we note that the graph of $x = y$ has slope 1 and the segment connecting $(5, 6)$ and $(6, 5)$ has slope -1. So, this segment is perpendicular to the graph of $x = y$. Moreover, the midpoint of the segment is $(5.5, 5.5)$, which is on the graph of $x = y$. So, the graph of $x = y$ is the perpendicular bisector of the segment connecting $(5, 6)$ and $(6, 5)$, which means that the image of $(5, 6)$ upon reflection over $x = y$ is $\boxed{(6, 5)}$.

(f) We wish to find the point (a, b) such that the graph of $2x + 3y = -5$ is the perpendicular bisector of the segment with endpoints (a, b) and $(5, 6)$. The line has slope $-2/3$, so the slope of the segment connecting (a, b) and $(5, 6)$ is $3/2$. Therefore, these two points are on the line $y - 6 = (3/2)(x - 5)$. Rearranging this equation gives $3x - 2y - 3 = 0$.

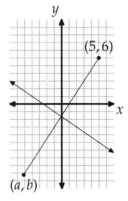

To find (a, b), we first find the point at which the lines $2x + 3y = -5$ and $3x - 2y - 3 = 0$ intersect, since this point is the midpoint of the segment with endpoints (a, b) and $(5, 6)$. Multiplying the first equation by 2 and the second by 3 gives $4x + 6y = -10$ and $9x - 6y - 9 = 0$. Adding these gives $13x - 9 = -10$, from which we find $x = -1/13$. Substituting this into either equation gives $y = -21/13$. So, the midpoint of (a, b) and $(5, 6)$ is $(-1/13, -21/13)$, which gives us

$$\left(\frac{a + 5}{2}, \frac{b + 6}{2} \right) = \left(-\frac{1}{13}, -\frac{21}{13} \right).$$

Solving for a and b, we have $(a, b) = \boxed{(-67/13, -120/13)}$.

17.6.2 The diagonals of a rectangle bisect each other and have the same length. So, we can draw a circle centered at the intersection point of the diagonals of a rectangle that passes through all four vertices of the rectangle (and has both diagonals as diameters). The midpoint of the given diagonal is the origin, and the length of the diagonal is $\sqrt{(-4 - 4)^2 + (-3 - 3)^2} = 10$. So, the center of the circle that passes through all four points of the rectangle is $(0, 0)$ and its radius is 5, *no matter what the other two vertices of the rectangle are*. This circle is the graph of the equation $x^2 + y^2 = 25$.

Now, we must find pairs of lattice points (points with integer coordinates) that are both on this circle and have the origin as their midpoint. So, we seek points (a, b) such that $a^2 + b^2 = 25$ and a and b are both integers. If $a = 0$, we have $b = \pm 5$. If we let $a = \pm 1$ or $a = \pm 2$, then b is not an integer. If $a = \pm 3$, then $b = \pm 4$. If $a = \pm 4$, then $b = \pm 3$, and if $a = \pm 5$, then $b = 0$. So, the only integer solutions are $(a, b) = (\pm 5, 0)$, $(0, \pm 5)$, $(\pm 3, \pm 4)$, and $(\pm 4, \pm 3)$. Going through these, we find $\boxed{5}$ pairs that could be the other two vertices: $(5, 0)$ and $(-5, 0)$, $(3, 4)$ and $(-3, -4)$, $(3, -4)$ and $(-3, 4)$, $(0, 5)$ and $(0, -5)$, $(4, -3)$ and $(-4, 3)$.

17.6.3

(a) Let W be the origin and X be $(6, 0)$. The circle centered at W is the graph of $x^2 + y^2 = 36$ and the circle centered at X is the graph of $(x - 6)^2 + y^2 = 36$. The intersection of these two circles with $y > 0$ is point A. Solving the first equation for y^2 gives us $y^2 = 36 - x^2$. Substituting this in the second equation gives $(x - 6)^2 + 36 - x^2 = 36$. Expanding the left side and simplifying the result gives us $-12x + 36 = 0$, so $x = 3$. (This is not surprising: we expect the x-coordinate of A to be 3.) From $y^2 = 36 - x^2$, we find $y = \pm 3\sqrt{3}$. We take the positive value of y so that A is above the x-axis. So, point A is $(3, 3\sqrt{3})$.

Similarly, the circle centered at Z is the graph of $x^2 + (y - 6)^2 = 36$. The intersection of this circle with the graph of $x^2 + y^2 = 36$ (with x positive) gives us D. Solving this system (which is the same as the earlier system, but with x and y reversed) gives us $(3\sqrt{3}, 3)$ as the coordinates of D. The distance formula then gives us

$$AD^2 = (3\sqrt{3} - 3)^2 + (3 - 3\sqrt{3})^2 = 2(3\sqrt{3} - 3)^2 = 2(36 - 18\sqrt{3}) = \boxed{72 - 36\sqrt{3}},$$

which is the area of $ABCD$.

(b) Since $\overline{WA}$ and $\overline{AX}$ are radii of circles with radius 6, we have $WA = AX = WX = 6$. Therefore, $\triangle WAX$ is equilateral, and A is $3\sqrt{3}$ from $\overline{WX}$. Similarly, C is $3\sqrt{3}$ from $\overline{ZY}$. The perpendicular bisector of $\overline{ZY}$ is also the perpendicular bisector of $\overline{WX}$, so this line contains both A and C. The distance between $\overline{ZY}$ and $\overline{WX}$ is 6, the distance from C to $\overline{ZY}$ is $3\sqrt{3}$, and the distance from A to $\overline{WX}$ is $3\sqrt{3}$, so the distance from A to C is $3\sqrt{3} + 3\sqrt{3} - 6 = 6\sqrt{3} - 6$ (since $3\sqrt{3} + 3\sqrt{3}$ gives us the distance from $\overline{ZY}$ to $\overline{WX}$ plus AC). Since $\overline{AC}$ is the diagonal of square $ABCD$, the area of $ABCD$ is

$$AC^2/2 = (6\sqrt{3} - 6)^2/2 = \boxed{72 - 36\sqrt{3}}.$$

17.6.4 Let the points where the x-axis intersects the two lines be A and B, so $AB = 5$. Let the vertical line through A hit the other line at C. Since the y-axis intersects the two lines at points that are 12 apart and the lines are parallel, every vertical line must hit the lines at points that are 12 apart. Therefore, $AC = 12$. Since $\overline{AB}$ is horizontal and $\overline{AC}$ is vertical, we know that $\triangle ABC$ is a right triangle with hypotenuse $BC = \sqrt{5^2 + 12^2} = 13$. The altitude from A to $\overline{BC}$ is the desired distance between the lines. The area of $\triangle ABC$ is $(5)(12)/2 = 30$. Letting the desired altitude be h, we must therefore have $13h/2 = 30$, which gives us $h = \boxed{60/13}$.

17.6.5 Because A and B are on the line, the altitude from C to the line is an altitude of the triangle. Because the triangle is equilateral, the length of this altitude is $\sqrt{3}/2$ times the length of each side of the triangle. We rearrange the equation as $2x - y + 7 = 0$ to find the length of the altitude (that is, the distance from $(3, -2)$ to the line). This length is

$$\frac{|2(3) - (-2) + 7|}{\sqrt{2^2 + (-1)^2}} = \frac{15}{\sqrt{5}} = 3\sqrt{5}.$$

So, we must have $AB(\sqrt{3}/2) = 3\sqrt{5}$, from which we find

$$AB = 3\sqrt{5} \cdot \frac{2}{\sqrt{3}} = \frac{6\sqrt{5}}{\sqrt{3}} = \boxed{2\sqrt{15}}.$$

17.6.6 Let A be the midpoint of $\overline{YZ}$, B be the midpoint of $\overline{XZ}$, and G be the centroid of $\triangle XYZ$. Let $AG = x$, so $GX = 2x$ and let $BG = y$, so $GY = 2y$. From right triangle XYG, we have $(2x)^2 + (2y)^2 = 64$. Therefore, we have $x^2 + y^2 = 16$. From right triangle BGX, we have $BX = \sqrt{BG^2 + GX^2} = \sqrt{y^2 + 4x^2}$, so $XZ = 2BX = 2\sqrt{y^2 + 4x^2}$. Similarly, right triangle AGY gives us $AY = \sqrt{x^2 + 4y^2}$, so $ZY = 2\sqrt{x^2 + 4y^2}$. Now, we apply the Pythagorean Theorem to $\triangle XYZ$ to find $64 + 4(y^2 + 4x^2) = 4(x^2 + 4y^2)$. Rearranging this gives us $y^2 - x^2 = 64/12 = 16/3$. Adding this to $x^2 + y^2 = 16$ gives us $y^2 = 32/3$, so $x^2 = 16/3$. Finally, we have $ZY = 2\sqrt{x^2 + 4y^2} = 2\sqrt{48} = \boxed{8\sqrt{3}}$.

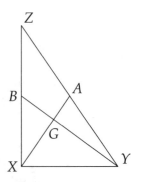

17.6.7   We might let $3x + 4y = k$, solve this equation for x in terms of y, then substitute the result into the given equation, then... That looks like a lot of work. Before we start plowing through all that algebra, we look for a less painful approach.

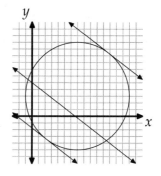

We recognize that the graph of the given equation is a circle. We rearrange the equation and complete the square to give $(x - 7)^2 + (y - 3)^2 = 64$. The graph of this equation is a circle with center $(7, 3)$ and radius 8. All points on the circumference of this circle satisfy the given equation. So, we want to find the largest value of $3x + 4y$ where (x, y) is a point on this circle.

The graphs of the equations $3x + 4y = k$ for different values of k form a family of parallel lines. As k increases, the resulting lines are farther and farther to the right. So, the largest possible k corresponds to the rightmost line that intersects the circle. This rightmost line must be tangent to the circle (otherwise, there would be a line to the right of it that still intersects the circle). Because the radius of the circle is 8, this tangent line must be 8 away from the center. Therefore, we seek the value of k such that the distance between $(7, 3)$ and $3x + 4y - k = 0$ is 8. So, we must have

$$\frac{|3 \cdot 7 + 4 \cdot 3 - k|}{\sqrt{3^2 + 4^2}} = \frac{|33 - k|}{5} = 8.$$

Multiplying by 5 gives us $|33 - k| = 40$, so we must have $33 - k = 40$ or $33 - k = -40$. The former equation gives $k = -7$, and the latter gives $k = 73$. These correspond to the two lines of the form $3x + 4y = k$ that are tangent to the circle. The one on the left (where $k = -7$) gives us our minimum value of $3x + 4y$ and the one on the right (where $k = \boxed{73}$) gives us our maximum value.

17.6.8   Let the origin be the center of the circle, so that the equation whose graph is the circle is $x^2 + y^2 = 64$. Let $\overline{AC}$ be parallel to the y-axis (so $\overline{BD}$ is parallel to the x-axis). We let $A = (a_1, a_2)$, so $a_1^2 + a_2^2 = 64$ (because A is on the circle) and $C = (a_1, -a_2)$ (because $\overline{AC}$ is vertical). Let $B = (b_1, b_2)$, so $b_1^2 + b_2^2 = 64$ and $D = (-b_1, b_2)$. Then $P = (a_1, b_2)$. Therefore, we have

$$
\begin{aligned}
PA^2 + PB^2 + PC^2 + PD^2 &= (a_2 - b_2)^2 + (b_1 - a_1)^2 + (b_2 + a_2)^2 + (a_1 + b_1)^2 \\
&= a_2^2 - 2a_2b_2 + b_2^2 + b_1^2 - 2a_1b_1 + a_1^2 + b_2^2 + 2a_2b_2 + a_2^2 + a_1^2 + 2a_1b_1 + b_1^2 \\
&= 2a_1^2 + 2a_2^2 + 2b_1^2 + 2b_2^2 \\
&= 2(a_1^2 + a_2^2) + 2(b_1^2 + b_2^2) \\
&= 2(64) + 2(64) = \boxed{256}.
\end{aligned}
$$

Review Problems

17.27   The graph of $5x - 2y + 7 = 0$ is a line with slope $-5/(-2) = 5/2$. Any line perpendicular to this line must have slope $-1/(5/2) = -2/5$. A point-slope form of the line through $(-5, 2)$ with slope $-2/5$ is $y - 2 = (-2/5)(x + 5)$. Multiplying both sides by 5 gives $5y - 10 = -2x - 10$. Rearranging this equation gives $\boxed{2x + 5y = 0}$. We could also have solved this problem by noting that because the graph of the desired equation is a line with slope $-2/5$, the equation has the form $2x + 5y = k$, for some constant k (such a line has slope $-(2)/(5) = -2/5$). Then, letting $(x, y) = (-5, 2)$, we find that $k = 0$.

17.28   The side with endpoints $(3, 4)$ and $(8, 4)$ is a horizontal segment with length 5. This side is on the line $y = 4$, which is 10 units from $(-6, -6)$, so the area of the triangle is $(5)(10)/2 = \boxed{25}$.

17.29

(a) The midpoint of $\overline{BC}$ is $(\frac{-7-1}{2}, \frac{1-5}{2}) = (-4, -2)$. The line through $(-2, 4)$ and $(-4, -2)$ has slope $(-2 - 4)/(-4 - (-2)) = 3$. So, a point-slope form of an equation whose graph contains the median from A to the midpoint of $\overline{BC}$ is $\boxed{y - 4 = 3(x + 2)}$. You might also have found other forms of this equation. For example, expanding the right side and rearranging gives the standard form $3x - y = -10$.

(b) Because the midpoint of $\overline{BC}$ is $(-4, -2)$, the distance formula gives the length of the median from A as $\sqrt{(-4 - (-2))^2 + (-2 - 4)^2} = \sqrt{40} = \boxed{2\sqrt{10}}$.

(c) The length of the altitude from A to $\overline{BC}$ is the distance from A to $\overleftrightarrow{BC}$. So, we find the equation whose graph is $\overleftrightarrow{BC}$. The slope of this line is $(-5 - 1)/(-1 - (-7)) = -1$, so a point-slope form of this equation is $y - 1 = -1(x + 7)$. Rearranging this equation gives us $x + y + 6 = 0$. The distance between $(-2, 4)$ and this line is $|-2 + 4 + 6|/\sqrt{1^2 + 1^2} = 8/\sqrt{2} = \boxed{4\sqrt{2}}$.

(d) The distance formula gives us $BC = \sqrt{(-7 - (-1))^2 + (1 - (-5))^2} = \sqrt{72} = 6\sqrt{2}$. We have already found that the altitude from A to $\overline{BC}$ has length $4\sqrt{2}$, so the area of $\triangle ABC$ is $(6\sqrt{2})(4\sqrt{2})/2 = \boxed{24}$.

17.30   The slope of $(2a - 3)x + (3a - 1)y = 3$ is $-(2a - 3)/(3a - 1)$. The segment with endpoints $(2, -4)$ and $(-1, 2)$ has slope $(2 - (-4))/(-1 - 2) = -2$. So, we must have $-(2a - 3)/(3a - 1) = -2$. Multiplying both sides by $3a - 1$ gives $-2a + 3 = -6a + 2$. Solving this equation gives $a = \boxed{-1/4}$.

17.31   Completing the square in both x and y gives us $x^2 - 6x + 9 + y^2 + 4y + 4 = -3 + 9 + 4$, so our equation is $(x - 3)^2 + (y + 2)^2 = 10$. The graph of this equation is a circle with center $(3, -2)$ and radius $\sqrt{10}$. The area of this circle therefore is $\boxed{10\pi}$. To graph the circle, we try to find some lattice points (points with integer coordinates) that are on the circle. We note that $3^2 + 1^2 = 10$, so we see that points that are 3 away from the center in one dimension and 1 away from the center in the other are on the circle. Some of these points are $(6, -3)$, $(0, -3)$, $(4, -5)$, $(4, 1)$. Others can be seen on the graph at right.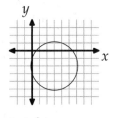

17.32   Because $\overline{PQ}$ is the diameter of the circle, the midpoint of $\overline{PQ}$ is the center of the circle. This midpoint is $(\frac{-2+8}{2}, \frac{5-3}{2}) = (3, 1)$. The distance between this point and P is $\sqrt{(-2 - 3)^2 + (5 - 1)^2} = \sqrt{41}$, so the radius of the circle is $\sqrt{41}$. Therefore, our desired equation is $\boxed{(x - 3)^2 + (y - 1)^2 = 41}$.

17.33

(a) The circle has radius $\sqrt{10}$ and center $(1, -2)$. Let the center be point Z. The distance formula gives us $XZ = 3\sqrt{10}$, so X is outside the circle. The point on the circle that is closest to the circle is the intersection of $\overline{XZ}$ with the circle. Let this point be Y. Since $XZ = 3\sqrt{10}$ and $YZ = \sqrt{10}$, we have $XY = XZ - YZ = \boxed{2\sqrt{10}}$.

(b) As noted in the previous part, point Y is the intersection of $\overline{XZ}$ and the circle. The line through X and Z has slope $1/3$, so a point-slope form of the equation of the line is $y - 1 = (1/3)(x - 10)$. Rearranging this equation gives us $x - 3y = 7$. Solving for x gives $x = 7 + 3y$. Substituting this into the equation of our circle gives us $(3y+6)^2 + (y+2)^2 = 10$. We notice that $y+2$ is a factor of $3y+6$, and we have $(3)^2(y+2)^2 + (y+2)^2 = 10$, so $10(y+2)^2 = 10$. Therefore, we have $(y+2)^2 = 1$, which means $y = -1$ or $y = -3$. We want the intersection point of the circle and $\overleftrightarrow{XZ}$ that is between X and Z, so we take $y = -1$. This gives us $x = 7 + 3y = 4$, so the point on the circle closest to X is $\boxed{(4,-1)}$. (Note that we also could have found Y by noting that that $XY = (2/3)XZ$.)

17.34 We rewrite the equation as $5x - 12y - 8 = 0$. The distance between $(a, 3)$ and this line is $|5a - 12 \cdot 3 - 8| / \sqrt{5^2 + (-12)^2} = |5a - 44|/13$. Since this must equal 7, we have $|5a - 44|/13 = 7$, so $|5a - 44| = 91$. Therefore, we must have $5a - 44 = 91$ or $5a - 44 = -91$. The first gives $a = \boxed{27}$ and the second gives $a = \boxed{-47/5}$. These are our two possible values of a.

17.35 *Solution 1: Find B and C.* Let B be (x_B, y_B). Since A is $(-3, 5)$ and the midpoint of $\overline{AB}$ is $(4, 2)$, we have $(\frac{x_B-3}{2}, \frac{y_B+5}{2}) = (4, 2)$. Solving for x_B and y_B gives us $(x_B, y_B) = (11, -1)$. Similarly, if we let C be (x_C, y_C), we must have $(\frac{x_C-3}{2}, \frac{y_C+5}{2}) = (3, 1)$. This gives us $(x_C, y_C) = (9, -3)$. So, the distance between B and C is $\sqrt{(11-9)^2 + (-1-(-3))^2} = \boxed{2\sqrt{2}}$.

Solution 2: The Midline Theorem. By the Midline Theorem, the length of $\overline{BC}$ is twice the length of the segment with the midpoints of $\overline{AB}$ and $\overline{AC}$ as endpoints. The segment connecting these midpoints has length $\sqrt{(4-3)^2 + (2-1)^2} = \sqrt{2}$, so $BC = \boxed{2\sqrt{2}}$.

17.36 *Solution 1: Algebra.* Let A be (x, y). Because $AO = 5$, point A is on the circle with center O and radius 5. So, we must have $x^2 + y^2 = 25$. Similarly, because $AB = 12$, point A is on the circle with center B and radius 12. This means we must have $x^2 + (y-13)^2 = 144$. Subtracting $x^2 + y^2 = 25$ from this equation eliminates x^2 and leaves $(y-13)^2 - y^2 = 144 - 25$. Simplifying both sides gives $-26y + 169 = 119$. Solving for y, we find that $y = 25/13$. From $x^2 + y^2 = 25$, we then find that $x = \sqrt{25 - y^2} = 60/13$. So, A is $\boxed{(60/13, 25/13)}$.

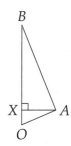

Solution 2: Geometry. We notice that $OB = 13$, so $\triangle AOB$ is a 5-12-13 right triangle. Let X be the foot of the altitude from A to OB, as shown at right. Then, because O is the origin and $\overline{OB}$ is on the y-axis, the x-coordinate of A is AX and the y-coordinate is OX. Because $\triangle AXO \sim \triangle BAO$, we have $AX/AO = AB/BO$, which gives us $AX = 60/13$. This similarity also gives us $XO/AO = AO/BO$, so $XO = 25/13$. Therefore, the coordinates of A are $\boxed{(60/13, 25/13)}$.

17.37

(a) The point $(5, -2)$ is 2 units below the x-axis, so its reflection is 2 units above the x-axis, at $\boxed{(5, 2)}$. (Note that the x-axis is the perpendicular bisector of the segment with endpoints $(5, -2)$ and $(5, 2)$.)

(b) The point $(5, -2)$ is 8 units to the right of the graph of $x = -3$, so its image upon reflection over $x = -3$ is 8 units to the left of $x = -3$, at $\boxed{(-11, -2)}$.

(c) Suppose the image is point P, with coordinates (a, b), and let $(5, -2)$ be point Q. Because P is the image of Q upon reflection over the graph of $2x - y = -1$, the graph of this line is the perpendicular bisector of $\overline{PQ}$. The slope of this line is $-2/(-1) = 2$, so the slope of $\overline{PQ}$ is $-1/2$. So, we must have $(b+2)/(a-5) = -1/2$. Multiplying both sides of this by $2(a-5)$ gives us $2b+4 = -a+5$, or $a + 2b = 1$.

Because the graph of $2x - y = -1$ bisects $\overline{PQ}$, it must pass through the midpoint of $\overline{PQ}$. The midpoint of $\overline{PQ}$ is $(\frac{a+5}{2}, \frac{b-2}{2})$. This point must be on the graph of $2x - y = -1$, so we must have

$$2\left(\frac{a+5}{2}\right) - \frac{b-2}{2} = -1.$$

Simplifying this equation gives us $2a - b = -14$. Adding twice this equation to $a + 2b = 1$ from above gives us $5a = -27$, which gives us $a = -27/5$. Substituting this in either equation gives $b = 16/5$. So, the desired image is $\boxed{(-27/5, 16/5)}$.

17.38 Substituting $x^2 + y^2 = 36$ into $x^2 + 2x + y^2 = 30$ gives us $2x + 36 = 30$, so $x = -3$. Substituting $x = -3$ into $x^2 + y^2 = 36$ gives $y^2 = 27$, so $y = \pm 3\sqrt{3}$. Therefore, the common chord of the two graphs has endpoints $(-3, 3\sqrt{3})$ and $(-3, -3\sqrt{3})$. This segment has length $\boxed{6\sqrt{3}}$.

17.39 We choose base $\overline{BC}$ to be along the x-axis such that the origin is the midpoint of $\overline{BC}$. We let B be $(b, 0)$, where $b > 0$, so that C is $(-b, 0)$. We let A be (c, a), where $a \neq 0$ so that A, B, and C are not collinear. Since we must have $AB = AC$, we must have

$$\sqrt{(c-b)^2 + (a-0)^2} = \sqrt{(c+b)^2 + (a-0)^2}.$$

Therefore, we must have $(c-b)^2 = (c+b)^2$, so $c^2 - 2bc + b^2 = c^2 + 2bc + b^2$. Simplifying this gives us $4bc = 0$. Since $b \neq 0$, we must have $c = 0$. So, our vertices are $A = (0, a)$, $B = (b, 0)$, and $C = (-b, 0)$. (Our work above shows that any isosceles triangle can be so described. Note that this work does not assume the result we are trying to prove!)

The midpoint of $\overline{BC}$ is the origin, so the median from A to $\overline{BC}$ is the segment with endpoints $(0, a)$ and $(0, 0)$. Because $\overline{BC}$ is horizontal, the altitude from A to $\overline{BC}$ must be vertical. Therefore, the foot of the altitude from A to $\overline{BC}$ must be the origin. So, the altitude and the median from A to $\overline{BC}$ are the same.

17.40 Let point A be $(-4, 5)$ and point B be $(-3, 7)$. Let C be another vertex of the square such that $\angle ABC = 90°$. To get from A to B, we go up 2 units and right 1 unit. Because $\overline{AB} \perp \overline{BC}$ and $AB = BC$, to get from B to C we must go either down 1 unit and right 2 units or up 1 unit and left 2 units. This ensures that $BC = AB$ and that the slope of $\overline{BC}$ is the negative of the reciprocal of the slope of $\overline{AB}$. This tells us that our two possible points for C are $(-1, 6)$ and $(-5, 8)$. These are marked C_1 and C_2, respectively, in the diagram.

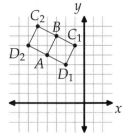

Suppose D is the final vertex of the square. Because A is 2 below and 1 to the left of B, we know that D must be 2 below and 1 to left of C, since we must have $CD = AB$, $\overline{CD} \parallel \overline{AB}$, and D on the same side of $\overline{BC}$ as A. So, if C is $(-1, 6)$, then D is $(-2, 4)$. If C is $(-5, 8)$, then D is $(-6, 6)$. Therefore, the possible points that could be vertices of the square are $\boxed{(-1, 6); (-5, 8); (-2, 4); (-6, 6)}$.

17.41 Let point P be (a, b) and Q be (b, a). The slope of $\overline{PQ}$ is -1, so $\overline{PQ}$ is perpendicular to the graph of $y = x$. Furthermore, the midpoint of $\overline{PQ}$ is $(\frac{a+b}{2}, \frac{a+b}{2})$, which is on the graph of $y = x$. So, the graph of $y = x$ is the perpendicular bisector of $\overline{PQ}$, which means that P and Q are images of each other upon reflection over the graph of $y = x$.

17.42 We first note that $XY = 4$, and that $\overline{XY}$ is a horizontal segment. If $\triangle XYP$ is equilateral, then the altitude from P to $\overline{XY}$ has length $XY\sqrt{3}/2 = 2\sqrt{3}$. So, point P must be $2\sqrt{3}$ above or below the line

$y = 3$. Moreover, because the altitude from P to $\overline{XY}$ must also be the median from P to $\overline{XY}$, point P must be $2\sqrt{3}$ directly above or below the midpoint of $\overline{XY}$, which is $(4,3)$. So, the two possible points P are $\boxed{(4, 3 + 2\sqrt{3}) \text{ and } (4, 3 - 2\sqrt{3})}$.

17.43 Let the parallelogram be $ABCD$, where $A = (0,0)$, $B = (b,0)$, $C = (c,d)$, and $D = (c-b,d)$. Note that $\overline{AB} \parallel \overline{CD}$ and $AB = CD$, so $ABCD$ is indeed a parallelogram. The squares of the lengths of the sides are:

$$\begin{aligned} AB^2 &= b^2, \\ BC^2 &= (c-b)^2 + d^2, \\ CD^2 &= b^2, \\ DA^2 &= (c-b)^2 + d^2. \end{aligned}$$

So, the sum of the squares of the lengths of the sides is

$$2(b^2 + (c-b)^2 + d^2) = 2(b^2 + c^2 - 2bc + b^2 + d^2) = 4b^2 - 4bc + 2c^2 + 2d^2.$$

The squares of the lengths of the diagonals are:

$$\begin{aligned} AC^2 &= c^2 + d^2, \\ BD^2 &= (c-2b)^2 + d^2. \end{aligned}$$

The sum of these is

$$c^2 + d^2 + c^2 - 4bc + 4b^2 + d^2 = 4b^2 - 4bc + 2c^2 + 2d^2,$$

which equals the sum of the squares of the lengths of the sides, as desired.

17.44 *Solution 1: The Long Way.* We first find the center of the circle by finding the intersection of the perpendicular bisectors of two of the sides of the triangle formed by connecting the three given points. The slope of the side with endpoints $(-1,5)$ and $(5,9)$ is $(9-5)/(5-(-1)) = 2/3$, so the slope of the perpendicular bisector of this side is $-3/2$. The midpoint of this side is $(2,7)$, so the perpendicular bisector is the graph of $y - 7 = (-3/2)(x-2)$.

The slope of the side with endpoints $(4,4)$ and $(5,9)$ is 5, so the slope of the perpendicular bisector of this side is $-1/5$. The midpoint of this side is $(9/2, 13/2)$, so the perpendicular bisector of this side is the graph of $y - 13/2 = (-1/5)(x - 9/2)$.

Solving each of the two equations of our perpendicular bisectors for y gives us

$$y = -\frac{3}{2}(x-2) + 7 = -\frac{1}{5}\left(x - \frac{9}{2}\right) + \frac{13}{2}.$$

We therefore have an equation for x. Multiplying this equation by 10 to get rid of the fractions gives us

$$-15(x-2) + 70 = -2x + 9 + 65.$$

Simplifying this equation gives us $13x = 26$, so $x = 2$. Substituting this in our equation for y above gives us $y = 7$, so the circle has center $(2,7)$. The distance between this and $(-1,5)$ is $\sqrt{(-1-2)^2 + (5-7)^2} = \sqrt{13}$, so the equation whose graph is the circle described in the problem is $\boxed{(x-2)^2 + (y-7)^2 = 13}$.

We can check our answer by noting that each of the three points does indeed satisfy this equation.

Solution 2: The Short Way. We could have avoided all that algebra in the first solution by checking if the triangle is special in any way. We note that the slope of the line through $(-1,5)$ and $(4,4)$ is $(5-4)/(-1-4) = -1/5$ and the slope of the line through $(4,4)$ and $(5,9)$ is $(9-4)/(5-4) = 5$. These two lines are perpendicular, so our triangle is a right triangle. The circumcenter of the triangle then is the midpoint of the hypotenuse, which is $(\frac{-1+5}{2}, \frac{5+9}{2}) = (2,7)$. The distance from the center to $(5,9)$ is $\sqrt{(5-2)^2 + (9-7)^2} = \sqrt{13}$, so the equation whose graph is the circle described in the problem is $\boxed{(x-2)^2 + (y-7)^2 = 13}$.

17.45 Since we are working with midpoints, we let $A = (0,0)$, $B = (2b, 0)$, and $C = (2c, 2d)$. Therefore, $D = (b+c, d)$, $E = (c, d)$, and $F = (b, 0)$.

(a) As shown in the text, the x-coordinate of the centroid of a triangle is the average of the x-coordinates of the vertices of the triangle, and the y-coordinate of the centroid of a triangle is the average of the y-coordinates of the vertices of the triangle. So, the centroid of $\triangle ABC$ is $\left(\frac{2b+2c}{3}, \frac{2d}{3}\right)$ and the centroid of $\triangle DEF$ is $\left(\frac{2b+2c}{3}, \frac{2d}{3}\right)$. Therefore, $\triangle ABC$ and $\triangle DEF$ have the same centroid.

(b) First, we find the coordinates of G_1, G_2, and G_3 by averaging the coordinates of the appropriate triangles. We have

$$G_1 : \quad \left(\frac{b+c}{3}, \frac{d}{3}\right),$$

$$G_2 : \quad \left(\frac{4b+c}{3}, \frac{d}{3}\right),$$

$$G_3 : \quad \left(\frac{b+4c}{3}, \frac{4d}{3}\right).$$

So, the centroid of $\triangle G_1 G_2 G_3$ is

$$\left(\frac{\frac{b+c}{3} + \frac{4b+c}{3} + \frac{b+4c}{3}}{3}, \frac{\frac{d}{3} + \frac{d}{3} + \frac{4d}{3}}{3}\right) = \left(\frac{6b+6c}{9}, \frac{6d}{9}\right) = \left(\frac{2b+2c}{3}, \frac{2d}{3}\right).$$

We found in the previous part that this point is also the centroid of $\triangle ABC$, so $\triangle G_1 G_2 G_3$ and $\triangle ABC$ have the same centroid.

17.46 Let points P and Q be the feet of the altitudes from X and Y to the x-axis, as shown. Because $\overline{YQ} \parallel \overline{XP} \parallel \overline{AO}$, we have $\triangle BYQ \sim \triangle BXP \sim \triangle BAO$. We have $BY : BX : BA = 1 : 2 : 3$, so our similarity gives us $BQ : BP : BO = YQ : XP : AO = 1 : 2 : 3$. Therefore, $XP = (2/3)AO = 6$ and $BP = (2/3)BO = 8/3$, so $OP = BO - BP = 4/3$, and the coordinates of X are $(4/3, 6)$. So, the slope of $\overline{OX}$ is $6/(4/3) = \boxed{9/2}$. Challenge: See if you can find a general formula for the coordinates of the point T on the segment with endpoints $A = (a_1, a_2)$ and $B = (b_1, b_2)$ such that $AT/BT = k$, for any positive constant k.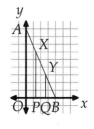

17.47 Because $\angle PQR = 30°$ and $\angle PRQ = 60°$, we have $\angle RPQ = 90°$. Because Q and R are on the x-axis and the y-coordinate of P is 2, we know that the altitude from P to the x-axis has length 2. Let this altitude be $\overline{PX}$ in the diagram, as shown. From 30-60-90 triangle PRX, we have $PR = 2PX/\sqrt{3} = 4/\sqrt{3}$. From 30-60-90 triangle PXQ, we have $PQ = 2PX = 4$. So, we have $[PRQ] = (PR)(PQ)/2 = 8/\sqrt{3} = \boxed{8\sqrt{3}/3}$.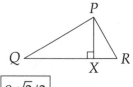

17.48 First, we note that $XY = 10$.

(a) The distance formula gives $XZ = \sqrt{(c-4)^2 + 36}$ and $YZ = \sqrt{(c+2)^2 + 4}$. If $XZ = XY$, then we have $\sqrt{(c-4)^2 + 36} = 10$. Squaring both sides gives $(c-4)^2 + 36 = 100$, so we must have $(c-4)^2 = 64$. Therefore, $c - 4 = 8$ or $c - 4 = -8$, which gives $c = 12$ and $c = -4$ as solutions. (Notice that this case corresponds to Z being on the circle with center X and radius XY.)

If $YZ = XY$, we have $\sqrt{(c+2)^2 + 4} = 10$. Squaring and rearranging gives $(c+2)^2 = 96$, so $c + 2 = 4\sqrt{6}$ or $c + 2 = -4\sqrt{6}$. This gives $c = -2 + 4\sqrt{6}$ and $c = -2 - 4\sqrt{6}$ as solutions. This case corresponds to Z being on the circle with center Y and radius XY.

If $XZ = YZ$, then Z is on the perpendicular bisector of $\overline{XY}$. Since $\overline{XY}$ has slope $-4/3$ and midpoint $(1, 1)$, the line through $(c, 3)$ and $(1, 1)$ must have slope $3/4$. Therefore, we have $(3-1)/(c-1) = 3/4$, so $3(c-1) = 8$, from which we find $c = 11/3$.

Combining all three cases, we have 5 possible values of c:

$$\boxed{-2 - 4\sqrt{6}, -4, 11/3, -2 + 4\sqrt{6}, 12.}$$

(b) We could use the distance formula and the Pythagorean Theorem, but instead we offer a simpler solution using slope.

Case 1: $\overline{XY} \perp \overline{XZ}$. The slope of $\overline{XY}$ is $-4/3$, so the slope of $\overline{XZ}$ must be $3/4$. Therefore, we must have $(3-(-3))/(c-4) = 3/4$. Cross-multiplying gives $3c - 12 = 24$, from which we find $c = 12$.

Case 2: $\overline{XY} \perp \overline{YZ}$. In this case, the slope of $\overline{YZ}$ must be $3/4$, which gives us $(3-5)/(c-(-2)) = 3/4$. Cross-multiplying gives $3c + 6 = -8$, from which we find $c = -14/3$.

Case 3: $\overline{XZ} \perp \overline{YZ}$. The slope of $\overline{XZ}$ is $6/(c-4)$ and the slope of $\overline{YZ}$ is $-2/(c+2)$. So, we must have

$$\left(\frac{6}{c-4}\right)\left(\frac{-2}{c+2}\right) = -1.$$

Multiplying both sides by $(c-4)(c+2)$ gives $-12 = -(c-4)(c+2) = -c^2 + 2c + 8$. Rearranging gives $c^2 - 2c - 20 = 0$. The quadratic formula then gives $c = 1 \pm \sqrt{21}$.

Combining all three cases gives 4 values of c such that $\triangle XYZ$ is a right triangle:

$$\boxed{1 - \sqrt{21}, -14/3, 1 + \sqrt{21}, 12}.$$

Challenge Problems

17.49 The x-intercept of the line is $(18, 0)$ and the y-intercept is $(0, -6)$. The circle passes through $(0, 0)$, $(18, 0)$, and $(0, -6)$. The triangle formed by connecting these three points has a right angle at the origin, so the circumcenter of the triangle is the midpoint of the hypotenuse, which is $(9, -3)$. The distance from this point to the origin is $\sqrt{81 + 9} = 3\sqrt{10}$, so our circle has center $(9, -3)$ and radius $3\sqrt{10}$. Therefore, an equation whose graph is this circle is $\boxed{(x-9)^2 + (y+3)^2 = 90}$.

17.50

(a) The graph of $3x - y = 0$ is a line through the origin with slope 3. Its image upon a $90°$ rotation about the origin is a line through the origin perpendicular to the original line. This image line therefore has slope $-1/3$. Since this line passes through the origin, it is the graph of $\boxed{y = -x/3}$.

(b) The graph of $3x - y = 6$ passes through $(2,0)$ and $(0,-6)$. Because $(2,0)$ is on the x-axis, its image upon $90°$ rotation about the origin is on the y-axis. Because the rotation is clockwise, the image of $(2,0)$ is $(0,-2)$. Similarly, the image of $(0,-6)$ is $(-6,0)$. So, the image of the graph of $3x - y = 6$ is the line that passes through $(0,-2)$ and $(-6,0)$. This line has slope $-2/6 = -1/3$ (note that this means the image line is perpendicular to the original line, as expected). A point-slope form of the equation of the image line is therefore $y + 2 = (-1/3)(x - 0)$. The standard form of this equation is $\boxed{x + 3y = -6}$.

17.51 The graph of each equation is a circle. We complete the square in both variables in both equations to give us the equations

$$(x - 6)^2 + (y - 3)^2 = 49 \qquad \text{and} \qquad (x - 2)^2 + (y - 6)^2 = k + 40.$$

The graph of the first is a circle with center $(6,3)$ and radius 7, and the graph of the second is the graph of a circle with center $(2,6)$ and radius $\sqrt{k + 40}$. The distance between the centers is $\sqrt{(6 - 2)^2 + (3 - 6)^2} = 5$. Therefore, the center of the second circle is inside the first circle. If the radius of the second circle is less than $7 - 5 = 2$, then the second circle is not large enough to intersect the first circle. So, we must have $\sqrt{k + 40} \geq 2$, which gives us $k \geq -36$. On the other hand, if the second circle has radius greater than $5 + 7 = 12$, then the first circle is completely inside the second, so they do not intersect. So, for the two circles to intersect, we must also have $\sqrt{k + 40} \leq 12$. Squaring this gives us $k + 40 \leq 144$, so $k \leq 104$. Combining our restrictions on k gives $-36 \leq k \leq 104$, so $b - a = 104 - (-36) = \boxed{140}$.

17.52 The x-axis, the y-axis, and the graph of $x + y = 2$ intersect to form a triangle, $\triangle ABC$ shown at right. The points that are equidistant from the axes form the line that contains the angle bisector of $\angle ACB$ (line ℓ_1), and the line through C perpendicular to ℓ_1 (line ℓ_2), which bisects exterior angles of $\triangle ABC$ at vertex C. Similarly, the points equidistant from the x-axis and the graph of $x + y = 2$ are lines m_1, which bisects $\angle ABC$, and m_2, which bisects the exterior angles of $\triangle ABC$ at B. Furthermore, the points equidistant from the y-axis and the graph of $x + y = 2$ are lines n_1, which bisects $\angle BAC$, and n_2, which bisects the exterior angles of $\triangle ABC$ at A.

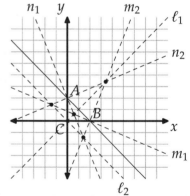

The lines ℓ_1, m_1, and n_1 meet at the incenter of $\triangle ABC$. Our diagram suggests that ℓ_1, m_2, and n_2 are concurrent. Indeed, since ℓ_1 consists of points equidistant from the axes and m_2 consists of points equidistant from the x-axis and the graph of $x + y = 2$, the intersection of ℓ_1 and m_2 is equidistant from the y-axis, the x-axis, and the graph of $x + y = 2$. Therefore, the intersection of ℓ_1 and m_2 must be on n_1 or n_2, which together consist of all points equidistant from the y-axis and the graph of $x + y = 2$. Our diagram shows that this intersection point is on n_2. Similarly, ℓ_2, m_2, and n_1 are concurrent, as are ℓ_2, m_1, and n_2. So, there are $\boxed{4}$ points that are equidistant from the four given lines. (The three points outside $\triangle ABC$ are the excenters of $\triangle ABC$. Just as the incenter is the center of the incircle, each excenter is the center of a circle that is tangent to the three lines that contain the sides of $\triangle ABC$.)

17.53 Let the radius of the circle be r and the center of the circle be the origin, so that the circle is the graph of the equation $x^2 + y^2 = r^2$. Next, we let the x-axis contain $\overline{PQ}$, so that we can let $P = (r, 0)$, $Q = (-r, 0)$, and $X = (a, 0)$. Since $\overline{AB}$ is a chord of the circle parallel to $\overline{PQ}$, if we let $A = (b, c)$ with $b \neq 0$, then $b^2 + c^2 = r^2$, so that $(-b, c)$ is also on the circle. Since the chord connecting (b, c) to $(-b, c)$ is parallel to $\overline{PQ}$ (both have slope 0), point B is $(-b, c)$. So, we have

$$
\begin{aligned}
XA^2 + XB^2 &= (a - b)^2 + (0 - c)^2 + (a + b)^2 + (0 - c)^2 \\
&= a^2 - 2ab + b^2 + c^2 + a^2 + 2ab + b^2 + c^2 \\
&= 2(a^2 + b^2 + c^2) \\
&= 2(a^2 + r^2).
\end{aligned}
$$

We also have

$$
\begin{aligned}
XP^2 + XQ^2 &= (a - r)^2 + 0 + (a + r)^2 + 0 \\
&= a^2 - 2ar + r^2 + a^2 + 2ar + r^2 \\
&= 2(a^2 + r^2).
\end{aligned}
$$

Therefore, we have the desired $XA^2 + XB^2 = XP^2 + XQ^2$.

17.54

(a) Point A is 3 units to the right of U, so its image A' upon $90°$ clockwise rotation about U will be 3 units below U, at $(-4, 0)$. Since A' is 6 units to the left and 1 unit above V, its image A'' upon $90°$ clockwise rotation about V is 6 units above and 1 unit to the right of V, at $\boxed{(3, 5)}$.

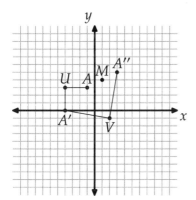

Figure 17.1: Diagram for Part (a)

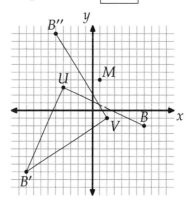

Figure 17.2: Diagram for Part (b)

(b) Point B is 11 units to the right and 5 units below U, so its image B' upon $90°$ clockwise rotation about U is 11 units below and 5 units to the left of U, at $(-9, -8)$. Since B' is 11 units to the left and 7 units below V, its image B'' is 11 units above and 7 units to the left of V, at $\boxed{(-5, 10)}$.

(c) The point $\boxed{(1, 4)}$ is the midpoint of both $\overline{AA''}$ and $\overline{BB''}$. This point is marked M in both our diagrams above.

(d) Point C is $a - (-4) = a + 4$ units to the right and $b - 3$ units above point U, so its image C' is $a + 4$ units below and $b - 3$ units to the right of U, at $(-4 + b - 3, 3 - (a + 4)) = (b - 7, -a - 1)$. Similarly, C' is $b - 7 - 2 = b - 9$ units to the right of V and $-a - 1 - (-1) = -a$ units above it, so its image C''

is $b - 9$ units below and $-a$ units to the right of V, at $(2 - a, -1 - (b - 9)) = (2 - a, 8 - b)$. So, the midpoint of $C = (a, b)$ and its image after the two rotations, $C'' = (2 - a, 8 - b)$, is always $(1, 4)$. In other words, the two rotations together are a $180°$ rotation about the point $(1, 4)$. We can use this to quickly find the result of the two rotations for any (a, b) by noting that $(1, 4)$ is the midpoint of the segment with endpoints (a, b) and $(2 - a, 8 - b)$.

Notice also that $\triangle UMV$ is a 45-45-90 triangle with $\angle UMV = 90°$. Is this a coincidence?

17.55 Squares are pretty easy to handle with coordinates, and the center of a square is simply the midpoint of the diagonals, so we set up the problem on the Cartesian plane. We let D be the origin, and define the axes so that $A = (0, 1)$, $B = (1, 1)$, and $C = (1, 0)$. Since P is on $\overline{CD}$, we have $P = (t, 0)$, where $0 \le t \le 1$. Since A is 1 above and t to the left of P, point Z is 1 to the left and t below A, at $(-1, 1 - t)$. Since P is 1 below and t to the right of A, point Y is 1 below and t to the right of Z, at $(t - 1, -t)$. Similarly, we determine that X is $(2, t)$ and W is $(t + 1, t - 1)$.

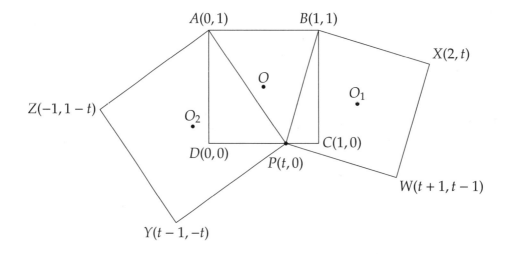

The center of each square is the midpoint of its diagonals. So, we have $O = (\frac{1}{2}, \frac{1}{2})$, $O_1 = (\frac{t}{2} + 1, \frac{t}{2})$, and $O_2 = (\frac{t}{2} - \frac{1}{2}, \frac{1}{2} - \frac{t}{2})$. Then the midpoint of $\overline{PO}$ is

$$\left(\frac{t + \frac{1}{2}}{2}, \frac{\frac{1}{2}}{2} \right) = \left(\frac{2t + 1}{4}, \frac{1}{4} \right),$$

and the midpoint of $\overline{O_1 O_2}$ is

$$\left(\frac{\frac{t}{2} + 1 + \frac{t}{2} - \frac{1}{2}}{2}, \frac{\frac{t}{2} + \frac{1}{2} - \frac{t}{2}}{2} \right) = \left(\frac{2t + 1}{4}, \frac{1}{4} \right).$$

The diagonals of OO_1PO_2 bisect each other, so quadrilateral OO_1PO_2 is a parallelogram.

17.56 The graph of $(x - 3)^2 + (y - 3)^2 = 6$ is a circle with center $(3, 3)$ and radius $\sqrt{6}$. If (x, y) satisfies the given equation, then the point (x, y) is on the circle. Furthermore, if $y/x = k$, then the point (x, y) is on the graph of the equation $y = kx$, which is a line through the origin with slope k. So, we seek the point on the circle such that the line through the point and the origin has the largest possible slope. The desired line then is the line through the origin that is tangent to the circle. Any line with larger slope will not intersect the circle, since we must rotate the tangent line counterclockwise about the origin (away from the circle) to produce a line with larger slope. Any other line that intersects the circle is the result of rotating the tangent line about the origin towards the center of the circle, reducing the slope of the line.

Since the point we seek is the only point that satisfies both $y = kx$ and $(x-3)^2 + (y-3)^2 = 6$ for the desired value of k, we seek the value of k such that this system of equations has exactly one solution. Substituting $y = kx$ in the second equation gives

$$(x-3)^2 + (kx-3)^2 = 6.$$

Expanding the left side and rearranging gives

$$(k^2 + 1)x^2 + (-6 - 6k)x + 12 = 0.$$

This quadratic equation in x has only one solution if its discriminant equals 0. The discriminant of this equation is

$$(-6 - 6k)^2 - 4(k^2 + 1)(12) = -12k^2 + 72k - 12 = -12(k^2 - 6k + 1).$$

The quadratic formula gives the two roots of this quadratic as

$$k = \frac{6 \pm \sqrt{36 - 4}}{2} = 3 \pm 2\sqrt{2}.$$

The two values of k are the slopes of the two tangents from the origin to the circle. We want the larger slope, $\boxed{3 + 2\sqrt{2}}$.

17.57 Let the center of the circle be the origin, so that the circle is the graph of $x^2 + y^2 = r^2$. We choose the axes so that $\overline{A_1 A_2}$ is parallel to the y-axis and $\overline{B_1 B_2}$ is parallel to the x-axis. This ensures that the chords are perpendicular. Let $A_1 = (a_1, a_2)$, so that $a_1^2 + a_2^2 = r^2$ and $A_2 = (a_1, -a_2)$. Let $B_1 = (b_1, b_2)$, so $b_1^2 + b_2^2 = r^2$ and $B_2 = (-b_1, b_2)$. Then, we have $P = (a_1, b_2)$, so $OP^2 = a_1^2 + b_2^2$. Using the distance formula, we now have

$$\begin{aligned}(A_1 A_2)^2 + (B_1 B_2)^2 &= 4a_2^2 + 4b_1^2 \\ &= 4(r^2 - a_1^2) + 4(r^2 - b_2^2) \\ &= 8r^2 - 4(a_1^2 + b_2^2) \\ &= 8r^2 - 4OP^2.\end{aligned}$$

Because point P is fixed, so is $8r^2 - 4OP^2$. Therefore, the quantity $(A_1 A_2)^2 + (B_1 B_2)^2$ is the same for all pairs of perpendicular chords that meet at P.

CHAPTER **18**

Introduction to Trigonometry

Exercises for Section 18.1

18.1.1 Our triangle is at right. The Pythagorean Theorem gives us

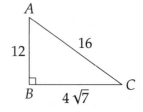

$$BC = \sqrt{AC^2 - AB^2} = \sqrt{256 - 144} = \sqrt{112} = 4\sqrt{7}.$$

So, we have $\sin A = BC/AC = \boxed{\sqrt{7}/4}$, $\cos A = AB/AC = \boxed{3/4}$, and $\tan A = BC/AB = \boxed{\sqrt{7}/3}$.

18.1.2 First, we tackle $\triangle PQR$. We have $\sin P = QR/PR$, so $PR = QR/(\sin P) = 16/(\sin 58°) \approx \boxed{18.9}$. We also have $\tan P = QR/PQ$, so $PQ = QR/(\tan P) \approx \boxed{10.0}$.

In $\triangle XYZ$, we have $\sin Z = XY/XZ$, so $XY = XZ \sin Z = 12 \sin 19° \approx \boxed{3.9}$. We also have $\cos Z = YZ/XZ$, so $YZ = XZ \cos Z = 12 \cos 19° \approx \boxed{11.3}$.

18.1.3 For any value of x between 0 and 1, there is a right triangle with hypotenuse of length 1 and a leg of length x. To construct such a triangle, let $AB = 1$. Draw the circle that has $\overline{AB}$ as diameter. Draw a circle with radius x and center B. Since $0 < x < 1$, these circles must meet at two points. Let either of these points be C, so that we have $BC = x$. Then, we have $\sin A = BC/AB = x$ and $\cos B = BC/AB = x$. So, for any value of x such that $0 < x < 1$, there are angles $\angle A$ and $\angle B$ such that $\sin A = \cos B = x$.

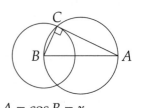

18.1.4 At right is right triangle $\triangle ABC$ with $\angle B = \theta$. We have $\cos\theta = BC/AB$. We also have $\angle A = 90° - \theta$, so

$$\sin(90° - \theta) = \sin A = \frac{BC}{AB}.$$

Therefore, we have $\sin(90° - \theta) = \cos\theta$ for any acute angle measure θ.

We also could have used the relationship $\sin x = \cos(90° - x)$ that we proved in the text for acute angle measures x. Specifically, if θ is acute, then $90° - \theta$ is acute, so we can let $x = 90° - \theta$ in this relationship to find

$$\sin(90° - \theta) = \cos(90° - (90° - \theta)) = \cos\theta.$$

18.1.5 Since $\angle B = 90°$, the hypotenuse of the triangle is $\overline{AC}$. So, we have $\sin A = BC/AC$. Since $BC/AC = \sin A = 5/7$, we have $BC = 5AC/7$. From the Pythagorean Theorem, we have

$$AB = \sqrt{AC^2 - BC^2} = \sqrt{AC^2 - \frac{25AC^2}{49}} = \sqrt{\frac{24AC^2}{49}} = \frac{2AC\sqrt{6}}{7}.$$

So, we have

$$\tan C = \frac{AB}{BC} = \frac{2AC\sqrt{6}/7}{5AC/7} = \boxed{\frac{2\sqrt{6}}{5}}.$$

18.1.6 First, we draw the diagram, which is shown at right. Because Y is on the angle bisector of $\angle XAB$, we have $\angle YAX = \angle YAB = 45°$. Since Z is on the bisector of $\angle YAB$, we have $\angle ZAP = 45°/2 = 22.5°$. Because $\angle ZAB = 22.5°$ and this angle is inscribed in $\widehat{ZB}$, we have $\widehat{ZB} = 2\angle ZAB = 45°$. So, we have $\angle ZOP = 45°$, which means that drawing altitude $\overline{ZP}$ will give us a 45-45-90 triangle ZOP. Because $OZ = 1$ as a radius of the circle, we have $ZP = OP = \sqrt{2}/2$. We also now have a right triangle with $\overline{AZ}$ as hypotenuse. We can use either the Pythagorean Theorem or trigonometry to find AZ. With trigonometry, we have $\sin \angle ZAP = ZP/AZ$, so $AZ = ZP/(\sin \angle ZAP) \approx \boxed{1.85}$. (We also could have drawn $\overline{BZ}$ and used right triangle $\triangle ABZ$ to find that $AZ = AB \cos \angle ZAB = 2\cos 22.5° \approx \boxed{1.85}$.)

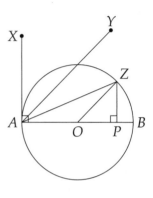

Extra challenge: Can you use the set-up of this problem to find an expression equal to $\sin 22.5°$?

18.1.7 We start with right triangle ABC with $\angle C = 90°$, $\angle A = 15°$, and $AB = 1$. We therefore have $\angle B = 90° - 15° = 75°$ and $\sin 15° = \sin A = BC/AB = BC$, so our problem is to find BC. We don't know much about 15° angles and 75° angles, but we know a lot about isosceles triangles and 60° angles. So, we choose point X on $\overline{AC}$ such that $\angle ABX = 15°$ and $\angle CBX = 60°$. Therefore, $\triangle ABX$ is isosceles with $AX = BX$, and $\triangle BCX$ is a 30-60-90 triangle. Letting $BC = x$, we therefore have $BX = 2x$, so $AX = 2x$. We also have $XC = x\sqrt{3}$. So, applying the Pythagorean Theorem to $\triangle ABC$ gives us

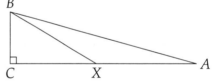

$$x^2 + (x\sqrt{3} + 2x)^2 = 1.$$

Expanding the left side gives

$$x^2 + 3x^2 + 4x^2\sqrt{3} + 4x^2 = 1,$$

and simplifying this gives us $8x^2 + 4x^2\sqrt{3} = 1$. So, we have

$$x^2 = \frac{1}{8 + 4\sqrt{3}}.$$

Taking the square root of both sides (and noting that x must be positive) gives us

$$x = \frac{1}{\sqrt{8 + 4\sqrt{3}}}.$$

While it's OK if you stopped here, we can simplify this expression more! We have $\sqrt{8 + 4\sqrt{3}} = \sqrt{6} + \sqrt{2}$ (read Art of Problem Solving's *Introduction to Algebra* to learn how to discover this), so we have

$$x = \frac{1}{\sqrt{6} + \sqrt{2}} = \frac{1}{\sqrt{6} + \sqrt{2}} \cdot \frac{\sqrt{6} - \sqrt{2}}{\sqrt{6} - \sqrt{2}} = \boxed{\frac{\sqrt{6} - \sqrt{2}}{4}}.$$

See if you can also find a solution by drawing the angle bisector of the 30° angle of a 30-60-90 triangle, then using the Angle Bisector Theorem!

Exercises for Section 18.2

18.2.1

(a) Because $90° < 120° < 180°$, the angle 120° is in the second quadrant. Since 120° is 60° clockwise from $(-1, 0)$, drawing altitude $\overline{BP}$ creates 30-60-90 right triangle BPO as shown, where $\angle BOP = 60°$. Since $BP = \sqrt{3}/2$, the y-coordinate of B is $\sqrt{3}/2$. Therefore, $\sin 120° = \boxed{\sqrt{3}/2}$.

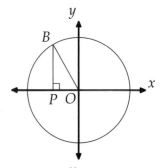

(b) Because $270° < 330° < 360°$, the angle 330° is in the fourth quadrant. Since 330° is 30° clockwise from $(1, 0)$, drawing altitude $\overline{BP}$ creates 30-60-90 triangle BPO as shown, where $\angle BOP = 30°$. Therefore, we have $BP = 1/2$. Since point B is below the x-axis, the y-coordinate of P is negative. So, we have $\sin 330° = -BP = \boxed{-1/2}$.

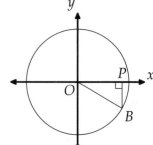

(c) Because $90° < 135° < 180°$, the angle 135° is in the second quadrant. Since 135° is 45° clockwise from $(-1, 0)$, drawing altitude $\overline{BP}$ creates 45-45-90 triangle BPO as shown, where $\angle BOP = 45°$. Therefore, we have $OP = BP = \sqrt{2}/2$. Since B is to the left of the y-axis and above the x-axis, its coordinates are $(-\sqrt{2}/2, \sqrt{2}/2)$, which means $\cos 135° = -\sqrt{2}/2$ and $\sin 135° = \sqrt{2}/2$. Therefore, we have $\tan 135° = (\sin 135°)/(\cos 135°) = \boxed{-1}$.

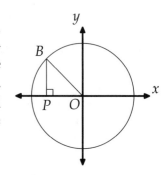

(d) To find the cosine of $-45°$, we find the x-coordinate of the point on the unit circle that is $45°$ *clockwise* from $(1,0)$. This point is in the fourth quadrant, as shown at right. Drawing altitude $\overline{BP}$ forms 45-45-90 triangle BOP, from which we find that $OP = \sqrt{2}/2$. So, the x-coordinate of B is $\sqrt{2}/2$, which means that $\cos(-45°) = \boxed{\sqrt{2}/2}$.

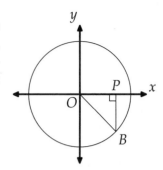

(e) Because $180° < 210° < 270°$, the angle $210°$ is in the third quadrant. Since $210°$ is $30°$ counterclockwise from $(-1,0)$, drawing altitude $\overline{BP}$ creates 30-60-90 triangle BPO as shown, where $\angle BOP = 30°$. We seek the x-coordinate of B. From $\triangle BPO$, we find $OP = \sqrt{3}/2$. Since point B is to the left of the y-axis, its x-coordinate is negative. So, we have
$$\cos 210° = -OP = \boxed{-\sqrt{3}/2}.$$

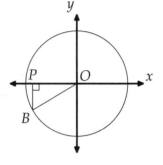

(f) Since $720°$ represents two full revolutions about a circle, the point on the unit circle that is $720°$ counterclockwise from $(1,0)$ is still $(1,0)$. Therefore, we have $\cos 720° = 1$ and $\sin 720° = 0$, so $\tan 720° = (\sin 720°)/(\cos 720°) = \boxed{0}$.

18.2.2 For any angle θ, the point $(\cos\theta, \sin\theta)$ is on the unit circle. Because this point is on the circle with center $(0,0)$ and radius 1, its coordinates satisfy the equation $x^2 + y^2 = 1$. So, for any angle θ, we have $\cos^2\theta + \sin^2\theta = 1$.

18.2.3 If $\sin\theta = 0.31$, then the point on the unit circle that is θ degrees counterclockwise from $(1,0)$ has y-coordinate 0.31. Therefore, this point must be on the line $y = 0.31$. The graph of $y = 0.31$ intersects the unit circle at two points, one in the first quadrant and the other in the second. Each corresponds to a value of θ such that $0 \le \theta < 360°$ and $\sin\theta = 0.31$. Therefore, there are $\boxed{2}$ such values of θ.

18.2.4 Let P be the point on the unit circle that is θ degrees counterclockwise from $(1,0)$. So, the coordinates of P are $(\cos\theta, \sin\theta)$. To find the point on the unit circle that is $360° + \theta$ degrees counterclockwise from $(1,0)$, we first go θ degrees counterclockwise, to P, then continue another $360°$ degrees counterclockwise. (If θ is negative, then our first step is going clockwise.) When we go $360°$ around a circle, we return exactly to our starting point. So, the point that is $360° + \theta$ counterclockwise from $(1,0)$ is also point P. Therefore, we can write the coordinates of P as $(\cos(360° + \theta), \sin(360° + \theta))$. These must be the same as our earlier coordinates for P, so we have $\cos(360° + \theta) = \cos\theta$ and $\sin(360° + \theta) = \sin\theta$.

18.2.5 Let A be $(1, 0)$ and B be $(-1, 0)$. Let C be $(\cos\theta, \sin\theta)$ and D be $(\cos(180° - \theta), \sin(180° - \theta))$. Therefore, point C is θ counterclockwise from A, and D is θ clockwise from B, as shown. We draw altitudes $\overline{CX}$ and $\overline{DY}$ to the x-axis, forming right triangles $\triangle OYD$ and $\triangle OXC$. Because $OD = OC = 1$, we have $\triangle OYD \cong \triangle OXC$ by SA Congruence for right triangles. This gives us $OY = OX$. Therefore, C and D are the same distance from the y-axis. Since they are on opposite sides of the y-axis, their x-coordinates must be opposites. So, we see that $-\cos\theta = \cos(180° - \theta)$.

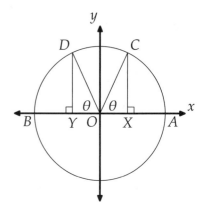

Exercises for Section 18.3

18.3.1 In both triangles, we have two side lengths and the measure of the angle between the sides. So, we can apply the Law of Cosines.

(a) We have $PR^2 = PQ^2 + QR^2 - 2(PQ)(QR)\cos Q \approx 58.83$. So, we have $PR \approx \boxed{7.7}$.

(b) We have $UV^2 = TU^2 + TV^2 - 2(TU)(TV)\cos T \approx 119.79$. Taking the square root of both sides gives $TU \approx \boxed{10.9}$.

18.3.2 For each triangle, we have a side length and two angle measures. So, we can find the third angle measure and use the Law of Sines to find the remaining side lengths.

(a) First, we find that $\angle C = 180° - 48° - 71° = 61°$. From the Law of Sines, we have

$$\frac{AB}{\sin C} = \frac{AC}{\sin B} = \frac{BC}{\sin A}.$$

So, we have

$$BC = \frac{AB}{\sin C} \cdot \sin A \approx \boxed{5.9}$$

and

$$AC = \frac{AB}{\sin C} \cdot \sin B \approx \boxed{7.6}.$$

(b) First, we find that $\angle X = 180° - 17° - 9° = 154°$. From the Law of Sines, we have

$$\frac{YZ}{\sin X} = \frac{XY}{\sin Z} = \frac{XZ}{\sin Y}.$$

So, we have

$$XZ = \frac{XY}{\sin Z} \cdot \sin Y \approx \boxed{16.8}$$

and

$$YZ = \frac{XY}{\sin Z} \cdot \sin X \approx \boxed{25.2}.$$

18.3.3

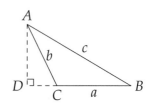

(a) We start by drawing altitude $\overline{AD}$ from A to the extension of $\overline{BC}$. From right triangle $\triangle ACD$, we have $\sin \angle ACD = AD/AC$ and $\cos \angle ACD = CD/AC$. Therefore, we have $AD = AC \sin \angle ACD = b \sin \angle ACD$ and $CD = AC \cos \angle ACD = b \cos \angle ACD$. However, we wish to write these in terms of $\angle ACB$, not $\angle ACD$.

We have $\angle ACD + \angle ACB = 180°$, so $\angle ACD = 180° - \angle ACB$. We showed in the text and Exercises that $\sin \theta = \sin(180° - \theta)$ and $\cos \theta = -\cos(180° - \theta)$ for all angles θ such that $0 \le \theta < 180°$. Therefore, we have $\sin \angle ACD = \sin \angle ACB$ and $\cos \angle ACD = -\cos \angle ACB$, which means $AD = b \sin \angle ACB$ and $CD = -b \cos \angle ACB$. Finally, we have $BD = DC + BC = a - b \cos \angle ACB$.

(b) Applying the Pythagorean Theorem to $\triangle ADB$ gives us $AB^2 = AD^2 + BD^2$, or

$$\begin{aligned}
c^2 &= (b \sin \angle ACB)^2 + (a - b \cos \angle ACB)^2 \\
&= b^2 \sin^2 \angle ACB + a^2 - 2ab \cos \angle ACB + b^2 \cos^2 \angle ACB \\
&= a^2 + b^2(\sin^2 \angle ACB + \cos^2 \angle ACB) - 2ab \cos \angle ACB \\
&= a^2 + b^2 - 2ab \cos \angle ACB.
\end{aligned}$$

18.3.4 We'll tackle right triangles first. Let $\triangle XYZ$ be a right triangle with $\angle Z = 90°$. Then, we have $\sin Z = \sin 90° = 1$, and we have $\sin Y = XZ/XY$, so $XZ/(\sin Y) = XY = XY/1 = XY/(\sin Z)$. Similarly, we have $\sin X = YZ/XY$, so $YZ/(\sin X) = XY = XY/(\sin Z)$. We have therefore proved that if $\triangle XYZ$ is right, we have

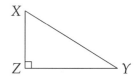

$$\frac{XZ}{\sin Y} = \frac{XY}{\sin Z} = \frac{YZ}{\sin X}.$$

Suppose $\triangle ABC$ is obtuse with $\angle C > 90°$. We draw altitude $\overline{AD}$ to the extension of $\overline{BC}$, as shown. From right triangle $\triangle ACD$, we have $\sin \angle ACD - AD/AC$, so $AD = AC \sin \angle ACD$. Since $\angle ACD = 180° - \angle ACB$ and $\sin \theta = \sin(180° - \theta)$, we have $AD = AC \sin \angle ACD = AC \sin(180° - \angle ACB) = AC \sin \angle ACB$. From right triangle ADB, we have $\sin \angle ABD = AD/AB$, so $AD = AB \sin \angle ABD$. Equating our two expressions for AD and noting that $\angle ABC = \angle ABD$, we have $AC \sin \angle ACB = AB \sin \angle ABC$, so $AC/(\sin \angle ABC) = AB/(\sin \angle ACB)$.

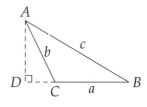

Similarly, we can draw the altitude from B to the extension of $\overline{AC}$ to show that $BC/(\sin \angle BAC) = AB/(\sin \angle ACB)$, so

$$\frac{AC}{\sin \angle ABC} = \frac{AB}{\sin \angle ACB} = \frac{BC}{\sin \angle BAC}.$$

18.3.5 We start with the diagram at right. The surveyor is at S, the base of the mountain is at B, and the top of the mountain is T. Point M is the foot of the altitude from T to $\overrightarrow{SB}$, so that MT is the height of the mountain. The surveyor measures $\angle TSB = 10°$ and we have $\angle TBM = 30°$. We also know that SB is 3 km. Our goal is to find MT.

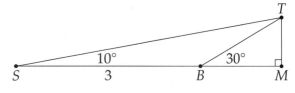

If we can find BT or ST, we can use right triangle $\triangle TBM$ or right triangle $\triangle TSM$ to find MT. Since we know SB, we focus first on $\triangle TBS$. First, we have $\angle SBT = 180° - \angle TBM = 150°$. So, we have $\angle BTS = 180° - 150° - 10° = 20°$. The Law of Sines then gives us

$$\frac{SB}{\sin 20°} = \frac{BT}{\sin 10°}.$$

So, we find $BT = (SB/(\sin 20°))\sin 10° \approx 1.52$. From 30-60-90 triangle $\triangle TBM$, we have $TM = TB/2 \approx$ $\boxed{0.76 \text{ km}}$.

18.3.6 We wish to relate the Law of Sines to the circumradius of $\triangle ABC$, so we draw $\triangle ABC$ and its circumcircle. We need to relate the triangle to the radius of this circle, so we draw diameter $\overline{AD}$ from a vertex of the triangle. We wish to relate the radius to the sine of the angles of the circle, so we connect D to B, forming right triangle ABD. However, this just gives us $\sin D = AB/AD = AB/2R$. We need the sine of one of the angles of the triangle. Since $\angle D$ and $\angle C$ are both inscribed in $\overparen{AB}$, we have $\angle D = \angle C$, so now we have $\sin C = \sin D = AB/2R$, which means $AB/(\sin C) = 2R$. Combining this with the Law of Sines gives us the Extended Law of Sines:

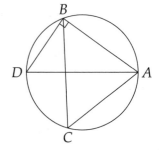

$$\frac{a}{\sin A} = \frac{b}{\sin B} = \frac{c}{\sin C} = 2R.$$

Notice that this proof holds even if $\triangle ABC$ is right or obtuse.

18.3.7 Since we know all three side lengths, we can use the Law of Cosines to find the cosines of the angles. From the Law of Cosines, we have

$$AB^2 = AC^2 + BC^2 - 2(AC)(BC)\cos C.$$

Substituting the side lengths we are given yields $\cos C = (31 - 25 - 36)/(-2 \cdot 5 \cdot 6) = 1/2$. There is only one point on the unit circle between $0°$ and $180°$ counterclockwise from $(1, 0)$ that has an x-coordinate of $1/2$, so there is only one possible value of $\angle C$. We know that $\cos 60° = 1/2$, so this is the angle measure we seek. Therefore, $\angle BCA = \boxed{60°}$.

Review Problems

18.24 First, we use the Pythagorean Theorem to find that $BC = \sqrt{39}$. Then, we have $\sin A = BC/AB = \boxed{\sqrt{39}/8}$, $\cos A = AC/AB = \boxed{5/8}$, and $\tan A = BC/AC = \boxed{\sqrt{39}/5}$.

18.25 We have $\sin P = QR/PR$, so $QR = PR\sin P \approx \boxed{15.1}$. We also have $\cos P = PQ/PR$, so $PQ = PR\cos P \approx \boxed{5.2}$.

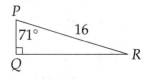

18.26 We have $\tan Y = XZ/XY$ and $\tan Z = XY/XZ$, so

$$(\tan Y)(\tan Z) = \left(\frac{XZ}{XY}\right)\left(\frac{XY}{XZ}\right) = 1.$$

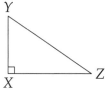

18.27 We start by noting that $\sin R = PQ/PR$. Because $\sin P = QR/PR = 1/4$, we have $QR = PR/4$. So, the Pythagorean Theorem gives us $PQ = \sqrt{PR^2 - QR^2} = \sqrt{PR^2 - PR^2/16} = PR\sqrt{15}/4$. So, we have $PQ/PR = \boxed{\sqrt{15}/4}$.

We also might have noted that $\sin R = \cos P$ and used the fact that $\sin^2 P + \cos^2 P = 1$ together with $0 < \cos P < 1$ to find the answer.

18.28

(a) Because $90° < 135° < 180°$, the angle $135°$ is in the second quadrant. Since $135°$ is $45°$ clockwise from $(-1, 0)$, drawing altitude $\overline{BP}$ creates 45-45-90 triangle BPO as shown, where $\angle BOP = 45°$. Therefore, we have $OP = BP = \sqrt{2}/2$. Since B is to the left of the y-axis and above the x-axis, its coordinates are $(-\sqrt{2}/2, \sqrt{2}/2)$, which means $\sin 135° = \boxed{\sqrt{2}/2}$.

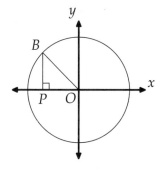

(b) Since $630° = 360° + 270°$, when we go $630°$ counterclockwise around the unit circle from $(1, 0)$, we end up at the same point as if we go $270°$ counterclockwise from $(1, 0)$. This point is $(0, -1)$, so $\sin 630° = \boxed{-1}$.

(c) Since the angle is negative, we seek the point that is $120°$ *clockwise* from $(1, 0)$. This point is in the third quadrant, as shown at right, and it is $60°$ counterclockwise from $(-1, 0)$. So, drawing altitude $\overline{BP}$ creates 30-60-90 triangle BPO as shown, where $\angle BOP = 60°$. We seek the x-coordinate of B. From $\triangle BPO$, we find $OP = 1/2$. Since point B is to the left of the y-axis, its x-coordinate is negative. So, we have $\cos(-120°) = -OP = \boxed{-1/2}$.

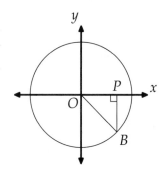

(d) Because $270° < 315° < 360°$, the angle $315°$ is in the fourth quadrant. Since $315°$ is $45°$ clockwise from $(1, 0)$, drawing altitude $\overline{BP}$ creates 45-45-90 triangle BPO as shown. Therefore, we have $OP = \sqrt{2}/2$, and $\cos 315° = OP = \boxed{\sqrt{2}/2}$.

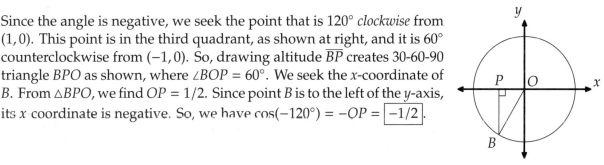

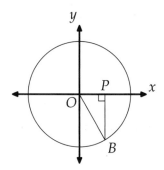

(e) Because $270° < 300° < 360°$, the angle $300°$ is in the fourth quadrant. Since $300°$ is $60°$ clockwise from $(1,0)$, drawing altitude $\overline{BP}$ creates 30-60-90 triangle BPO as shown, where $\angle BOP = 60°$. Therefore, we have $OP = 1/2$ and $BP = \sqrt{3}/2$. Since point B is below the x-axis, the y-coordinate of B is negative. So, we have $\sin 300° = -\sqrt{3}/2$ and $\cos 300° = 1/2$, which means $\tan 300° = \boxed{-\sqrt{3}}$.

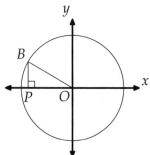

(f) Because $90° < 150° < 180°$, the angle $150°$ is in the second quadrant. Since $150°$ is $30°$ clockwise from $(-1,0)$, drawing altitude $\overline{BP}$ creates 30-60-90 right triangle BPO as shown, where $\angle BOP = 30°$. So, the coordinates of B are $(-\sqrt{3}/2, 1/2)$, from which we find $\tan 150° = (1/2)/(-\sqrt{3}/2) = -1/\sqrt{3} = \boxed{-\sqrt{3}/3}$.

18.29

(a) In the diagram at right, the surveyor is at S, the top of the building is T, and the base of the building is B. We have $\tan S = BT/SB$, so $SB = BT/(\tan S) \approx \boxed{5715 \text{ ft}}$.

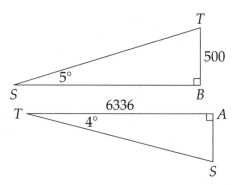

(b) In the diagram, the top of the taller mountain is T, the top of the smaller mountain is S, and A is directly above S such that $\overline{TA}$ is horizontal. Our map shows T and A as 1 inch apart, which corresponds to 1.2 miles. Since there are $1.2(5280) = 6336$ feet in 1.2 miles, we have $TA = 6336$ ft. So, we have $\tan T = AS/AT = AS/6336$. Therefore, $AS = 6336 \tan T \approx 443$ ft. So, the smaller mountain is 443 feet smaller than the taller mountain, which means it is $14000 - 443 = \boxed{13557 \text{ feet}}$ tall.

(c) In the diagram, the bee is at B, and the side of the building is $\overline{GT}$, so $GT = 400$. We seek BX. We have $XT/BX = \tan \angle TBX = \tan 4°$, so $XT = BX \tan 4°$. Similarly, we have $XG/BX = \tan \angle GBX = \tan 2°$, so $XG = BX \tan 2°$. Because $XT + XG = 400$, we have $BX \tan 4° + BX \tan 2° = 400$. Factoring out BX gives $BX(\tan 4° + \tan 2°) = 400$, so $BX = 400/(\tan 4° + \tan 2°) \approx \boxed{3815 \text{ ft}}$.

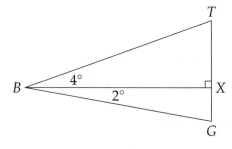

18.30   Let θ_1 be the angle such that $0° \le \theta < 90°$ and $\sin \theta_1 = 0.48$. (This angle corresponds to the point in the first quadrant where the graph of $y = 0.48$ intersects the unit circle.) Since $\sin(180° - x) = \sin x$ for all acute x, we have $\sin(180° - \theta_1) = \sin \theta_1 = 0.48$, as well, so $\theta_2 = 180° - \theta_1$, which means $\theta_1 + \theta_2 = \boxed{180°}$.

18.31   We can start by drawing altitude $\overline{AX}$ from A to $\overline{BC}$, thus forming 45-45-90 triangle $\triangle AXB$. This altitude then has length $AB/\sqrt{2} = 7/\sqrt{2}$, from which we have $[ABC] = (1/2)(AX)(BC) = (1/2)(7/\sqrt{2})(8) =$

$28/\sqrt{2} = \boxed{14\sqrt{2}}$.

After doing this, we might remember that this is essentially how we proved that for any triangle ABC, we have $[ABC] = (1/2)(AB)(BC)\sin\angle ABC$. Using this formula gives us $(1/2)(7)(8)(\sqrt{2}/2) = 14\sqrt{2}$, as before.

18.32 If $\tan\theta$ is negative, then $\sin\theta$ and $\cos\theta$ have opposite signs. This only occurs when θ is in $\boxed{\text{quadrant II or quadrant IV}}$.

18.33 We have $\tan^2 x + 1 = \dfrac{\sin^2 x}{\cos^2 x} + \dfrac{\cos^2 x}{\cos^2 x} = \dfrac{\sin^2 x + \cos^2 x}{\cos^2 x} = \dfrac{1}{\cos^2 x} = \sec^2 x$.

18.34 The shaded region is a parallelogram, so its area equals the product of the length of a side and the height drawn between that side and the opposite side. We build right triangle $\triangle ABC$ as shown, so the height of the parallelogram between its horizontal sides is AB. We have $AB = 1$ and $\sin\alpha = AB/AC = 1/AC$. Therefore, $AC = 1/\sin\alpha$. Similarly, we can show that all four sides of the shaded region have length $1/\sin\alpha$, so the parallelogram is a rhombus. The area of this rhombus then is $(1/\sin\alpha)(AB) = 1/(\sin\alpha)$, as desired.

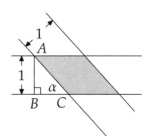

18.35 At right is the diagram for the problem. From right triangle $\triangle OAB$, we have $\tan\theta = AB/OA$. Since $\overline{OA}$ is a radius of a unit circle, we have $OA = 1$, so $\tan\theta = AB$.

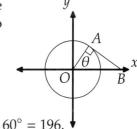

18.36

(a) The Law of Cosines gives us
$$PQ^2 = RQ^2 + RP^2 - 2(RQ)(RP)\cos R = 100 + 256 - 2(10)(16)\cos 60° = 196.$$
Taking the square root of both sides gives $PQ = \boxed{14}$.

(b) The Law of Cosines gives us
$$TU^2 = VT^2 + VU^2 - 2(VT)(VU)\cos V = 64 + 49 - 2(8)(7)\cos 46° \approx 35.2.$$
Taking the square root gives $TU \approx \boxed{5.9}$.

18.37

(a) The Law of Sines gives us
$$\frac{YZ}{\sin X} = \frac{XZ}{\sin Y} = \frac{XY}{\sin Z}.$$
We also have $\angle Y = 180° - 103° - 31° = 46°$. So, we find $XZ = (\sin 46°)(XY/(\sin 31°)) \approx \boxed{12.6}$, and $YZ = (\sin 103°)(XY/(\sin 31°)) \approx \boxed{17.0}$.

(b) The Law of Sines gives us
$$\frac{AB}{\sin C} = \frac{AC}{\sin B} = \frac{BC}{\sin A}.$$
We also have $\angle C = 180° - 88° - 52° = 40°$. So, we find $AC = (\sin 52°)(8/(\sin 40°)) \approx \boxed{9.8}$, and $BC = (\sin 88°)(8/(\sin 40°)) \approx \boxed{12.4}$.

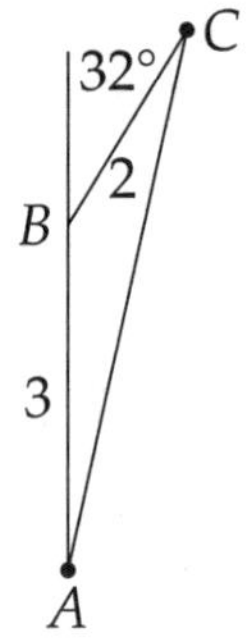

18.38   In the diagram at right, I start at point A, walk north 3 miles, turn $32°$ at point B, and then walk another 2 miles to point C. We have $\angle ABC = 180° - 32° = 148°$, $AB = 3$, and $BC = 2$, so we can use the Law of Cosines to find that

$$AC^2 = AB^2 + BC^2 - 2(AB)(BC)\cos \angle ABC \approx 23.2.$$

Taking the square root gives us $AC \approx \boxed{4.8 \text{ miles}}$.

18.39   In the text, we used the Pythagorean Theorem to prove the Law of Cosines. In the 'proof' offered in this problem, we then use the Law of Cosines to prove the Pythagorean Theorem. Because we used the Pythagorean Theorem to prove the Law of Cosines, we cannot use the Law of Cosines to prove the Pythagorean Theorem. (If we do, then we have essentially used the Law of Cosines to prove the Law of Cosines, which is clearly absurd!) We say this proof is guilty of 'circular reasoning,' and is therefore not a valid proof.

18.40

(a)   We have three side lengths and we seek the value of the cosine of one of the angles. The Law of Cosines has three side lengths and the cosine of an angle, so we try that. We have

$$QR^2 = PQ^2 + PR^2 - 2(PQ)(PR)\cos P.$$

We know all three side lengths, so we can substitute these into our equation to find

$$25 = 9 + 36 - 2(3)(6)\cos P.$$

Solving for $\cos P$, we have $\cos P = \boxed{5/9}$.

(b)   It's not as clear how to find $\sin Q$. If we knew the sine of one of the other angles, we could use the Law of Sines. Since we know $\cos P$, we can find $\sin P$ because $\sin^2 P + \cos^2 P = 1$. Therefore, we have $\sin^2 P = 1 - \cos^2 P = 56/81$. Since $\angle P$ is between $0°$ and $180°$, we know that $\sin P$ is nonnegative. So, taking the square root of $\sin^2 P = 56/81$ gives us $\sin P = 2\sqrt{14}/9$. Now, we can use the Law of Sines! Specifically, we have

$$\frac{QR}{\sin P} = \frac{PR}{\sin Q}.$$

Solving for $\sin Q$ gives us

$$\sin Q = \frac{\sin P}{QR} \cdot PR = \frac{2\sqrt{14}/9}{5} \cdot 6 = \boxed{\frac{4\sqrt{14}}{15}}.$$

See if you can find another solution by first using the Law of Cosines as in part (a) to find $\cos Q$.

Challenge Problems

18.41   In the text, we showed that $[ABC] = \frac{1}{2}ab\sin C$. As an Exercise, we proved the Extended Law of Sines, which gives us $c/(\sin C) = 2R$. Solving this for $\sin C$ gives us $\sin C = c/2R$. Substituting this into our expression for $[ABC]$ gives the desired $[ABC] = abc/4R$.

18.42 We are interested in OB, and we know both AB and $\angle O$. We can relate OB to AB and $\angle O$ through the Law of Sines, which gives us

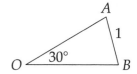

$$\frac{OB}{\sin A} = \frac{AB}{\sin O} = \frac{1}{\sin 30°} = 2.$$

Therefore, we have $OB = 2\sin A$. We wish to maximize OB, so we must maximize $\sin A$. The largest $\sin A$ can be occurs when $A = 90°$, which makes $\sin A = 1$ and $OB = \boxed{2}$. This occurs when $\triangle AOB$ is a 30-60-90 triangle with $AO = \sqrt{3}$, $AB = 1$, and $OB = 2$.

18.43 The base of the triangle with endpoints $(5,0)$ and $(-5,0)$ has length 10. Because the area of the triangle is 10, the height to this base must be 2. So, the final vertex of the triangle is either 2 units above the y-axis or 2 units below the y-axis. Therefore, this vertex is on the line $y = 2$ or the line $y = -2$.

We are told that this last vertex is $(5\cos\theta, 5\sin\theta)$ for some value of θ. Since the last vertex is on the graph of $y = 2$ or the graph of $y = -2$, we must have either $5\sin\theta = 2$ or $5\sin\theta = -2$. Therefore, we have either $\sin\theta = 2/5$ or $\sin\theta = -2/5$. There are two points on the unit circle with y-coordinate $2/5$ (the intersections of the graph of $y = 2/5$ with the unit circle) and two others with y-coordinate $-2/5$. So, there are $\boxed{4}$ possible triangles.

18.44 We must find the area of $ABCD$ and of the shaded region. Because the triangles are 30-60-90 triangles with hypotenuse 2, their legs have lengths $AW = 1$ and $WX = \sqrt{3}$. We can now find the area of the inner square and the right triangles, but we don't yet have the shaded triangles. We do have $\angle AXB = 180° - 30° = 150°$. Here are two different ways to finish the problem from this point:

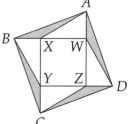

Solution 1: Law of Cosines. The shaded triangles are all congruent, from which we can determine that $ABCD$ is a square. The Law of Cosines gives us $AB^2 = AX^2 + BX^2 - 2(AX)(BX)\cos\angle AXB = 5 - 4\cos 150° = 5 + 2\sqrt{3}$. Therefore, the area of $ABCD$ is $AB^2 = 5 + 2\sqrt{3}$. Each right triangle has area $(1/2)(1)(\sqrt{3}) = \sqrt{3}/2$ and the small square has area $(\sqrt{3})^2 = 3$, so the unshaded portion of $ABCD$ has area $3 + 4(\sqrt{3}/2) = 3 + 2\sqrt{3}$. Therefore, the shaded region has area $(5 + 2\sqrt{3}) - (3 + 2\sqrt{3}) = 2$, which means it is $\boxed{2/(5 + 2\sqrt{3})}$ of the area of $ABCD$. We can write this fraction with a rational denominator by multiplying the numerator and denominator by $5 - 2\sqrt{3}$:

$$\frac{2}{5 + 2\sqrt{3}} = \frac{2}{5 + 2\sqrt{3}} \cdot \frac{5 - 2\sqrt{3}}{5 - 2\sqrt{3}} = \boxed{\frac{10 - 4\sqrt{3}}{13}}.$$

Solution 2: Find the shaded areas directly. We have $[AXB] = \frac{1}{2}(AX)(XB)\sin AXB = \frac{1}{2}(2)(1)\sin 150° = \frac{1}{2}$, so the four shaded regions together have area $4(\frac{1}{2}) = 2$. Adding the right triangles and the inner square to these gives us $[ABCD] = 5 + 2\sqrt{3}$, as before. So, the desired fraction is $2/(5 + 2\sqrt{3}) = (10 - 4\sqrt{3})/13$.

18.45 We start with the diagram at right. We focus first on $\sin\theta$, and find that $\sin\theta = AB/OB$. Next, we turn to OC and the fact that $\overline{BC}$ bisects $\angle OBA$. The Angle Bisector Theorem gives us $OC/OB = AC/AB$. Rearranging this gives us $AC/OC = AB/OB = \sin\theta$. Now, all we have to do is express AC in terms of OC or $\sin\theta$. Fortunately, this is easy. Because the circle has radius 1, we have $AC = 1 - OC$, so we have $(1 - OC)/OC = \sin\theta$. Multiplying both sides by OC gives $1 - OC = OC\sin\theta$. Adding OC to both sides then dividing by $1 + \sin\theta$ gives $OC = 1/(1 + \sin\theta)$.

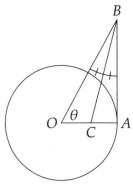

18.46 Substituting $\sin x = 3\cos x$ into $\sin^2 x + \cos^2 x = 1$ gives us $9\cos^2 x + \cos^2 x = 1$, from which we find $\cos x = \pm 1/\sqrt{10}$. If $\cos x = 1/\sqrt{10}$, then $\sin x = 3\cos x = 3/\sqrt{10}$. If $\cos x = -1/\sqrt{10}$, then $\sin x = 3\cos x = -3/\sqrt{10}$. In both cases, we have $(\sin x)(\cos x) = \boxed{3/10}$.

18.47 There are essentially two possible configurations. These are shown at right below. Any other possibility is the result of reflecting the shown diagrams over some vertical line. In all these other cases, the distance PQ between the x-intercepts is the same as one of the two cases shown.

For our first case, let the lines intersect at T as shown at right. The y-axis is not included because it is not relevant to the problem. We draw the altitude from T to the x-axis, which both forms right triangles and allows us to use the information that the lines meet at a point with y-coordinate 10. Therefore, we have $TU = 10$. From right triangle $\triangle TPU$, we have $TU/PU = \tan 40°$, so $PU = TU/(\tan 40°) \approx 11.9$. From right triangle $\triangle TQU$, we have $TU/QU = \tan 50°$, so $QU = TU/(\tan 50°) \approx 8.4$. So, we have $PQ = PU - QU \approx 3.5$.

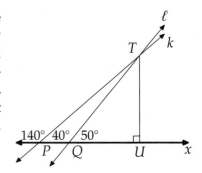

For our second case, let the lines again meet at T, and let the foot of the altitude from T to the x-axis be U. Once again, we use right triangles $\triangle TPU$ and $\triangle TQU$ to find that $PU = TU/(\tan 40°) \approx 11.9$ and $QU = TU/(\tan 50°) \approx 8.4$, so $PQ = PU + QU \approx 20.3$.

These two cases give us our two possibilities for the desired distance, $\boxed{3.5 \text{ and } 20.3}$.

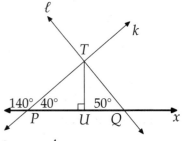

18.48 Let our triangle be $\triangle ABC$ at right, with $\angle B = 90°$. Because $\sin A = 12/13$, we have $BC/AC = 12/13$, so $BC = 12AC/13$. Therefore, the Pythagorean Theorem gives us

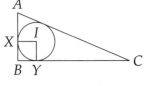

$$AB = \sqrt{AC^2 - BC^2} = \sqrt{AC^2 - \frac{144AC^2}{169}} = \frac{5AC}{13}.$$

We might also have recognized $\triangle ABC$ as similar to a 5-12-13 triangle.

We draw inradii to $\overline{AB}$ and $\overline{BC}$. This forms square $IXBY$. Because the inradius is 1, we have $BX = 1$. But we also have $BX = (AB + BC - AC)/2$. Substituting our expressions for AB and BC above gives us $BX = 2AC/13$. So, we have $AC = 13BX/2 = \boxed{13/2}$.

18.49 We cannot use the Law of Sines to prove AA Similarity because our very use of sine assumes that

AA Similarity is true. For example, to show that every right triangle with a 23° angle has the same ratio

$$\frac{\text{leg opposite the 23° angle}}{\text{hypotenuse}},$$

we used AA Similarity. In other words, we use AA Similarity to show that sine 'works'. So, we can't turn around and then use sine to show that AA Similarity works.

18.50   At first, all we can determine is that $\angle BAC = 36°/3 = 12°$. To find any other angles, we have to focus on the other given piece of information, that $BX = CX$. We might guess that this means that $\angle BXD = \angle DXC$, but we have to prove it. To do so, we focus on what is special about X: it is on diameter $\overline{AD}$. We know that X is not the center of the circle because $\angle BXC \neq 2\angle BAC$. Because $\angle BXC > 2\angle BAC$, we know that the center is between A and X.

We connect B and C to O, the center of the circle, and we have $\triangle BOX \cong \triangle COX$ by SSS Congruence. Therefore, $\angle BXO = \angle CXO$, which means $\angle BXD = \angle CXD$ and $\angle BXA = \angle CXA$. So, we also have $\triangle AXB \cong \triangle AXC$ by SAS Congruence, which means that $\angle XAB = \angle XAC$. Therefore, diameter $\overline{AD}$ bisects both $\angle BAC$ and $\angle BXC$. So, we have $\angle BXD = \angle DXC = (\angle BXC)/2 = 18°$, $\angle BXA = \angle CXA = 162°$, and $\angle BAD = \angle CAD = 6°$. So, we have $\angle ABX = \angle ACX = 180° - 162° - 6° = 12°$.

Turning to the length we wish to find, AX, we apply the Law of Sines to find

$$\frac{AX}{\sin \angle ABX} = \frac{AB}{\sin \angle AXB},$$

so we have

$$AX = \frac{AB}{\sin 162°} \cdot \sin 12°.$$

We're getting close, but we have to deal with AB. We also know we would like a final expression with a 6° angle in it. We can take care of both by drawing $\overline{BD}$. Because $\overline{AD}$ is a diameter, we have $\angle ABD = 90°$, so $\cos \angle BAD = AB/AD = AB$ (because $AD = 1$). So, we have $AB = \cos 6°$, and

$$AX = \frac{(\cos 6°)(\sin 12°)}{\sin 162°}.$$

Finally, because $\sin 162° = \sin(180° - 162°) = \sin 18°$, we have the desired

$$AX = \frac{(\cos 6°)(\sin 12°)}{\sin 18°}.$$

18.51   We can write AC in terms of α using the Law of Cosines, which gives

$$AC^2 = 25 + 64 - 2(5)(8)\cos \alpha = 89 - 80 \cos \alpha.$$

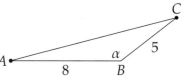

We wish to know the values of α for which $AC < 7$. If $AC < 7$, then $AC^2 < 49$, and we have $89 - 80 \cos \alpha < 49$. Rearranging this inequality gives us $\cos \alpha > 1/2$. So, $AC < 7$ if and only if $\cos \alpha > 1/2$. When $\cos \alpha = 1/2$, we have $\alpha = 60°$, since $0° < \alpha < 180°$.

We have $\cos \alpha > 1/2$ when the x-coordinate of the point on the unit circle corresponding to α is greater than $1/2$. We have $\cos 60° = 1/2$, and all smaller positive angles α correspond to points on the unit circle to the right of the point corresponding to $60°$ (point B in the diagram at right). So, the x-coordinates of these points are larger than $1/2$, which means $\cos \alpha > 1/2$ if $\alpha < 60°$. Similarly, if $\cos \alpha < 1/2$, then we have $\alpha > 60°$. So, because $AC < 7$ if $0 < \alpha < 60°$, and $AC \geq 7$ for $60° \leq \alpha < 180°$, the probability that $AC < 7$ is $60°/180° = \boxed{1/3}$.

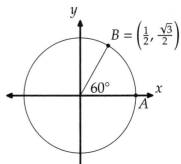

18.52 Because $BA' = 1$ and $A'C = 2$, we have $BC = BA' + A'C = 3$. Since $\triangle ABC$ is equilateral, we have $AB = AC = BC = 3$, and we have $\angle B = \angle A = \angle C = 60°$. Furthermore, because $\angle PA'Q$ is the image of $\angle A$ upon reflection over $\overline{PQ}$, we have $\angle PA'Q = \angle A = 60°$. So, if we have AP and AQ, we can use the Law of Cosines on $\triangle APQ$ or $\triangle A'PQ$ to find PQ. Letting $AP = x$ and $AQ = y$, we apply the Law of Cosines to $\triangle APQ$ to find

$$PQ^2 = x^2 + y^2 - 2xy \cos 60° = x^2 + y^2 - xy.$$

Since $A'P = AP = x$, $BA' = 1$, and $BP = AB - AP = 3 - x$, applying the Law of Cosines to $\triangle A'BP$ gives us

$$x^2 = 1^2 + (3 - x)^2 - 2(1)(3 - x) \cos 60° = x^2 - 5x + 7.$$

Solving $x^2 = x^2 - 5x + 7$ for x gives $x = 7/5$. Similarly, we have $CQ = 3 - y$, $A'Q = y$, and $A'C = 2$, so applying the Law of Cosines to $\triangle A'QC$ gives

$$y^2 = 4 + (3 - y)^2 - 2(2)(3 - y) \cos 60° = y^2 - 4y + 7.$$

Solving $y^2 = y^2 - 4y + 7$ for y gives us $y = 7/4$. We can now substitute $x = 7/5$ and $y = 7/4$ into our equation for PQ^2 to find

$$PQ^2 = \frac{49}{25} + \frac{49}{16} - \frac{49}{20} = \frac{49 \cdot 16 + 49 \cdot 25 - 49 \cdot 20}{400} = \frac{49 \cdot 21}{400}.$$

Taking the square root gives us $PQ = \boxed{7\sqrt{21}/20}$.

19

Problem Solving Strategies in Geometry

Exercises for Section 19.1

19.1.1 Let P be the center of the small circle nearest A and Q be the center of the other small circle. Since a diagonal of the square is a diameter of $\odot O$, the square has diagonal length 2, and therefore has side length $2/\sqrt{2} = \sqrt{2}$. Since OH is half the length of a side of the square, we have $GH = OG - OH = 1 - \sqrt{2}/2$. Therefore, $GP = GH/2 = 1/2 - \sqrt{2}/4$, so $OP = OG - GP = 1/2 + \sqrt{2}/4$. Since $\triangle OPQ$ is a 45-45-90 triangle, we have $PQ = OP\sqrt{2} = \sqrt{2}/2 + 1/2$. From 45-45-90 triangle $\triangle APH$, we have $AP = PH\sqrt{2} = GP\sqrt{2} = \sqrt{2}/2 - 1/2$. Similarly, we find $BQ = AP = \sqrt{2}/2 - 1/2$. Finally, we have $AB = PQ - AP - BQ = \boxed{3/2 - \sqrt{2}/2}$.

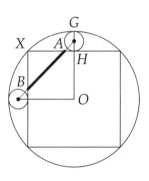

We also could have noted that $AH = PH = GP = 1/2 - \sqrt{2}/4$, so $XA = \sqrt{2}/2 - AH = 3\sqrt{2}/4 - 1/2$, from which 45-45-90 triangle $\triangle XAB$ gives us the answer.

19.1.2 Because the length of a side of the square is 4, diagonal $\overline{AD}$ of the square has length $4\sqrt{2}$ and the quarter-circle has radius 4. Therefore, $AE = AD - DE = 4\sqrt{2} - 4$. Letting the small circle have radius r and center B, we have $AE = AB + BE = BC\sqrt{2} + r = r\sqrt{2} + r$. Therefore, we have $r\sqrt{2} + r = 4\sqrt{2} - 4$, so

$$r = \frac{4\sqrt{2} - 4}{\sqrt{2} + 1} = \frac{4(\sqrt{2} - 1)(\sqrt{2} - 1)}{(\sqrt{2} + 1)(\sqrt{2} - 1)} = \boxed{12 - 8\sqrt{2}}.$$

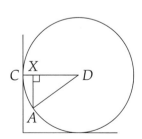

19.1.3 We build a right triangle by connecting A to the center of the circle, D, drawing the radius to the point of tangency of the circle and the wall on the left, and then drawing altitude $\overline{AX}$ to this radius. Since A is 2 inches from the wall, we have $DX = CD - CX = 37 - 2 = 35$. Since $\overline{DA}$ is a radius of the circle, we have $DA = 37$. Therefore, from right triangle $\triangle ADX$, we have $AX = \sqrt{DA^2 - DX^2} = 12$. Since A is 12 inches from radius $\overline{CD}$, and this radius is parallel to the other wall and 37 inches from it (since the table is tangent that wall as well), we find that A is $37 - 12 = \boxed{25 \text{ inches}}$ from the other wall.

19.1.4 Extend $\overline{AB}$ past B to hit $\overrightarrow{ZY}$ at P. Since $\overline{WZ} \parallel \overline{BY}$, we have $\angle PBY = \angle PWZ$ and $\angle PYB = \angle PZW$, so $\triangle PWZ \sim \triangle PBY$ by AA Similarity. Since $YB/XY = 1/2$ and $XY = WZ$, we have $BY = WZ/2$. Therefore, our triangle similarity gives us $PY = PZ/2$, so Y is the midpoint of $\overline{PZ}$. Since Y is the midpoint of hypotenuse $\overline{PZ}$ of right triangle APZ, the median to this hypotenuse has length half the hypotenuse. Therefore, $AY = PZ/2 = PY = YZ$.

Exercises for Section 19.2

19.2.1 Let $\angle C = x$ and $\angle D = y$. From isosceles triangles $\triangle ABD$ and $\triangle ABC$, we have $\angle BAD = y$ and $\angle ABC = x$. From right triangle BEC, we have $\angle EBC = 90° - x$, so $\angle ABD = \angle ABC - \angle EBC = x - (90° - x) = 2x - 90°$. Therefore, from $\triangle ABD$, we have $\angle ABD + \angle ADB + \angle DAB = 180°$, so $(2x - 90°) + y + y = 180°$. We therefore find $2x + 2y = 270°$, so $x + y = \boxed{135°}$.

19.2.2 Let X be the point described in the problem. We draw radii to the points of tangency nearest X. We then draw altitudes from X to these two radii. Since $OABC$ is a square (because $\angle OAB = \angle OCB = \angle ABC = 90°$ and $OA = OC$), we see that $OEXD$ is a rectangle (since $\angle EOD = \angle ODX = \angle OEX = 90°$). Let the radius of the circle be r. Since X is 1 and 2 units from the sides of the square, we have $EA = 1$ and $CD = 2$ (because $\overline{XE} \parallel \overline{AB}$ and $\overline{XD} \parallel \overline{BC}$). Therefore, $OE = r - 1$ and $OD = r - 2$. $\overline{OX}$ is a radius of the circle and a diagonal of rectangle $OEXD$, so the diagonals of rectangle $OEXD$ have length r. Therefore, we have $(r-1)^2 + (r-2)^2 = r^2$. Expanding and rearranging gives us $r^2 - 6r + 5 = 0$, so $(r-1)(r-5) = 0$. Therefore, 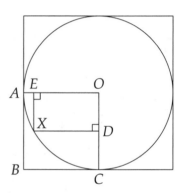 $r = 1$ or $r = 5$. Since $r = 1$ is impossible (make sure you see why), we have $r = 5$. The side length of the square is twice the radius of the circle, so the area of the square is $10^2 = \boxed{100}$.

19.2.3 Because $AB < AD$, when we make our fold, the fold must run from a point on $\overline{AD}$ to a point on $\overline{BC}$. Let these points be F and G, as shown, and let A be at point E after folding. Let $AG = x$, so $EG = x$ as well, and $GD = AD - AG = 12 - x$. We also have $ED = AB = 8$, because $\overline{AB}$ after the fold is $\overline{ED}$. From right triangle GED, we have $8^2 + x^2 = (12 - x)^2$, from which we find $x = 10/3$. Since $\angle DEG = \angle DCF$, $ED = DC$, and $\angle CDF = 90° - \angle ADF = \angle EDF - \angle ADF = \angle EDA$, we have $\triangle DFC \cong \triangle DGE$ by ASA Congruence. Therefore, $FC = EG = 10/3$, so $BF = BC - FC = 26/3$.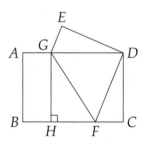

We draw altitude $\overline{GH}$ from G to $\overline{BF}$ to build right triangle $\triangle GHF$. Since $AGHB$ is a rectangle, we have $BH = AG = 10/3$, so $HF = BF - BH = 16/3$. We also have $GH = AB = 8$. Therefore, from right triangle $\triangle GHF$, we have $GF = \sqrt{HF^2 + GH^2} = \sqrt{(16/3)^2 + 8^2} = \sqrt{(8/3)^2(2^2 + 3^2)} = \boxed{8\sqrt{13}/3}$.

19.2.4 Let $BE = x$. Therefore, $AE = 3/2 - x$, and right triangle $\triangle ABE$ gives us $1 + x^2 = (3/2 - x)^2$. Solving, we find that $x = 5/12$, so $AE = 13/12$ and $EC = 7/12$. Let G be the foot of the altitude from F to $\overleftrightarrow{AE}$ of $\triangle AEF$. Since $AF = AF$, $\angle GAF = \angle DAF$, and $\angle D = \angle FGA$, we have $\triangle FGA \cong \triangle FDA$ by AAS Congruence. Therefore, $AG = AD = 1$, so $AG < AE$, which means G is on $\overline{AE}$, as shown. Let $FG = y$, so $FD = y$ from our congruent triangles. Therefore, $FC = 1 - y$, so right triangle $\triangle EFC$ gives us $EF = \sqrt{FC^2 + EC^2} = \sqrt{(1-y)^2 + (7/12)^2}$. Since $EA = 13/12$ and $AG = AD = 1$, we have $EG = 1/12$, so right triangle $\triangle EFG$ gives us $EF = \sqrt{EG^2 + GF^2} = \sqrt{y^2 + (1/12)^2}$. We thus have $EF = \sqrt{(1-y)^2 + (7/12)^2} = \sqrt{y^2 + (1/12)^2}$. Solving this equation by first squaring both sides then doing some rearranging, we find $y = \boxed{2/3}$.

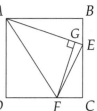

Exercises for Section 19.3

19.3.1 Because $\angle CBE$ and $\angle CAE$ are inscribed in the same arc, we have $\angle CAE = \angle CBE$. Since $\overline{AD}$ bisects $\angle CAB$, we have $\angle CBE = \angle CAE = \angle BAE$. We also have $\angle AEB = \angle DEB$, so we have $\triangle DBE \sim \triangle BAE$ by AA Similarity. Therefore, we have $BE/DE = AE/BE$, or $BE^2 = (AE)(DE)$.

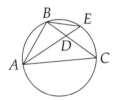

19.3.2 Since $\angle CYB = 2\angle AYB$ and $\angle BZC = 2\angle AZC$, triangles $\triangle YCZ$ and $\triangle YBZ$ give us

$$
\begin{aligned}
\angle YCZ + \angle YBZ &= (180° - \angle CYZ - \angle YZC) + (180° - \angle BZY - \angle BYZ) \\
&= 360° - \angle CYB - \angle BYZ - \angle YZC - \angle BZC - \angle CZY - \angle BYZ \\
&= 360° - 2\angle AYB - 2\angle BYZ - 2\angle YZC - 2\angle AZC \\
&= 2(180° - \angle AYB - \angle BYZ - \angle YZC - \angle AZC) \\
&= 2(180° - \angle AYZ - \angle AZY)
\end{aligned}
$$

Since $180° - \angle AYZ - \angle AZY = \angle YAZ$ from $\triangle AYZ$, we have the desired $\angle YCZ + \angle YBZ = 2\angle YAZ$.

19.3.3 We have $\angle ZXL = \widehat{ZL}/2 = \angle ZAL = \angle LAY = \widehat{YL}/2 = \angle YXL$. (We used the fact that $\overline{AL}$ bisects $\angle ZAY$.) Combining this with $XL = XL$ and $XZ = XY$ (from $\triangle CYX \cong \triangle BZX$, which we proved in the text), we have $\triangle XZL \cong \triangle XYL$ by SAS Congruence.

Since $XZLY$ is a cyclic quadrilateral, we have $\angle XZL + \angle XYL = 180°$. We also have $\angle XZL = \angle XYL$ from our triangle congruence. Therefore, $\angle XZL = \angle XYL = 90°$, so $\triangle XZL$ and $\triangle XYL$ are right triangles. The hypotenuse of these triangles, $\overline{XL}$, is therefore a diameter of the circumcircle of $XZLY$. This is the same circle as the circumcircle of $\triangle ALM$, so $\overline{XL}$ is a diameter of the circumcircle of $\triangle ALM$.

19.3.4 Since $\angle YOX = \angle YPZ$ and $\angle OXY = 90° - \angle OYX = \angle ZYX - \angle OYX = \angle ZYP$, we have $\triangle ZYP \sim \triangle YXO$ by AA Similarity. Therefore, we have $OY/OX = ZP/YP$, or $(OY)(YP) = (ZP)(OX)$. Similarly, we have $\triangle ZWP \sim \triangle WXO$, so $WP/ZP = OX/OW$, or $(WP)(OW) = (ZP)(OX)$. Therefore, we have $(OY)(YP) = (WP)(OW)$. Since $OY = YP - OP$ and $WP = OW - OP$, we have $(YP - OP)(YP) = (OW)(OW - OP)$. Therefore, we have $YP^2 - OW^2 + (OW)(OP) - (YP)(OP) = 0$. A little factoring then gives us $(YP - OW)(YP + OW) + (OP)(OW - YP) = 0$, so $(YP - OW)(YP + OW - OP) = 0$. Clearly $YP + OW - OP > 0$, so we must have $YP - OW = 0$, or $YP = OW$.

19.3.5

(a) Suppose we complete the circle with diameter $\overline{AC}$ and continue $\overrightarrow{DB}$ to meet this circle again at E. Since diameter $\overline{AC}$ is perpendicular to chord $\overline{ED}$, it bisects it, so $BD = BE$. The power of point B then gives $(AB)(BC) = (BD)(BE) = (BD)^2$. Letting the radius of the semicircle with diameter $\overline{BC}$ be s and the radius of the semicircle with diameter $\overline{AB}$ be t, we have $BD^2 = (2t)(2s) = 4st$. Therefore, the area of the circle with $\overline{BD}$ as diameter is $BD^2\pi/4 = st\pi$. The area of the arbelos is the area of the semicircle with diameter $\overline{AC}$ minus the areas of the other two semicircles. The diameter of the large semicircle is the sum of the diameters of the two smaller semicircles, so the radius of the large semicircle is $s + t$. Therefore, the area of the arbelos is $(s + t)^2\pi/2 - s^2\pi/2 - t^2\pi/2 = st\pi$. Hence, the area of the arbelos equals the area of the circle with $\overline{BD}$ as a diameter.

(b) We let $AB = 2t$ and $BC = 2s$ as before, so the radius of the large semicircle is $s + t$. Let the center of the circle on the right be X and the center of the smallest semicircle (which is the midpoint of $\overline{BC}$) be P. Let the center of the large semicircle (which is the midpoint of $\overline{AC}$) be O. Let the feet of the altitudes from X to $\overline{AC}$ and $\overline{BD}$ be Y and Z, respectively. Let r be the radius of the small circle tangent to the semicircle with diameter $\overline{BC}$.

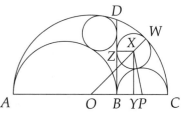

We find XY in two ways in order to show that $r = st/(s + t)$. First, if we continue $\overrightarrow{OX}$, it will meet the point where $\odot X$ is tangent to the large semicircle. We call this point W. Therefore, $OX = OW - XW = s + t - r$. Since $XZBY$ is a rectangle, we have $BY = ZX = r$. We also have $OB = OC - BC = t - s$. Therefore, $OY = OB + BY = t - s + r$. Right triangle OXY then gives us

$$XY = \sqrt{OX^2 - OY^2} = \sqrt{(s + t - r)^2 - (t - s + r)^2} = \sqrt{(2t)(2s - 2r)} = 2\sqrt{t(s - r)},$$

where we used the difference of squares factorization to simplify our algebra. We can also use right triangle $\triangle XYP$ to find XY, since $XP = r + s$ and $YP = PB - BY = s - r$. We find $XY = \sqrt{XP^2 - YP^2} = 2\sqrt{rs}$. These two expressions for XY must be equal, so we have $2\sqrt{rs} = 2\sqrt{t(s - r)}$. A little rearranging gives $r = st/(s + t)$. We notice that our expression for r is symmetric in s and t (meaning if we switch s and t, we get the same expression). Therefore, it's unsurprising that following pretty much the same steps, we find that the radius of the other little circle is also $st/(s + t)$. Therefore, our two little circles have the same radius.

Challenge Problems

19.11 Let X' be the point where X lands on $\overline{YZ}$ when the triangle is folded over $\overline{YP}$. Therefore, $\angle YX'P$ is the image of $\angle YXP$ upon reflection over $\overline{YP}$, so $\angle YX'P = \angle YXP = 90°$. We also similarly have $\angle X'YP = \angle XYP$. Since $PY = PZ$, $X'P = X'P$, and $\angle PX'Y = \angle PX'Z = 90°$, we have $\triangle PX'Y \cong \triangle PX'Z$ by HL Congruence. Therefore, $\angle Z = \angle X'YP$. Since $\angle Z + \angle X'YP + \angle XYP = 180° - \angle X = 90°$, and $\angle Z = \angle X'YP = \angle XYP$, we have $\angle Z = \angle X'YP = \angle XYP = \boxed{30°}$.

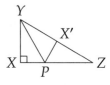

19.12 We have $\angle LPA = \angle MNC = 90°$ and $\angle ALP = 180° - \angle MLP - \angle MLB = 90° - \angle MLB = \angle LMB = 180° - \angle LMN - \angle NMC = 90° - \angle NMC = \angle NCM$, so $\triangle ALP \sim \triangle MCN$ by AA Similarity. Therefore, $AP/LP = MN/NC$, so $(AP)(NC) = (LP)(MN) = PN^2 = \boxed{36}$.

19.13 By symmetry, the center of the small circle in the middle must be equidistant from the vertices of the square. Therefore, the center of the small circle is the intersection of the diagonals of the square. Letting the radius of each of the large circles be R and the radius of the little circle be r, diagonal $\overline{BD}$ therefore has length $2R + 2r$. The side length of the square is double the radius of an outer circle, or $2R$. We are given that the side length of the square is 4, so $R = 2$. Therefore, the diagonal of the square has length $4\sqrt{2}$. We can equate our expressions for the diagonal to find $4 + 2r = 4\sqrt{2}$. Therefore, $r = \boxed{2\sqrt{2} - 2}$.

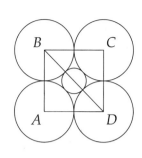

19.14 Because our four interior regions have equal area, each has area 1/4 the area of $\triangle ABC$. Specifically, $[PQR] = [ABC]/4$. Since the two triangles are equilateral, they are similar. The ratio of their sides is the square root of the ratio of their areas, so the sides of $\triangle PQR$ are half the sides of $\triangle ABC$. Therefore, $PQ = AB/2 = 10$. The outer trapezoids all have the same height because their bases are the same and their areas are the same.

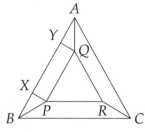

We draw altitudes $\overline{PX}$ and $\overline{QY}$ to $\overline{AB}$ from P and Q, respectively. $PXYQ$ is a rectangle because $\overline{PQ} \parallel \overline{AB}$. Therefore, we have $XY = PQ = 10$ and $PX = QY$. By HL Congruence, we have $\triangle AQY \cong \triangle BPX$, so $BX = AY$. Since $AY + BX = AB - XY = 10$, we have $AY = BX = 10/2 = 5$. Since $\overline{PB}$ bisects $\angle ABC$, we have $\angle PBC = \angle PBA = (\angle ABC)/2 = 30°$. Therefore, $\triangle BXP$ is a 30-60-90 triangle and $PB = BX(2/\sqrt{3}) = \boxed{10\sqrt{3}/3}$.

19.15 We draw altitudes from P to the sides of $WXYZ$ as shown. Since $\overline{PX}$ and $\overline{AD}$ are perpendicular to $\overline{AB}$, they are parallel. Since $\overline{AD} \perp \overline{CD}$ and $\overline{PX} \parallel \overline{AD}$, we know that $\overleftrightarrow{PX} \perp \overline{CD}$. Since Z is the foot of the perpendicular from P to $\overline{CD}$ and $\overleftrightarrow{PX} \perp \overline{CD}$, XPZ is a straight line. Similarly, WPY is a straight line as well. Hence, $AXZD$ and $BXZC$ are rectangles, so $AX = DZ$ and $BX = ZC$. We then use the Pythagorean Theorem to find

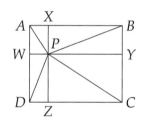

$$\begin{aligned}
AP^2 &= AX^2 + PX^2 \\
CP^2 &= CZ^2 + ZP^2 \\
BP^2 &= BX^2 + PX^2 \\
DP^2 &= DZ^2 + PZ^2
\end{aligned}$$

Adding the first two equations and the last two equations and noting $AX = DZ$ and $CZ = BX$, we have

$$AP^2 + CP^2 = AX^2 + PX^2 + CZ^2 + PZ^2 = DZ^2 + PX^2 + BX^2 + PZ^2 = (BX^2 + PX^2) + (DZ^2 + PZ^2) = BP^2 + DP^2.$$

19.16 Let diagonals $\overline{WY}$ and $\overline{XZ}$ meet at P. Since $ZM/ZW = ZO/ZY$ and $\angle MZO = \angle WZY$, we have $\triangle WZY \sim \triangle MZO$ by SAS Similarity. Therefore, $\angle ZOM = \angle ZYW$ and we have $\overline{WY} \parallel \overline{MO}$. Hence, $\triangle ZPY \sim \triangle ZNO$, so $ZN/ZP = ZO/ZY = 1/2$. Because the diagonals of a parallelogram bisect each other, we have $ZP = ZX/2$, so $ZN = ZX/4$, from which we find $ZN/NX = \boxed{1/3}$.

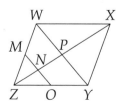

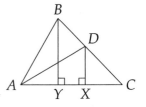

19.17 We have $\angle B = 180° - \angle BAC - \angle C = 75°$ and $\angle ADB = 180° - \angle BAD - \angle B = 180° - (\angle BAC)/2 - \angle B = 75°$. Therefore, $\triangle ABD$ is isosceles with $AB = AD$. We draw altitudes $\overline{BY}$ and $\overline{DX}$ from B and D to $\overline{AC}$ as shown. From 30-60-90 triangle $\triangle ABY$, we have $BY = AB(\sqrt{3}/2) = 9\sqrt{3}$. From 30-60-90 triangle $\triangle DXA$, we have $DX = AD/2 = 9$ and $XA = 9\sqrt{3}$. From 45-45-90 triangle $\triangle DXC$, we have $CX = DX = 9$. Therefore, we have $AC = AX + XC = 9 + 9\sqrt{3}$, so $[ABC] = (AC)(BY)/2 = \boxed{(243 + 81\sqrt{3})/2}$.

19.18 Since the area outside the square but inside the circle equals the area outside the circle but inside the square, the areas of the circle and the square must be the same. Therefore, letting the radius of the circle be r and the side length of the square be s, we have $s^2 = \pi r^2$, so $s^2/r^2 = \pi$, and $s/r = \boxed{\sqrt{\pi}}$.

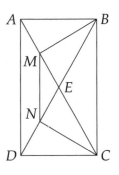

19.19 Since $\overline{CN}$ and $\overline{CE}$ are angle trisectors of $\angle BCD$, we have $\angle BCE = \angle ECN = \angle NCD$. Therefore, median $\overline{CN}$ of $\triangle ECD$ is also an angle bisector of $\triangle ECD$. The Angle Bisector Theorem then gives us $EC/CD = EN/ND = 1$, so $\triangle ECD$ is isosceles with $EC = CD$, and $\overline{CN}$ is also an altitude of $\triangle ECD$. Similarly, $\triangle ABE$ is isosceles with $AB = BE$ and $\overline{BM} \perp \overline{AE}$. Since $\angle BME = \angle CNE$ and $\angle MEB = \angle CEN$, we have $\triangle BEM \sim \triangle CEN$ by AA Similarity. Therefore, $\angle ECN = \angle EBM$, so $\angle ABC = 3\angle MBE = 3\angle ECN = \angle BCD$. Since $\overline{AB} \parallel \overline{CD}$, we have $\angle ABC + \angle BCD = 180°$, as well. Therefore, $\angle ABC = \angle BCD = 90°$. Our angle trisectors give us $\angle EBC = \angle ECB$, so $BE = CE$. Isosceles triangles $\triangle ECD$ and $\triangle ABE$ give us $BA = BE = CE = CD$. Since $\overline{AB} \parallel \overline{CD}$ and $AB = CD$, $ABCD$ is a parallelogram. Since we also have $\angle ABC = 90°$, $ABCD$ is a rectangle.

19.20 Let the radius of the base of the cone be r, the slant height of the cone be l, and the angle of the sector in the given diagram be θ. The lateral surface area of the cone is $\pi r l = \pi(2l/3)l = 2\pi l^2/3$. The area of the sector is $(\pi l^2)(\theta/360°)$. When this sector is rolled up, it becomes the lateral surface of the cone. Therefore, we have $(\pi l^2)(\theta/360°) = 2\pi l^2/3$, so $\theta = 240°$. Hence, each of our little circles is inscribed in a $240°/4 = 60°$ sector.

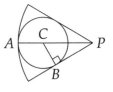

Shown in our diagram is one such circle with the sector in which it is inscribed. Let C be the center of the circle, B be the point where the circle touches a radius of the sector, and A be the point where the circle is tangent to the arc of the sector. Since C is equidistant from the sector radii at the sides of the sector, C is on the angle bisector formed by these radii. Therefore, $\angle CPB = 60°/2 = 30°$. Letting the radius of the little circle be s, we have $CP = 2CB = 2s$ from 30-60-90 triangle CBP, so $AP = AC + CP = 3s$, and the ratio of the radius of the sector to the radius of each circle is $\boxed{3}$.

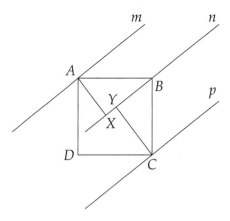

19.21 We draw altitudes from A and C to n, as shown. Since $AB = BC$, $\angle AXB = \angle BYC$, and $\angle YCB = 90° - \angle YBC = \angle ABC - \angle YBC = \angle ABX$, we have $\triangle ABX \cong \triangle BCY$ by AAS Congruence. Therefore, $BY = AX = 7$, so $BC^2 = BY^2 + CY^2 = \boxed{130}$.

19.22 Let the chords intersect at N. Since the chords bisect each other and have the same length, N is equidistant from X, Y, T, and U. Therefore, the circle with center N and radius NX passes through all four of these points. Our given $\odot O$ passes through all four of these points as well. For any three noncollinear points in a plane, there is exactly one circle that passes through all three points: the circumcircle of the triangle formed by connecting the three points. Since $\odot O$ and $\odot N$ pass through the same four points, the two circles must therefore be the same circle. Thus, the point where $\overline{XY}$ and $\overline{TU}$ intersect is the center of $\odot O$.

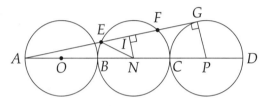

19.23 We draw the altitude from N to $\overline{AG}$, and we draw radius $\overline{GP}$ of $\odot P$. Since $\overline{AG}$ is tangent to this circle at G, we have $\overline{PG} \perp \overline{AG}$. Therefore, $\overline{PG} \parallel \overline{NI}$, so we have $\triangle AIN \sim \triangle AGP$ by AA Similarity. Since $AN/AP = 3/5$, we have $NI/PG = AN/AP = 3/5$, so $NI = (3/5)(PG) = 9$. From right triangle NIE we have $IE = 12$ (since $NE = 15$). Similarly, $IF = 12$ (or we note that $\overline{NI} \perp \overline{EF}$ means I is the midpoint of $\overline{EF}$), so $EF = 2EI = \boxed{24}$.

19.24 By symmetry, the sphere must be tangent to the octahedron at the centers of the faces of the octahedron. The centers of the faces of a regular octahedron are the vertices of a cube. A space diagonal of the cube is a diameter of the sphere, so our problem now is to find the length of a space diagonal of the cube.

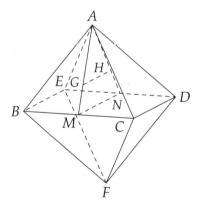

Let ABC and ADE be faces of the octahedron, and let G and H be the centers of these two faces. Let M be the midpoint of $\overline{BC}$ and N be the midpoint of $\overline{DE}$. We have $MN = 12$, because $\overline{MN}$ connects the midpoints of opposite sides of square $BCDE$. Since G and H are the centroids of ABC and ADE, respectively, we have $AG/AM = AH/AN = 2/3$. Combining this with $\angle MAN = \angle GAH$, we have $\triangle MAN \sim \triangle GAH$ by SAS Similarity. Therefore, we have $GH/MN = AG/AM = 2/3$, so $GH = (2/3)(MN) = 8$. $\overline{GH}$ is a face diagonal of our cube, so our cube has side length $8/\sqrt{2} = 4\sqrt{2}$. Therefore, it has diagonal length $(4\sqrt{2})(\sqrt{3}) = 4\sqrt{6}$. Finally, this means the sphere has radius $2\sqrt{6}$, so the sphere has volume $(4/3)(\pi)(2\sqrt{6})^3 = \boxed{64\pi\sqrt{6}}$.

19.25 Our shaded region equals the area of sector AOC minus $[AOM]$. Since M is the midpoint of $\overline{OC}$, $[AOM] = [AOC]/2$. $\triangle AOC$ is an isosceles triangle with $OA = OC = 12$ and $\angle OAC = \angle OCA = (180° - 120°)/2 = 30°$. We can find the area either by drawing an altitude from A to $\overleftrightarrow{OC}$, or by drawing an altitude from O to $\overline{AC}$. Let X be the foot of the altitude from A to $\overleftrightarrow{OC}$, extended as shown. From 30-60-90 triangle AXO, we have $AX = AO(\sqrt{3}/2) = 6\sqrt{3}$. Therefore, $[AOM] = [AOC]/2 = ((AX)(OC)/2)/2 = 18\sqrt{3}$. Sector AOC is 1/3 of a circle with radius 12, so it has area $(12)^2\pi/3 = 48\pi$. Therefore, the shaded region has area $\boxed{48\pi - 18\sqrt{3}}$.

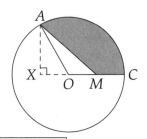

19.26 Let D be the foot of the altitude from A to $\overline{BC}$ and E be the foot of the altitude from B to $\overline{AC}$. Since $\angle BEC = \angle ADC = 90°$ and $\angle DCA = \angle BCE$, we have $\triangle DAC \sim \triangle EBC$ by AA Similarity. Therefore, $DC/AC = CE/BC$, so $(DC)(BC) = (AC)(CE)$. The dimensions of the rectangle with area R_2 are DC and BC (since one side of the rectangle is a side of the square with side $\overline{BC}$), so $R_2 = (BC)(DC)$. Similarly, $R_3 = (AC)(CE)$, so our $(DC)(BC) = (AC)(CE)$ gives us $R_2 = R_3$. In a similar fashion, we have $R_1 = R_6$ and $R_4 = R_5$.

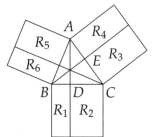

When $\angle ACB$ is a right angle, we have $R_2 = R_3 = 0$, $R_5 = R_4 = AC^2$, $R_6 = R_1 = BC^2$, and $R_5 + R_6 = AB^2$. Therefore, we have $AC^2 + BC^2 = AB^2$, and another cool proof of the Pythagorean Theorem!

19.27 By the Pythagorean Theorem, we have $AC^2 = BA^2 + BC^2$, $CF^2 = BC^2 + BF^2$, and $FA^2 = BF^2 + BA^2$, so

$$AC^2 + CF^2 - FA^2 = (BA^2 + BC^2) + (BC^2 + BF^2) - (BF^2 + BA^2) = 2BC^2 > 0.$$

Therefore, $AC^2 + CF^2 > FA^2$, so $\angle ACF$ is acute.

19.28

(a) Let X be the midpoint of $\overline{AD}$ and Y be the midpoint of $\overline{BC}$. We have $DX/DA = DM/DB$ and $\angle ADB = \angle XDM$, so $\triangle DAB \sim \triangle DXM$. Therefore, $\angle DXM = \angle DAB$, so $\overline{XM} \parallel \overline{AB}$. The line through X that is parallel to $\overline{AB}$ contains the median of trapezoid $ABCD$, so M must be on the median. Similarly, we can show that N is on the median. Since M and N are both on the median of the trapezoid, and the median of a trapezoid is parallel to its bases, we have $\overline{MN} \parallel \overline{AB} \parallel \overline{CD}$.

(b) Let the diagonals meet at E. Because $\overline{MN} \parallel \overline{AB} \parallel \overline{CD}$, we have $\triangle ABE \sim \triangle NME \sim \triangle CDE$. Therefore, $NE/AE = MN/AB = 3/4$, so $EN = (AE)(3/4)$. Since N is the midpoint of $\overline{AC}$, we have $NC = NA = AE + EN = 7AE/4$. Therefore, $EN/EC = EN/(EN + NC) = (3AE/4)/(10AE/4) = 3/10$. From similar triangles $\triangle EMN$ and $\triangle EDC$, we thus have $MN/DC = EN/EC = 3/10$, so $DC = MN(10/3) = \boxed{20}$.

19.29 Let $\overline{AB}$ be the chord in the given diagram of which M is the midpoint. Since M is the midpoint of $\overline{AB}$, we have $\overline{OM} \perp \overline{AB}$ because $\overline{OM}$ is part of a radius that bisects chord $\overline{AB}$. Therefore, $\angle OMP = 90°$. Similarly, $\angle ONP = 90°$. Therefore, the midpoint of $\overline{OP}$ is the circumcenter of both $\triangle ONP$ and $\triangle MOP$. $\overline{OP}$ is a diameter of this circle, and $\overline{MN}$ is a chord of the circle. No chord of a circle can be longer than the circle's diameter, so $MN \leq OP$. The only way we can have $MN = OP$ is if MN is a diameter of the circle with diameter $\overline{OP}$. If this is the case, then $\angle MPN$ is inscribed in a semicircle and is therefore right. Hence, we have $MN = OP$ if and only if our initial chords are perpendicular.

19.30 Let $x = [SPC]$ and $y = [RPC]$. Triangles SPC and BPC have bases $\overline{SP}$ and $\overline{BP}$, respectively, and the same altitude, so

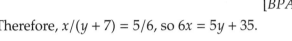

$$\frac{[SPC]}{[BPC]} = \frac{x}{y+7} = \frac{SP}{BP}.$$

Similarly, triangles SPA and BPA also have bases $\overline{SP}$ and $\overline{BP}$, respectively, and the same altitude, so

$$\frac{[SPA]}{[BPA]} = \frac{5}{6} = \frac{SP}{BP}.$$

Therefore, $x/(y+7) = 5/6$, so $6x = 5y + 35$.

By the same argument,

$$\frac{[APC]}{[RPC]} = \frac{x+5}{y} = \frac{AP}{RP} = \frac{[APB]}{[RPB]} = \frac{6}{7}.$$

Therefore, $6y = 7x + 35$.

Solving this system of equations, we get $x = 385$ and $y = 455$. Therefore, the total area of triangle ABC is $5 + 6 + 7 + 385 + 455 = \boxed{858}$.

19.31 $\angle ACB = \widehat{AB}/2 = \theta/2$. Since ℓ bisects $\angle ACB$, we have $\angle ACM = \angle BCM = \theta/4$. Because P is on the perpendicular bisector of $\overline{AC}$, it is equidistant from A and C, so $AP = PC$. Therefore, $\triangle APC$ is isosceles with $\angle PAC = \angle PCA = \angle MCA$. Since $\angle APM$ is an exterior angle of $\triangle APC$, we have $\angle APM = \angle PCA + \angle PAC = 2\angle PCA = \theta/2$, as desired.

19.32 By symmetry, the sphere touches each edge length at the midpoint of the edge. From here, we can either grind out a solution with lots of cross-sections, or very cleverly choose just one cross-section. We'll show the long way first.

Solution 1: In the text, we showed that a regular tetrahedron with side length 6 has volume $18\sqrt{2}$. We can follow exactly the same steps to show that a regular tetrahedron with side length 12 has volume $144\sqrt{2}$. (We could also note that since a regular tetrahedron with side length 12 is similar to one with side length 6, its volume is $(12/6)^3 = 8$ times that of the smaller tetrahedron.) Let O be the center of $ABCD$. By symmetry, O is the center both of the sphere inscribed in $ABCD$ and the center of $\mathcal{Z}$. We can find the radius of the inscribed sphere by noting that the volume of $ABCD$ equals $r[ABC]/3 + r[ABD]/3 + r[ACD]/3 + r[BCD]/3$ (see Problem 15.49). Since the faces of $ABCD$ are all congruent, the volume of $ABCD$ equals $(4/3)r[ABC]$. Since $[ABC] = 12^2\sqrt{3}/4 = 36\sqrt{3}$ and the volume of the tetrahedron is $144\sqrt{2}$, we find that the radius of the sphere inscribed in the tetrahedron is

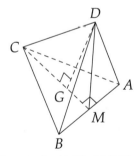

$$r = \frac{\text{Volume}}{(4/3)[ABC]} = \frac{(144\sqrt{2})(3/4)}{(36\sqrt{3})} = \sqrt{6}.$$

This inscribed sphere is tangent to face ABC at the center, G, of the face. The distance from G to the midpoint, M, of $\overline{AB}$ is $1/3$ the length of median $\overline{CM}$. From 30-60-90 triangle CMA, we have $CM = AM\sqrt{3} = 6\sqrt{3}$, so $GM = 2\sqrt{3}$. Finally, right triangle OMG gives us $OM = \sqrt{OG^2 + GM^2} = \boxed{3\sqrt{2}}$.

Solution 2: Let W, X, Y, and Z be the midpoints of $\overline{AB}$, $\overline{AC}$, $\overline{CD}$, and $\overline{BD}$, respectively. Since $AW/AB = AX/AC$ and $\angle WAX = \angle BAC$, we have $\triangle WAX \sim \triangle BAC$, so $\angle AWX = \angle ABC$ and $\overline{WX} \parallel \overline{BC}$. Similarly, $\overline{YZ} \parallel \overline{BC}$. We also have $YZ = WX = BC/2 = 6$. Since $\overline{WX} \parallel \overline{BC} \parallel \overline{YZ}$ and $WX = YZ$, we know $WXYZ$ is a parallelogram. Since by symmetry, we must have $\angle ZWX = \angle WXY$ and $WX = XY$ because the tetrahedron is regular, we conclude that $WXYZ$ is a square. Moreover, by symmetry, a diagonal of this square is a diameter of the sphere. Hence, the diameter of the sphere is $WX\sqrt{2} = 6\sqrt{2}$, so the radius is $\boxed{3\sqrt{2}}$.

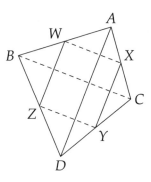

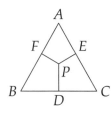

19.33 Let D, E, and F be the feet of the perpendiculars from P to sides $\overline{BC}$, $\overline{CA}$, and $\overline{AB}$, respectively. We must prove that $PD + PE + PF$ is a constant. The area of triangle PBC is $\frac{1}{2}BC \cdot PD = 3PD$. Similarly, the area of triangle PCA is $\frac{1}{2}CA \cdot PE = 3PE$, and the area of triangle PAB is $\frac{1}{2}AB \cdot PF = 3PF$. The sum of their areas is the total area of the triangle, which is $6^2\sqrt{3}/4 = 9\sqrt{3}$. Therefore, $3PD + 3PE + 3PF = 9\sqrt{3}$, so $PD + PE + PF = 3\sqrt{3}$ for any point P inside $\triangle ABC$.

19.34 Draw $\overline{MC}$. Triangles MUC and NUC have bases $\overline{MU}$ and $\overline{NU}$, respectively, and equal heights to these bases, so $[MUC]/[NUC] = MU/NU = s/6$. Therefore, $[MUC] = [NUC](s/6)$, so

$$[NCM] = [MUC] + [NUC] = \frac{s}{6}[NUC] + [NUC] = \frac{s+6}{6}[NUC].$$

Also, triangles HCM and NCM have bases $\overline{HC}$ and $\overline{NC}$, respectively, and equal heights to these bases, so $[HCM]/[NCM] = HC/NC = s/20$. We thus have

$$[HCM] = \frac{s}{20}[NCM] = \frac{s}{20} \cdot \frac{s+6}{6}[NUC] = \frac{s^2+6s}{120}[NUC].$$

Therefore,

$$[MUCH] = [MUC] + [HCM] = \frac{s}{6}[NUC] + \frac{s^2+6s}{120}[NUC] = \frac{s^2+26s}{120}[NUC].$$

Since $[MUCH] = [NUC]$, we have $s^2 + 26s = 120$, or $s^2 + 26s - 120 = (s+30)(s-4) = 0$. Since s is positive, $s = \boxed{4}$.

19.35 Let N be the midpoint of $\overline{O_1O_2}$. Since $\overline{O_1P}$ and $\overline{O_2Q}$ are radii to the points where $\overline{PQ}$ is tangent to the circles, we have $\overline{O_1P} \perp \overline{PQ}$ and $\overline{O_2Q} \perp \overline{PQ}$. Therefore, we have $\overline{O_1P} \parallel \overline{O_2Q}$. Hence, $\overline{MN}$ is the median of trapezoid PO_1O_2Q, and therefore has length $(O_1P + O_2Q)/2$. So, we have $2MN = O_1P + O_2Q$. $\overline{O_1O_2}$ passes through the point of tangency of C_1 and C_2, so $O_1O_2 = O_1P + O_2Q = 2MN$. Hence, $MN = O_1O_2/2$. $\overline{MN}$ is also a median of $\triangle O_1O_2M$. Since its length is half the length of the side to which it is drawn, $\triangle O_1O_2M$ must be a right triangle with right angle $\angle O_1MO_2$.

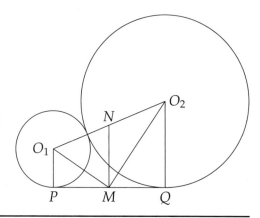

19.36 Let A, B, C, and D be the centers of the four balls. Since the four balls are tangent and each has radius 1, $AB = AC = AD = BC = BD = CD = 2$. Hence, tetrahedron $ABCD$ is regular with side length 2.

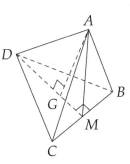

Let O be the center of the tetrahedron. Extend $\overrightarrow{OA}$ to meet the ball centered at A at P. Then the radius of the smallest sphere that encloses all four balls is OP.

Let M be midpoint of $\overline{BC}$, and let G be the centroid of triangle BCD. Then $AB = 2$, $BM = 1$, and $\angle AMB = 90°$, so triangle ABM is a 30-60-90 triangle. Therefore, $AM = \sqrt{3}$. Similarly, $DM = \sqrt{3}$.

Since G is the centroid of triangle BCD, G divides median $\overline{DM}$ such that $GM/GD = 1/2$. Therefore, $GM = DM/3 = \sqrt{3}/3$. We also have $\angle AGM = 90°$, so by the Pythagorean Theorem, $AG = \sqrt{AM^2 - GM^2} = \sqrt{(\sqrt{3})^2 - (\sqrt{3}/3)^2} = \sqrt{3 - 1/3} = \sqrt{8/3} = 2\sqrt{6}/3$.

OG is the distance from O to face BCD, so it is the inradius of tetrahedron $ABCD$. The volume of tetrahedron is

$$V = \frac{1}{3}[BCD](AG) = \frac{1}{3} \cdot [BCD] \cdot 2 \cdot \frac{\sqrt{6}}{3} = \frac{2\sqrt{6}}{9}[BCD].$$

As we showed in Problem 15.49, the inradius of the tetrahedron is

$$r = \frac{3V}{[ABC] + [ABD] + [ACD] + [BCD]} = \frac{3(2\sqrt{6}/9)[BCD]}{4[BCD]} = \frac{\sqrt{6}}{6}.$$

Hence,

$$OP = OA + AP = (AG - OG) + AP = \frac{2\sqrt{6}}{3} - \frac{\sqrt{6}}{6} + 1 = \boxed{\frac{\sqrt{6}}{2} + 1}.$$

Question: What is the radius of the largest ball that is externally tangent to all four?

19.37 We rotate the heptagon about D such that the image of D is on $\overrightarrow{AC}$. (Our goal is to find a segment that is both clearly equal to AD, and clearly equal to $AB + AX$.) Since D' is the image of D upon this rotation, we have $AD' = AD$. Since $\overline{BC} \parallel \overline{AD}$, we have $\overline{B'C'} \parallel \overline{AD'}$. Since $AD = AE$ and $\angle DAE = \angle CAD$ (consider the circumcircle of $ABCDEFG$; $\overset{\frown}{CD} = \overset{\frown}{DE}$, so $\angle CAD = \angle DAE$), a rotation of an angle of $\angle CAD$ counterclockwise about A maps E to D as shown. Since $\overline{CD} \parallel \overline{BE}$, we have $\overline{C'D'} \parallel \overline{B'E'}$. Since $\angle ADB = \angle AEB$ (we can see this from the circumcircle of $ABCDEFG$) and $\angle AEB = \angle AE'B'$, we have $\angle ADB = \angle AE'B' = \angle ADB'$, so B' is on $\overrightarrow{DB}$ and X is on $\overline{B'D}$. Therefore, we have $\overline{C'D'} \parallel \overline{B'X}$ and $\overline{B'C'} \parallel \overline{XD'}$, so $B'C'D'X$ is a parallelogram. Finally, we have $AX + XD' = AD'$, $XD' = B'C' = AB' = AB$, and $AD' = AD$, so $AB + AX = AD$.

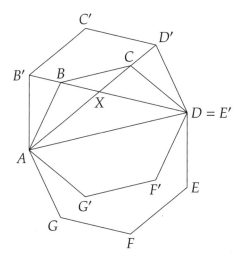

www.artofproblemsolving.com

The Art of Problem Solving (AoPS) is:

- ## Books

    For over 15 years, the classic *Art of Problem Solving* books have been used by students as a resource for the American Mathematics Competitions and other national and local math events.

    > *Every school should have this in their math library.*
    > – Paul Zeitz, past coach of the U.S. International Mathematical Olympiad team

    Our new Introduction series of textbooks – covering Algebra, Counting & Probability, Geometry, and Number Theory – constitutes a complete curriculum for outstanding math students in grades 6-10.

    > *The new book [Introduction to Counting & Probability] is great. I have started to use it in my classes on a regular basis. I can see the improvement in my kids over just a short period.*
    > – Jeff Boyd, coach of the 2005 MATHCOUNTS National Championship team from Texas

- ## Classes

    The Art of Problem Solving offers online classes on topics such as number theory, counting, geometry, algebra, and more at beginning, intermediate, and Olympiad levels.

    > *All the children were very engaged. It's the best use of technology I have ever seen.*
    > – Mary Fay-Zenk, coach of National Champion California MATHCOUNTS teams

- ## Forum

    As of July 2009, the Art of Problem Solving Forum has over 64,000 members who have posted over 1,490,000 messages on our discussion board. Members can also participate in any of our free "Math Jams".

    > *I'd just like to thank the coordinators of this site for taking the time to set it up... I think this is a great site, and I bet just about anyone else here would say the same...*
    > – AoPS Community Member

- ## Resources

    We have links to summer programs, book resources, problem sources, national and local competitions, scholarship listings, a math wiki, and a LATEX tutorial.

    > *I'd like to commend you on your wonderful site. It's informative, welcoming, and supportive of the math community. I wish it had been around when I was growing up.*
    > – AoPS Community Member

- ## ...and more!

Membership is **FREE**! Come join the Art of Problem Solving community today!